Progress in Systems and Control Theory
Volume 14

Series Editor
Christopher I. Byrnes, Washington University

Essays on Control: Perspectives in the Theory and its Applications

H. L. Trentelman

J. C. Willems

Editors

Springer Science+Business Media, LLC

H. L. Trentelman
University of Groningen
Mathematics Institute
The Netherlands

J. C. Willems
University of Groningen
Mathematics Institute
The Netherlands

Library of Congress Cataloging In-Publication Data

Essays on control : perspectives in the theory and its applications /
H. L. Trentelman, J. C. Willems, editors.
    p.   cm. -- (Progress in systems and control theory : v. 14)
    Includes bibliographical references.
    ISBN 978-1-4612-6702-7   ISBN 978-1-4612-0313-1 (eBooK)
    DOI 10.1007/978-1-4612-0313-1
    1. Control theory. I. Trentelman, H. L. II. Willems, Jan C.
III. Series.
QA402.3.E87   1993   93-1980
003'.5--dc20   CIP

Printed on acid-free paper
© Springer Science+Business Media New York 1993
Softcover reprint of the hardcover 2nd edition 1993
Originally published by Birkhäuser Boston in 1993
Second printing 1994

ISBN 978-1-4612-6702-7

Typeset by the Authors in TEX.

9 8 7 6 5 4 3 2

Contents

Contents

Preface

This book contains the text of the plenary lectures and the mini-courses of the European Control Conference (ECC'93) held in Groningen, the Netherlands, June 28–July 1, 1993. However, the book is not your usual conference proceedings. Instead, the authors took this occasion to take a broad overview of the field of control and discuss its development both from a theoretical as well as from an engineering perpective.

The first essay is by the key-note speaker of the conference, A.G.J. MacFarlane. It consists of a non-technical discussion of information processing and knowledge acquisition as the key features of control engineering technology.

The next six articles are accounts of the plenary addresses.

The contribution by R.W. Brockett concerns a mathematical framework for modelling motion control, a central question in robotics and vision.

In the paper by M. Morari the engineering and the economic relevance of chemical process control are considered, in particular statistical quality control and the control of systems with constraints.

The article by A.C.P.M. Backx is written from an industrial perspective. The author is director of an engineering consulting firm involved in the design of industrial control equipment. Specifically, the possibility of obtaining high performance and reliable controllers by modelling, identification, and optimizing industrial processes is discussed.

One of the most active areas of control theory research goes under the somewhat cryptic name of H_∞-control. The fundamental problem is to obtain feedback control algorithms such that the performance, measured in terms of the induced L_2-norm from the disturbance input to the to-be-controlled output, is minimized. The paper by A.J. van der Schaft gives an overview of the H_∞-problem for nonlinear systems.

In the article by M. Gevers an overview of an emerging area of control theory research is presented: the problem of obtaining identification algorithms for use in the design of robust controllers.

Distributed parameter systems form the topic of Y. Yamamoto's essay, in which the issue of learning control is addressed.

In addition to these plenary lectures and several hundreds of contributed papers, there were three mini-courses given at the ECC'93. These three-hour courses were based on tutorial exposé's of recent developments in control, illustrating advances in both theoretical as well as the engineering aspects of the field. The first mini-course was based on the article by M. Fliess and S.T. Glad, explainii.g the use of differential algebra as a mathematical framework for describing systems by means of models involving differential equations.

The second mini-course was based on the article by I. Postlethwaite and S. Skogestad, in which the methodology in designing robust industrial controllers using H_∞-methods is explained.

The third mini-course at the ECC'93 was devoted to the emerging field of neural networks. It was based on the article by E.D. Sontag on neural networks in control, and the article by J.-J. Slotine and R.M. Sanner on neural networks for adaptive control and identification.

We would like to take this occasion to thank all those who have contributed to this book. A special word of thanks goes to our colleague Rein Smedinga who has lend us his LaTeX expertise for formatting this book.

Groningen, April 6, 1993

The editors:

Harry L. Trentelman
Jan C. Willems.

1. Information, Knowledge and Control

A.G.J. MacFarlane*

Abstract

The importance of control has led to the development of one of the modern world's key technologies, and one whose use will increase rather than diminish. As such it needs a coherent philosophy - a world view, a conceptual framework which will allow the fundamental problems which confront its developers to be posed, which will allow its nature to be explained to non- specialists in an illuminating way, which will ensure that its fundamental concerns are not overlooked in an ever-increasing welter of specialisation and diversity of application, and which will ensure that it maintains a proper balance between theory and practice.

The basic concepts of : data and process, information, complexity and interaction are used to indicate the potential importance of information- theoretic concepts for such an undertaking. Much of the exposition draws on Kolmogorov's measure of algorithmic complexity, and analogies are drawn between the properties of effective interaction and the fundamental Laws of Thermodynamics.

Control is regarded as a technology which must necessarily draw on a range of cognate fields. Any resolutions of the fundamental problems which it faces in further developing as a technology will, it is claimed, share certain key features : they will use information-theoretic concepts; they will harness an emerging technology of knowledgeable machines to help manage the complexity of the data generated by theory; they will draw widely on, and interact with, a widening range of cognate fields; and they increasingly will adopt a pragmatic and empirical approach.

1 Introduction

Control is effective interaction : a bird hovering and an aircraft landing are both examples of objects interacting with their environments in a controlled

---

*Lord Balerno Building, Heriot-Watt University, Riccarton, Edinburgh EH14 4AS, Scotland, UK

way. That one of these examples is biological and one is technological shows why the idea of control is so important : it is a concept of the widest ranging applicability. The importance of control has led to the development of one of the modern world's key technologies, and one whose use will increase rather than diminish. As such it needs a coherent philosophy - a world view, a conceptual framework which will allow the fundamental problems which confront its developers to be posed, which will allow its nature to be explained to non-specialists in an illuminating way, which will ensure that its fundamental concerns are not overlooked in an ever-increasing welter of specialisation and diversity of application, and which will ensure that it maintains a proper balance between theory and practice.

It is the purpose of this overview to address this difficult task; not in any sense by offering a fully worked out philosophy of control - that would be a huge and complex undertaking, involving many points of view and so requiring the concerted efforts of many people - but rather by considering a few key aspects. Subsidiary aims are to identify a number of key sources in cognate fields which can provide access points for further study of major relevant topics and concepts, and to consider future development.

Major, coherent domains of knowledge are characterised by a fundamental, extensively used and essentially unifying concept : for example the cell in biology. Control is such a coherent domain, and its fundamental unifying concept is that of *feedback*. This concept is so widely useful that it has generated one of the few terms associated with modern technology which have passed into the currency of everyday speech. A synoptic overview of feedback and its many uses has been given by MacFarlane (1976), [16]. Feedback is the principal weapon in the armoury of the control system designer. At the most basic level of description however, the terms interaction, feedback and control are equivalent and may be used interchangeably. Consider for example a simple interconnection of a spring and a mass. Such a dynamical system can be represented, in a block diagram showing the interrelationship of the key dynamical variables, as a feedback system. We can, and indeed often do, speak of the spring as controlling the position of the mass. Thus we may, usefully and as appropriate, regard the same system in terms of a pair of interacting objects, in feedback terms, or in terms of control.

Control is delimited as a subject by:

- its focus on a clear and simply stated problem - the achievement of effective interaction

- by the fact that this problem generates clean interfaces with cognate fields

- by the rich resonances which its basic concern of the achievement of effective interaction sets up with mathematics, dynamics, information theory, physiology, neurophysiology, computational science and artificial intelligence.

An acid test of any claim for the coherency and wide utility of a subject is that its central purposes, and the fundamental nature of its limitations, should be communicable to a lay audience in terms of a few fundamental ideas. What is being attempted here is such a synoptic overview in terms of : data and process, information, complexity and interaction.

2 Data and process

H.A. Simon, in his analysis of artefacts and systems and their uses, distinguishes between two fundamentally different types of description :

- state descriptions, and

- process descriptions.

In his words (Simon (1981), [28]):

> These two modes of apprehending structures are the warp and weft of our experience. Pictures, blueprints, most diagrams, and chemical structural formulas are state descriptions. Recipes, differential equations, and equations for chemical reactions are process descriptions. The former characterise the world as sensed; they provide the criteria for identifying objects, often by modelling the objects themselves. The latter characterise the world as acted upon; they provide the means for producing or generating objects having the desired characteristics.

Since the term state is used in dynamics and control in a very well established but very different sense, we will use the terms *data* description and *process* description to make Simon's distinction. For example in a computer we can store a picture of a circle as a bit map - *a data description* - and print from that, or as a program to generate points lying on the circle - *a process description* - and output the same picture in a different way, with the same end result. However the *amounts* of information involved in the two cases will not be the same; in general the amount of information required for the process description can be much less than that for the data description. Indeed, as we store data for more and more points, the information required for the data description can become unbounded while the information required for the process description remains small.

Any computation carried out by an information processor will involve a mixture of data and process. Explanatory processes underpin all our understanding - to understand is to be able to explain. Explanation in essence arises from our ability to replace data by process. The working life of the scientist and engineer is a constant battle to strike a working balance between the management of data and the management of process, and to replace data by process to the largest possible extent.

2.1 Fractals and Chaos

Good examples of the nature of the difference between data and process can be found in a consideration of fractals and chaos (Peitgen, Jurgens and Saupe (1992), [21]).

Fractals are sets, usually exhibited in a plane, which are generated by processes, usually of modest complexity. There is a striking contrast between the apparently rich structure of the generated fractal data sets and the simple nature of the generating process. The chief feature of the structure of the generated data sets is one of *self-similarity*. When one looks at subsets of the fractal set, each has essentially a similar form regardless of scale. All the subsets appear to contain all the detail of the complete set. The computer- generated bit maps which create pictures of fractals generate unbounded amounts of data, and it appears that one could explore the details of such sets for an unbounded length of time. Thus a fractal is an example of a simple process which can apparently generate an unbounded amount of data.

The defining feature of chaos is that simple deterministic processes can generate unpredictable behaviour. The most usually demonstrated examples are deterministic dynamical systems which exhibit an extreme sensitivity to their initial conditions. If we think of such systems as described by a mixture of data (the initial conditions) and process (the dynamical equations), then exploration of the output behaviour in arbitrary detail requires an unbounded amount of data in the initial conditions. The output data sets generated by chaotic systems require an unbounded amount of information to describe them.

Fractals and chaos exhibit different aspects of data and process. They are linked in that some of the subsets of the output behaviour description of chaotic systems (such as attractors) can be characterised in fractal terms. In chaos *data dominates process*; the complexity of chaos is real - a chaotic system is associated with unbounded data. In fractals process dominates data; the complexity of a fractal data set is only apparent - the process imposes self-similarity on the generated data and binds it into a tightly self- referential structure. In a way, fractals and chaos provide an instructive parable for Theory and Practice. Theory, like fractals, can only generate limited amounts of information, although it may seem overwhelmongly complicated. Experiment, drawing on the unbounded amount of information in the real universe can, like chaos, generate unbounded amounts of information. All scientific and technical endeavour is an unending struggle to resolve the conflicting demands of process and data, and to manage the trade- offs between them. Fractals are increasingly used for *data compression* - to generate complicated objects (representations of clouds, mountains, richly textured surfaces) using only modestly complex processes. This is analogous to theory, which compresses experimental data into modestly complex formal structures. Chaos, like our experimental encounters with reality, reminds us of the vast complexity needed for the accurate prediction of real system behaviour. Fractals, like theory, provides us with visions of dazzling beauty; chaos, like practice, gives us a grim reminder of reality. Fractals and chaos lie at opposite ends of a spectrum of the relative dominance of process and data in system characterisation.

3 Information, complexity and interaction

Kolmogorov introduced a characterisation of *complexity* which gives an intuitively appealing measure of the amount of information required to obtain or reconstruct an object by any effective procedure. In general terms, one can think of the Kolmogorov complexity (Kolmogorov (1968), [14]) of a process as the length (in bits) of the *shortest possible* computer program which can instantiate it, and so the complexity of a system can be considered as being the Kolmogorov complexity of a computer simulation of it. In what follows the term complexity is to be understood in this Kolmogorovian sense. Kolmogorov's approach has been used by Solomonoff (1964), [29], to characterise induction and the testing of scientific theories, and has been extensively developed by Chaitin (1990), [4].

3.1 Information

Information has been described, together with energy and matter, as the third fundamental concept of modern science (Gitt (1989), [11]). As a fundamental concept it cannot be defined, but must be comprehended by its use and justified by its usefulness.

Information is a quantitative measure of pattern or order, a way of ascribing a quantitative measure to a symbol or set of symbols. In the simplest case, when we are concerned with the presence or absence of one symbol, the corresponding unit of information is one bit. By defining information in this way we have of course simply transferred the burden of definition to finding one for pattern and order. Any attempt to find firm ground in definitions will lead to an endless regression, which is best avoided axiomatically by using fundamental concepts which are illuminated and justified by their use. A valuable discussion of the problems and pitfalls of definitions and their use and misuse can be found in Popper (1972), [23].

Two theories of information have evolved : one developed from an analysis of communication and signals, and the other arising out of the analysis of models and formal structures. These are:

- that which is appropriate to communication theory, and was developed by Shannon (1948), [27]. This is based on information as a measure of improbability or *uncertainty*. In this approach a process is characterised by its *entropy*. Entropy is a measure of the average uncertainty in a random variable; it is the number of bits on average required to describe the random variable (Cover and Thomas (1991), [7]).

- that which is appropriate to modelling and algorithms, and was developed by Kolmogorov (1968), [14] and Chaitin (1990), [4]. This is based on information as a measure of complexity. In this approach a process is characterised by its complexity. In all cases the information is measured in bits.

The Kolmogorov-Chaitin theory is usually called *algorithmic information theory* to distinguish it from Shannon's earlier theory. Since information theory has two aspects, the corresponding measures - entropy and complexity - will appear in complementary roles. There is a huge technical literature on information theory and its applications. Excellent general references are Cover and Thomas (1991), [7], which deals with both forms of the theory, and Resnikoff (1989), [24], which looks at an interesting range of applications.

3.2   Complexity

Chaitin makes many and subtle uses of the concept of complexity, and the interested reader is strongly urged to consult his collected papers (Chaitin (1990), [4]). For our present purposes it is useful to give one of the characterisations which he uses in his own words, and then use the term in the spirit of this characterisation:

> The complexity of a binary string is the information needed to define it, that is to say, the number of bits of information which must be given to a computer in order to calculate it, or in other words, the size in bits of the shortest program for calculating it. It is understood that a certain mathematical definition of an ideal computer is being used, but it is not given here, because as a first approximation it is sufficient to think of the length in bits of a program for a typical computer in use today.

Chaitin characterises Solomonoff's approach in the following way :

> Solomonoff represented a scientist's observations as a series of binary digits. The scientist seeks to explain these observations through a theory, which can be regarded as an algorithm capable of generating the series and extending it, that is, predicting future observations. For any given series of observations there are always several competing theories, and the scientist must choose among them. The model demands that the smallest algorithm, the one consisting of the fewest bits, be selected. Stated another way, this rule is the familiar formulation of Occam's Razor : given differing theories of apparently equal merit, the simplest is to be preferred.
>
> Thus in Solomonoff's model a theory that enables one to understand a series of observations is seen as a small computer program that reproduces the observations and makes predictions about possible future observations. The smaller the program, the more comprehensive the theory and the greater the degree of understanding. Observations that are random cannot be reproduced by a small program and therefore cannot be explained by a theory. In addition the future behaviour of a random system cannot be predicted. For random data the most compact way for the scientist to communicate his observations is for him to publish them in their entirety.

In general, a program will consist of a mixture of process and data. In limiting cases, of course, it may be wholly process or wholly data. The shortest program to generate wholly random data is the data itself; the shortest program to generate a sequence of identical symbols is the shortest instruction set which will achieve this end. The shortest program to generate a circle will, in fact, not be a pure process but will be the shortest instruction set to generate points so located plus the shortest specification of the data needed to fix the centre and radius required. Note that data, process, and any mixture of data and process can all be allocated a complexity measure; the idea of complexity is not confined to process.

Cover and Thomas (1991), [7] give an illuminating discussion of Kolmogorov complexity and some of its uses. In particular they make the observation that one should use Kolmogorov complexity "as a way to think" rather than as an immediately useable computational tool. A main contention of this paper is that Kolmogorov complexity is a very helpful way to think about the general aspects of modelling, systems and control. Cover and Thomas make the following observation about the relationship between Kolmogorov complexity and entropy :

[The Kolmogorov] definition of complexity turns out to be universal, that is, computer independent, and is of fundamental importance. Thus Kolmogorov complexity lays the foundation for *the* theory of descriptive complexity. Gratifyingly, the Kolmogorov complexity K is approximately equal to the Shannon entropy H if the sequence is drawn at random from a distribution that has entropy H. So the tie-in between information theory and Kolmogorov complexity is perfect. Indeed, we consider Kolmogorov complexity to be more fundamental than Shannon entropy.

3.3   Complexity and models

Models are built from information. If a model is stored in a digital computer, then the information involved can be exactly quantified in bits. For any model, we have the following :

> **Bounds on model complexity**: The complexity of a model is bounded above by the complexity of the data from which it was constructed (and below by zero).

Good models use small amounts of information to relate large amounts of input-output data; that is they have substituted small amounts of process information for large amounts of data information (Maciejowski (1978), [19]). There are striking differences between engineering and econometric models in this regard. When constructing dynamical models for predictive purposes, one should not be misled into thinking that complexity implies effectiveness. For example, too high an order of dynamical model may simply arise from some of the degrees of freedom in the model being used to replicate the noise behaviour in the data. An optimal dynamical order can be generated by testing the predictive power against new data rather than by merely testing its replicative or simulating power.

It is useful, when deploying complexity arguments, to think of generating a model by "squeezing down from" the raw data into the shortest program which will reproduce the data. This program will consist of a process component and a data component, neither of which can be further compressed. If the data is wholly random, no process component will be squeezed out. In the case of very regular data, little or no data component will be left after the squeezing. The complexity of the original data will (by definition) equal the complexity of the squeezed-out process-component plus the complexity of the squeezed-out data-component. We will call the squeezed-process-component complexity the *P-complexity*, and the squeezed- data-component complexity the *D-complexity* respectively.

Engineers use a multiplicity of models in their work. For example in studying the behaviour of an aircraft engine one could use : thermodynamic models, mechanical stressing models, dynamic models, fatigue and crack propagation models, and no doubt others. Furthermore, in addition to the many different types of model used for an engine, one would have a whole range of models for its various subsystems : for example electrical, mechanical and hydraulic. These models are usually used independently and for very specific purposes. A key point is that they are mostly used for their *explanatory* purposes in design. It is of course true that such models also have in some sense a predictive role, but this is only verified under rigorously controlled experimental conditions, which isolate the reduced set of circumstances under which the particular model is appropriate. Hence their *predictive* roles are seldom used or required in the way in which atmospheric or econometric models are ostensibly used for predictive purposes in weather and economic forecasting. In engineering work theory and experiment, that is process and data, always go together hand in hand. Design leads to prototypes which are rigorously tested, leading to a refinement of design until satisfactory behaviour is achieved. The emergence of a successful artefact is a vivid example of the interplay between process and data, and of the continuing struggle to replace data by process.

The following question is of fundamental importance : Why is it that we can, for example, design and operate complex mechanisms like an aircraft and its associated engine and flight control systems when we cannot, in any absolute sense, predict the behaviour of very simple objects like a spherical pendulum (because they exhibit chaotic behaviour)? The resolution of this apparent paradox lies in the fact that *the uses of models for explanatory and for predictive purposes can require vastly different amounts of complexity.*

A collection of models each of modest complexity, and each associated with one aspect of a machine's behaviour, may be used to make a prediction in the limited sense that one may say that the machine will work in the way the designer intended. An altogther different magnitude of complexity is required in order to make specific predictions about the detailed future behaviour of any specific machine at some specific time. In the field of meteorology and weather forecasting, one can construct a climatic model which will accurately reproduce the main qualitative features of the Earth's climate in considerable detail, when one cannot produce a model which is guaranteed to predict specific future weather behaviour at a given place and future time with great accuracy. The reason is that the former task simply requires vastly less information. Indeed the latter task may require, for prediction beyond a certain time limit, impossibly large amounts of information, for the meteorological system equations are chaotic. To effectively interact with anything to the extent that we can accurately predict its future behaviour requires much more information than interacting at a level which allows us to reproduce the main qualitative features of its behaviour.

3.4   Management of complexity

Two *systems* will be said to be *effectively interacting* if their *inter*dependent behaviour, *perceived externally as that of a single composite system*, is significantly different from their independent behaviour.

To consider how complexity is related to effective interaction, we first establish how the complexities of two interacting systems are related to an externally perceived complexity. The simple but fundamentally important result which we need is :

Bounds on externally perceived complexity: The externally perceived complexity of a composite system consisting of two subsystems is bounded above by the sum of their individual complexities and bounded below by zero.

The upper bound is established by noting that one of the ways of regarding the whole system is simply as the sum of its parts, and so the sum of their complexities must bound the complexity of the whole system. The lower bound is established by making the two systems identical and then, by an appropriate further arrangement, making the output of one system wholly cancel the output of the other, so that the externally perceived complexity then becomes zero. Starting from this situation of complete interdependence we could modify one of the systems "bit by bit" until we reached a condition in which the two systems would be fully independent, and the upper bound of externally perceived complexity had been attained. In principle then, we could arrange that, as we passed through this "bit by bit" procedure of incremental modification of one of the associated systems, the externally perceived complexity of the composite system would move monotonically from its lower to its upper bound.

The upper bound holds when the two subsystems are completely *independent*, and the lower bound holds when the two subsystems are completely *interdependent*. The externally perceived complexity of the whole is a function of the degree of interdependence of the two subsystems. The effectiveness of the action of one of the subsystems on the other is measured by their degree of interdependence and so by the way in which their individual complexities are related to the externally perceived complexity of the composite system. *Thus interdependence can reduce externally perceived complexity.* This is, for example, what a controller does : by effectively interacting with the system which it is controlling it achieves a reduction in the externally perceived complexity - that is a simpler pattern of behaviour - of the system which is being controlled. The essence of an effectively controlled system is that the whole of the externally perceived system-plus-its-controller must be less complex than the sum of its parts.

3.4.1   Aggregation and hierarchisation

Typical procedures for the management of complexity are aggregation and hierarchisation. To be able to lump a set of systems together in order to treat them as a simpler composite single system, the sub-systems must effectively interact. The reason that we can treat a metal rod, say for certain purposes of dynamical analysis, as a single mass is that the interactions between the myriad atoms which comprise it are so effective in holding it together that it can be externally perceived as a single object of simple complexity. The various sub-systems comprising an engine must interact together effectively in order for the engine to be externally perceived as a single entity, performing a simply describable task, and, for the purposes of external description, of a complexity much less than the sum of the complexities of its sub-systems.

The best example of the implications of effective interaction for the management of complexity is in hierarchical organisation. Consider the external perception of a complicated object such as an automobile engine. Such an external perception can be described in hierarchical terms. At the highest level of the hierarchy of possible forms of interaction, which is also the lowest level of complexity for the perceived object with which we are interacting, we perceive and describe it, say, as a power unit responding to a throttle input. At the next level down we perceive it in terms of an *interacting set of units*, each performing functions which interact in such a way as to give an object - the engine-as-a-power-unit - at the higher level of description. We can only effectively describe how it works at the lower level *if all the units at the lower level of description effectively interact*. In describing how these units at the lower level of description work together we will use similar amounts of complexity in describing each of their individual functions - how the function of the carburettor relates to the function of the cylinders and valves, how the gearbox relates to the transmission, and so on. At the next level down we might be concerned, for example, in discussing how the various bits of a piston - sealing rings, connecting rods and bearings relate. And so we could go on right down to the molecular level, when we might be considering metallurgical properties of a particular part, and how this relates to fatigue and crack propagation, and so on.

So long as the engine is functioning well, we need only consider it, from an external point of view, as an object of low complexity. When we are interacting with it when it no longer functions as it should, then we have to move down through the hierarchy of complexity in descriptions to locate the particular part which is not functioning properly. In so doing we are facing a much higher level of complexity than we needed to use when it was functioning normally. This "breakdown", which requires an expansion of the level of complexity with we have to interact with an object, exactly mirrors the notion of breakdown in Heideggerian philosophy

(Dreyfus (1992), [8]), when an individual's wholeness of being-in-the-world is disrupted by an extraneous and unforseen event. Nature, in the aeons of experiment over which evolution has taken place, has discovered that aggeregation, modularisation and hierarchisation are effective ways of managing complexity, building systems capable of highly effective interaction with their environment by assembling the total complexity necessary by an incremental approach. In an effectively functioning hierarchy, the interactions between systems or units at lower levels is such as to create a reduced level of complexity at the level perceived above. This reduction of externally perceived complexity proceeds up the hierarchy till the top level perceives the aggregated effect of all the lower levels in terms of a single entity with a manageable level of perceived complexity.

The price to be paid for a reduction of externally perceived complexity is two-fold:

- A loss of detailed information.

- An appropriate degree of complexity in the sub-systems.

3.5   Complexity and control

Consider a region of space and time where a process is taking place, due to the action of the rest of the universe on it, and that this process has a given entropy. Now create an apparatus, influencing that region of space and time. This apparatus can be associated with another process having a different entropy, and the combination of apparatus and the original region can be regarded as a *joint* process with an entropy which is a function of the two processes. Standard texts on information theory, such as Cover and Thomas (1991), [7], show that the entropy of the joint process is bounded above by the sum of the individual process entropies with equality if and only if the random process variables are *independent*. Now assume that the apparatus is interacting effectively with that region of space and time to *reduce* the entropy of the joint process characterising the controlled situation. This reduction in joint entropy arises from the *mutual information* between the two processes, and for such a reduction it is necessary that the two processes are interdependent (Cover and Thomas (1991), [7]).

Since the entropy of the combined process has been reduced, the complexity of the externally perceived combined system must have been reduced, for which, as we have seen, it is necessary that the two interacting subsystems be *interdependent*. So the reduction in *overall* complexity can only have been effected by the creation of a necessary amount of complexity in the controlling system. For the entropy of the combined system to be reduced to zero, the complexity of the controller would have to match exactly the complexity of a model capable of exactly reproducing the effect of the rest of the universe on that region of space and time which is being controlled. In practice just enough complexity is created in the controlling system in order to achieve the degree of control - that is the reduction in entropy - which is sought for the practical purposes in hand. Complexity is the price we pay to reduce entropy.

Mutual information between processes has emerged as a useful measure in information theory for the estimation of the reduction of the entropy of a joint process from its upper bound. So the idea of "mutual complexity" between system substructures, as a measure of the reduction in complexity of a combined system from its upper bound, might be useful for control considerations. A useful name for this analogue of mutual information might be *mutual organisation*. In these terms we would say that effective interaction between systems implies a degree of mutual organisation between them, and the greater the effective interaction the greater the necessary degree of mutual organisation.

An instructive example of the price paid in complexity to reduce uncertainty is given by the technology of time keeping. As time is measured ever more accurately we go through a progression of ever more complex apparatus for the purpose, starting off with simple mechanical clocks, and progressing through elaborate chronometers to things like atomic clocks, which fill racks with equipment. We can think of the time-read-out device as a simple controlled object generating a process associated with the value of present time. As the entropy of the controlled time-keeping process is reduced, squeezing out the uncertainty in time values, so the complexity of the controlling process must be increased.

An interesting general question which arises is : to what extent could Kolmogorov complexity and entropy measures be used for an *empirical, experimental* approach to the determination of the necessary complexity of controllers to effect a variety of effective controls? In a totally different subject area - an investigation of the implications of algorithmic information theory for Arithmetic - Chaitin (1987), [3] has given a virtuoso example of how the relevant computational issues might be handled.

A further example of the price paid in complexity to reduce uncertainty is given by large-scale experiments in particle physics. As greater and greater control is exerted to reduce the uncertainties in an ever smaller region of space and time, so does the complexity of the physical apparatus which is necessary monotonically increase. This is because, in order to achieve the effective interdependence between the apparatus and the region of space being controlled, the entropy squeezed out of the region of space by reducing the effect of the rest of the universe on it must be paid for by the complexity of the apparatus which is being used to generate the effective control. Exploration of ever more simple basic particles necessarily requires ever more complex apparatus.

In the same way that we can understand the mechanisms of the weather while only having a limited ability to predict it (most short term weather prediction depends heavily on the *empirical* device of looking at what is *actually happening* via satellite photography), so we can understand the qualitative aspects of how complex systems work while only being able to control them to a limited extent. For example, compare the complexity of the models normally available for a petrochemical refining plant with the complexity of the controllers normally used to control it. It is commonplace to use plant models having hundreds or even thousands of state variables while using controllers which have only tens of state variables. Indeed I was once shown over a large research establishment where there were three adjacent offices all full of dynamical specialists. It was explained to me that the first office contained modellers building very large plant models, the middle office contained people creating reduced order models of the plant - from these same very high order models - for the people in the end office to use in designing the control systems to be used on the plant. I have always considered this to be a deeply instructive parable. All these activities were necessary - the first to understand it, and to devise a point of departure for further modelling work on the process; the second to produce a related, simpler system which reflected only those aspects of plant behaviour whose complexity could be successfully matched by a feasible controller design, which would in turn achieve an attainable level of effective interaction with the plant. (Anyone who thinks that this point of view is exaggeratedly pessimistic should try designing and testing effective controllers for some simple mechanical systems such as a quadruple inverted pendulum.)

Using the terms introduced in Section 3.3, we can say that the petrochemical example plant controllers were of low complexity, compared to the complexity of the available plant models, because the P-complexity of data available from the real plant is small. The large models of the plant would be largely theoretically constructed and aggregated in a piecemeal fashion, and not generated from real, experimentally obtained data. The complexity of a useful controller for a plant will normally match the P-complexity of a plant model derived from plant data. This is because the squeezed-process-component extracts the maximum amount of useful predictive machinery from the plant data, and we can only control something to the extent that we can predict its future. Any attempt to exert more control on the plant than the P- complexity merits will result in instability. In the aerospace field, when the objects controlled are essentially ballistic in nature, the P-complexities of models being used for control purposes will be significantly higher than for messy earth-bound industrial plants.

Information and complexity concepts then, and in particular the distinction between the information associated with data and process, can provide a powerful unifying thread when considering complicated matters of modelling, prediction and control. For discrete-time dynamical systems, or sampled data systems, the incidence of chaotic behaviour increases. A one-dimensional discrete-time system can exhibit chaotic behaviour, and it has been observed that in digital control systems the use of a reduced word length (that is the presence of amplitude as well as time quantization) increases the incidence of chaotic behaviours. Loss of information also reduces the effectiveness of observations for feedback - digital systems have less tolerance of feedback than analogue systems. Hence loss of information reduces our ability to combat uncertainty. If a control system cannot perform well enough on measured information by a direct feedback of observed variables, then one can seek to incorporate a model of the system which is being controlled. This extra information, in the form of a process, can be thought of as tradeable for measured data. That is we can have more data, or better processes, or trade off between some combination of them. This is an interesting practical example of *information trading*, that is of the interchange of data information for process information. Techniques of this sort are used in high integrity systems which substitute model-derived information for measured information to cope with certain sensor failure conditions.

The predictive role of models is used in engineering for control purposes, and engineers have learned that the problem of controlling a dynamical system is a mixture of being able to predict its behaviour and to observe its behaviour, that is a mixture of process and data. The sophisticated control systems used, for example, in current aerospace systems contain models of the systems which they are controlling. These models are however simulating rather than explanatory models. A controller thus uses two kinds of information : data, in the form of direct measurements, and process in the form of system models. The key weapon in the armoury of the control engineer is feedback, using measured information. Feedback is also Nature's way, in biological systems of reducing entropy or uncertainty. Given that there is a limit to the predictive capability of a dynamic model, then the use of feedback is the only route remaining for the reduction of uncertainty.

4   The laws of interaction

Complexity provides a useful way of thinking about many general aspects of systems : how we can create simplified representations, the price necessary for good control, whether machines can be created with human levels of functionality, and so on. It is also useful for the explanation of such general aspects of system behaviour without getting bogged down in the complicated arguments which are necessarily deployed when drawing on highly specific fields of knowledge. The generality and usefulness of complexity for such purposes is reminiscent of that of entropy in Thermodynamics, and we have noted that modern Information Theory has established the closest links between these two concepts. For this reason it is instructive to formulate the way in which consideration of some general properties of systems might be informed by complexity considerations in terms of an analogy with the formulation of the basic Laws of Thermodynamics. (The analogy must not, of course, be taken literally - a random process, after all, has an entropy but not a temperature.) Pursuing such an analogy, a set of Laws of Interaction can be set out:

> **First Law**: Complexity enables interaction, and interaction determines perceived complexity.

This states an equivalence between complexity and interaction, and is analogous to the First Law of Thermodyamics in the form which states the equivalence of heat and work.

> **Second Law**: The complexity of an interacting system must increase if its interaction is to become more effective.

The Second Law of Thermodynamics tell us that entropy is the price paid for work, and so as work is done entropy increases. The Second Law of interaction tells us that complexity is the price paid for effective interaction, and so necessarily increases as we seek more and more effective interaction.

> **Third Law**: No interaction can be arbitrarily effective. It is impossible to manage arbitrarily large amounts of complexity.

The Third Law of Thermodynamics states that it is impossible to attain an absolute zero temperature, and so sets conceptual limits to what may be contemplated in systems concerned with the flow of heat. The Third Law of Interaction sets an analogous conceptual limit to what may be accomplished in systems concerned with the flow of information.

> **Zeroth Law**: There is a Zeroth Law of Themodynamics (Pippard (1959), [22]) which states that : if A is in thermal equilibrium with B, and B with C, then C is in thermal equilibrium with A. It states an obvious assumption made in the application of the other Laws. Analogous forms could be devised for a Zeroth Law of Interaction : for example, if A and B effectively interact so that they can be considered as an externally perceived composite system of reduced complexity, and so do B and C, then so does the composite system formed by A, B and C.

These three "laws" are put forward as no more than a convenient summary of the roles played by complexity and interaction, and because one can then use terms like Second Law in a rough analogy with the Second Law of Thermodynamics. The Second Law is reminiscent of Ashby's Law of Requisite Variety : that only variety can destroy variety (Ashby (1953), [1]).

4.1   Data catastrophes and process catastrophes

In trying to reduce uncertainty by creating complexity, we can be overtaken by two kinds of catastrophe, which we may call a *data catastrophe* and a *process catastrophe* respectively. A data catastrophe arises when we need an unbounded amount of data to resolve uncertainty, and a process catastrophe arises when we need an unbounded amount of process to resolve uncertainty.

Examples of data catastrophe are provided by the field of chaos (Peitgen *et al* (1992), [21]), which deals with dynamical systems which have an unbounded sensitivity in their initial conditions. As one tries to predict their behaviour at a given point in time more and more precisely, the precision required in the specification of the initial conditions increases without bound. In a data catastrophe, process is overwhelmed by data. It is impossible to achieve more than a modest level of reduction in the uncertainty of predicted behaviour because of the catastrophic growth in the amount of data which is necessary.

Examples of process catastrophes are provided by undecidable propositions in formal systems. The uncertainty as to whether they are true or not cannot be resolved because, as increasingly serious attempts are made to resolve the proposition, more and more complex theories are generated. From the point of view of algorithmic complexity theory it is not a question as to whether or not a proposition in some formal system is true or not true, but rather the degree of complexity of the proof which is required (Chaitin (1990), [3]). An undecidable proposition is one for which the complexity of the process required to resolve it is unbounded.

In this context it is interesting to note the complexity characterisation of Godel's theorem (Shanker (1988), [27]) that any formal system will necessarily have associated with it propositions which may be formulated validly in terms of the system but which cannot be shown to be either true or false by using it. Stewart (1988), [30] has succintly stated this in information-theoretic terms as : the theorems deducible from an axiom system cannot contain more information than the axioms themselves do. A process catastrophe arises in this context when the extra axioms required to resolve a proposition involve an unbounded amount of information; the complexity of the processes which specify the formal system needed becomes unbounded.

These catastrophes illustrate the Third Law - that it is impossible to manage arbitrarily large amounts of complexity. As we struggle to create and manage systems which require ever increasing amounts of complexity, we are eventually undone by either the data requirements of our representations or their process requirements becoming unbounded.

5   Knowledgeable machines and intelligent machines

An excellent overview of epistemology (the theory of knowledge) is given in the extensive collection of sources and commentaries assembled by Moser and Vander Nat (1987), [20]. Very little of this is relevant to technology (because of its emphasis on subjective knowledge), but the approach adopted by C.I. Lewis (1926), [15] is illuminating and helpful in developing a characterisation appropriate for our context.

In Lewis's original treatment knowledge is taken as having three elements :

- given data

- concepts

- acts which interpret data in terms of concepts.

In his characterisation of knowledge, and in his own words:

- Knowledge is an interpretation driven by need or interest and tested by its consequences in action.

- Order or logical pattern is the essence of understanding. Knowledge arises when some *conceptual pattern* of *relationships* is imposed on the given data by interpretation.

- The business of learning is not a process in which we have passively absorbed something which experience has presented to us. It is a process of trial and error in which we have attempted to impose on experience one interpretation or conceptual pattern after another and, guided by our own practical success or failure, have settled down to that mode of constructing it which accords best with our purposes and interests of action.

- Knowledge is the active interpretation which unites the concept and the data.

In considering the technology of control, we are concerned with both machines acting in isolation and in conjunction with humans. We therefore need a means of handling the concepts of knowledge and intelligence which can be applied equally well to human and machine. This can be achieved by regarding intelligence and knowledge as necessary components of *effective action*.

- **Intelligence**: is a *capacity* for effective action. It is the dynamic component of effective action. We will think of it as a *process*.

- **Knowledge**: is a *potential* for effective action. It is the static component of effective action. We will think of it primarily as *data*; in Lewis's terms : as a highly organised and interactively accessible *structure of relationships*.

5.1 Limitations on achievable levels of artificial intelligence

The human brain is a system of the most enormous complexity. One estimate of the number of distinct neural states has been given as 10 to the power 100,000,000,000,000 whereas the total number of elementary particles in the universe has been estimated as a relatively modest 10 to the power 87 (Flanagan (1992), [9]). This complexity has evolved over billions of years in order that we may interact effectively with the universe in which we live. That is to say : the degree of complexity of animal nervous systems is neccessary in order that they may be interdependent with their surroundings to the extent required to survive and flourish. Chaitin (1990), [3] has characterised biological systems by their unity, that is by the huge degree of interdependence among their constituent parts. As their individual parts, such as cells, are very complex, this degree of interdependence is, in Chaitin's words "either monstrously impossible or is the result of long and gradual evolution."

Flanagan (1992), [9] regards evolutionary processes as having produced a nervous system which has two components : a fixed structure and a plastic, or adaptable, structure. In his words:

> The synaptic connections that evolve in a brain over time are the complex causal outcome of genotypic instructions, endogenous biochemical processes, plus vast amounts of individually unique interactions between organism and environment . The key to our magnificent abilities as anticipation machines involves fixing gross architecture while leaving the development of connections at the microstructural level undedicated and adaptable ... individual brains are extraordinarily diverse in terms of structure and connectivity.

The awesome complexity of the human brain reflects our degree of effective control over our immediate environment, that is to say our ability to interact effectively with our surroundings. This complexity has two components : an inherited component whose complexity gives our endowed innate capability, and a component whose complexity is generated by a long, intensive and effective interaction with the environment. These are augmented by a third component, an objective knowledge component, whose complexity has been generated by the creation of artefacts such as books, machines, computers, etc.

There is a large literature on artificial intelligence, enlivened by vigorous polemic and battles between opposing camps. A good introduction to the central controversies which are involved can be found in Dreyfus (1992), [8] and Searle (1984), [25]. The central point of contention is whether we can build (meaning in practical terms and in the reasonably near future) machines which will replicate human functionality in problem solving and task execution over a range of tasks and to a level at least equal to skilled human performance levels. The view that we can do this may be called the Strong AI view. An answer to the question of whether the *Strong AI* view is correct obviously has great implications for the field of Control. Since it is now a demonstrable fact that we can build a range of machines which are capable of independent effective action, and which can autonomously carry out a range of precisely specifiable tasks, there is no argument against what one might call a Weak AI view; such machines are possible, and what is at issue is their achievable levels of functionality.

Dreyfus (1992), [8] generates a powerful attack on the claims of Strong AI from the concepts of Heideggerian philosophy, and makes a crucial distinction between two levels of functionality which he calls know-how and know-that. Know-how can be thought of, for example, as associated with the exercise of a complex acquired skill with human levels of insight, adaptivity and sensitivity such as playing the piano with the virtuosity associated with great musicianship. In contrast, know-that can be thought of as captured in rule- based descriptions of some activity as achieved, for example, by an AI expert system. Dreyfus argues that human know-how is essentially *different* from machine know-that; he argues that, in the exercise of their everyday skills, humans do not simply exercise rule bases, however complicated - that the range and extent of any conceivable rule base will not be able to replicate human levels of functionality. In our terms, using complexity arguments, we would say simply that human know-how requires processes of vastly more complexity than those of machine know-that. From the point of view adopted here there is a necessity to distinguish between human and machine, but the difference which is acknowledged does not lie in any mystical indescribable property, or in any mysterious mind-body d-uality, but simply reflects a *pragmatic* concession to the fact that the levels of complexity required in our know-how and a machine's know-that makes it practically useful to treat human and machine as different forms of entity. This argument implicitly assumes that the complexity of the plastic component of the brain's functionality is very large. This seems entirely reasonable : the overall complexity of the whole brain is so vast that even if the plastic component is a tiny fraction of it, it will still be of many orders of magnitude in complexity greater that any achievable machine complexity.

In my view, complexity arguments deliver a conclusive and instructive resolution of this AI argument. The case against Strong AI is not that it is logically impossible, but simply that the Second Law tells us that the complexity of any system able to interact with the physical environment at a level of human functionality must approximate the complexity of the human nervous system. But an animal nervous system is of such vast complexity that it is utterly impractical to consider building a machine with anything remotely approaching it. In other words, we take the neurophysiologists' experimentally established complexity of the human brain as an empirical measure of the complexity of a system required for effective interaction with the universe to our levels of functionality. The Second Law then tells us that we *must* achieve this level of complexity to justify the Strong AI view, and this is patently not feasible. Thus the key arguments about the limitations on achievable control, and about the limitations on achievable machine intelligence both hinge on the Second Law : for effective interaction between two systems the complexity of one must mirror the complexity of the other. In thermodynamic terms we struggle against the necessity for entropy to increase; in effective interaction terms we struggle against the necessity for complexity to increase.

5.2   Knowledgeable machines and intelligent machines

Knowledge evolves from interactions: between people, society and their physical surroundings. As a result of such social and experimental inter-actions, there have been developed a variety of *formal systems* - languages and theories, and *artefacts* - models, machines, computers, books, libraries, and so on. The essence of formal systems and artefacts is that they are *objective* . Hence in addition to our individual subjective knowledge there is objective knowledge, existing independently of individuals and available in a variety of media. The philosophical implications of such objective knowledge have been discussed by Popper (1972), [23]; his views however have been challenged by Churchland (1989), [5]. The most important implication from our point of view is that objective knowledge can be *instantiated* in an information processor. Instantiated knowledge allows us to create intelligent and knowledgeable machines. There is nothing mysterious (*pace* Popper) about objective knowledge; information is just information regardless of whether it is stored in a brain, in a book or in a computer. What matters is its complexity, and how this complexity is distributed between data and process, and between human and machine.

An intelligent machine is a machine which has a capability for independent, efective action. By calling it an intelligent machine, we are emphasising the *process* aspect of its functioning. All machines fitted with a control system are, by this definition, intelligent machines. In a wide sense control téchnology is the technology of the intelligent machine.

A knowledgeable machine is one which makes instantiated knowledge interactively acessible, for example for the purposes of supporting the design function. By calling it a knowledgeable machine we are emphasising the *data* aspect of its functioning. It allows a user to access both a structured knowledge base and appropriate simulations of reality, interrogating the knowledge base, and receiving advice from it, and manipulating the simulated reality and perceiving the result. As the knowledge base of control technology expands it will become ever more pressing to make systematic use of knowledgeable machines for design purposes. A discussion of the essential features of such systems in a control engineering context has been given by MacFarlane, Grubel and Ackermann (1989), [18], and detailed descriptions of an implementation of a design support system along these lines has been given by Grubel (1992), [13].

6 The future of control

Control is essentially a *technology*: it addresses an important practical problem - that of achieving effective interaction. As such it *necessarily* draws on all possible, relevant and cognate fields from which it can gain useful inputs. The vitality of any technology derives from the number, nature and intensity of these links, and so the future development of Control is inextricably bound up with its interrelationships with these fields.

Control will continue to develop as a major technology because it addresses a fundamental, general and omnipresent need - the need for effective interaction. This generality of applicability allows it to draw on, and in return to contribute to, a wide range of other fields of activity in science and technology. Control would lose its vitality as a subject area if it ceased to interact strongly with its cognate fields. This is unlikely however as the drive to create controllers of high functionality and great flexibility will force continuing interactions with : mathematics and dynamics, information theory, computational and cognitive science, artificial intelligence, physiology and neurophysiology, and with all aspects of the technology of knowledgeable machines and intelligent machines. The following brief summaries seek to do no more than point out a number of useful references.

6.1 Mathematics and dynamics

A very comprehensive survey of the links between mathematics and control theory, with a special emphasis on dynamics, has been given in the report of a SIAM Panel on Future Directions in Control Theory (Fleming (1989), [10]). It noted that: control theory is an interdisciplinary field; the field has an excellent record for transferring knowledge from mathematics to applications; that mathematical and engineering advances have been closely intertwined at every stage of development; and that advances arise from a combination of mathematics, modelling, computation and experimentation. As discussed in detail in this SIAM report, a striking range of mathematical tools have been employed in control studies : complex function theory, linear and abstract algebra, ordinary and partial differential equations, variational methods, stochastic processes, functional analysis, operator theory, differential geometry, Lie algebras, algebraic geometry, discrete mathematics, optimization and numerical analysis. Indeed, as noted in the report, work on control problems has had, in certain areas, a significant impact on mathematical research.

Mathematics and dynamics will surely remain the key cognate areas for control technology for the forseeable future because they bring to bear on the welter of results and data from other cognate fields *the organising power of abstraction*. In common with other branches of science, such as physics, control will draw on an ever more abstract range of mathematical disciplines in order to bring the necessary level of abstraction to bear. Some areas, like complex function theory's use for problems of linear multivariable feedback system stability, and H-infinity techniques in multivariable feedback system design, will be able to find made-to-measure tools of great power. Other areas may show a need for new mathematical tools. Fields of particular difficulty - and therefore, from the mathematical point of view of great promise - are : control of nonlinear and stochastic systems; adaptive control (see here the excellent survey by Astrom (1987), [2]); control of distributed parameter systems; computational methods for control - particularly distributed control, and for the use of parallel programming; use of symbolic methods; control of chaotic systems; decentralised control; discrete-event and hybrid dynamical system control; and intelligent and learning control.

6.2 Information theory

The way in which information theory has been extended by Kolmogorov, Chaitin and Solomonoff is potentially of great importance for the development of control technology. Control is such an all-pervasive, enabling technology that it deserves a framework of the same fundamental nature as that enjoyed by Thermodynamics. Complexity theory seems to hold out such a possibility. What is particularly enticing is the prospect of developing quantitative assessments of controller complexity, and means of assessing controller effectiveness, which are comparable with the ways in which information theory is used with great effect to characterise the ideal performance of communication systems.

6.3   Computational and cognitive sciences

If the arguments of Section 5.1 are accepted then it will be necessary to make an absolute distinction betwen machine and human capability. Cognitive science will thus be a relevant cognate field, and controller/human interface design will need to draw on it for guidance in achieving effective interaction between human and machine. The development of cognitive science has been summarised by Gardner (1987), [12]. A discussion of the use of cognitive science in design is given by Winograd and Flores (1986), [31]; this is of particular interest because it uses concepts from Heideggerian philosophy (which was briefly alluded to in Section 5.1). For reasons suggested in Section 6.5 below, it is likely that there will be an increasing emphasis on parallel processing in controller implementation.

6.4   Artificial intelligence

Although it has been argued here that, because of limits to attainable levels of machine complexity, there are severe limits to the extent to which AI systems can mimic human capacities, it still remains the case that AI is a cognate field of significant importance to the future development of control technology. Many applications of controllers will involve the use of rule bases. An important use of AI in a control context will be in the support of control design function, in the development and use of knowledgeable machines.

6.5   Physiology and neurophysiology

Physiological feedback mechanisms are discussed in MacFarlane (1976), [16] where some further references are given. The range and subtlety of such mechanisms would repay careful study by control technologists.

Major advances are taking place in neurophysiology which will surely provide important pointers to possible ways of developing high functionality controllers. The development of a combination of fixed and flexible structures in nervous systems is of particular relevance. An excellent survey of current knowledge in this field is given by Churchland and Sejnowski (1992), [6], and discussions of some wider implications are in Churchland (1989), [5]. As the implications of work in this field are absorbed by control technologists, it is likely that there will be an increasing use of parallel processing and neural network approaches in the implementation of control strategies, and it may well be that control technology will see some of the most important and widespread uses of these computing techiques.

6.6   Future development

Increasingly, control technologists will have to take a broad view of relevant work in all these cognate fields. As the amount of knowledge of the subject inexorably increases, it will be necessary to use knowledgeable machines to create intelligent machines, in a sort of evolutionary bootstrap. It will become increasingly important to devise general, and easily communicable, approaches to the fundamental aspects of Control of the "thermodynamic type" attempted in a sketchy form here, which can be used to give a synoptic overview of the subject, to convey its essential features to non-specialists and to facilitate interaction with cognate fields. The difficulties we face are all describable in terms of complexity : its creation, management and deployment into effective interaction.

Stand in any big library, say before the stacks of control-related journals. You are surrounded by information, but it is not, to put it mildly, in an immediately useful form. You are surrounded by data but what you need is process. The knowledge with which the library is filled is very difficult to access, and even more difficult to turn into either illuminating insight or to practical use. This - the difficulty of harnessing such forms of objective knowledge to useful effect - poses a very real and a very formidable problem. The biggest single challenge facing Control, or indeed any other modern technology, is how to:

- organise and make interactively accessible the large and growing amount of objective formal knowledge which is available

- develop a coherent, communicable, synoptic overall view of the subject which can be used to illuminate and organise this knowledge

- make the knowledge useable by practitioners - to provide them with a powerful toolkit

- relate the knowledge to reality, that is to experimental and practical investigations.

Control has largely adhered to the rationalist tradition rather than the empirical, that is to say it has been more concerned with representational, formalist approaches (Winograd and Flores (1986), [31]) than with measurement or experimental approaches. The overwhelming preponderence of theoretical papers over experimental papers in the major jounals of the subject is striking. For a healthy future development both approaches are needed. Control has clung too long to formalism, and its long term health now requires an injection of pragmatism and empiricism.

The drive to create, and an increasing use of, knowledgeable machines should lead to more coherence in theoretical approaches. Better ways of displaying and using structured knowledge should lead to a less fragmented use of the huge range of techniques and knowledge which is now available, and this will, hopefully, be accompanied by a revival of empirical and experimental investigations. Thus a use of knowledgeable machines will make applicable theories more widely available and useable, and so break down the gap between theory and practice. The width of this much discussed gap is an experimental demonstration of the difficulty of manageing complex knowledge structures. As such it will remain the best quantitative measure of our progress.

In this overview, little more has been done than to point at the problems which we face in moving Control forward as a technology, and to sketch out how information-theoretic concepts, such as complexity, might be deployed. However it is claimed that any significant resolutions of these problems will share certain key features : they will use information-theoretic concepts; they will harness an emerging technology of knowledgeable machines to help manage the complexity of the data generated by theory; they will draw widely on, and interact with, a widening range of cognate fields; and they increasingly will adopt a pragmatic and empirical approach.

References

[1] Ashby, W.R., "Requisite variety and its implications for the control of complex systems," *Cybernetica*,1, 1953, pp. 83-99.

[2] Aström, K.J., "Adaptive feedback control,"*IEEE Proceedings*,75, (2), 1987, pp.185 -217.

[3] Chaitin, G.J., *Algorithmic information theory*, Cambridge University Press, Cambridge UK, 1987.

[4] Chaitin, G.J., *Information, randomness and incompleteness* (Second edition), World Scientific, Singapore, 1990.

[5] Churchland, P.M., *Neurophilosophy*, MIT Press, Cambridge MA, 1989.

[6] Churchland, P.S., and Sejnowski, T.J., *The computational brain*, MIT Press, Cambridge MA, 1992.

[7] Cover, T.M., and Thomas, J.A., *Elements of information theory*, Wiley, New York, 1991.

[8] Dreyfus, H.L., *What computers still can't do*, MIT Press, Cambridge MA, 1992.

[9] Flanagan, O. *Consciousness reconsidered*, MIT Press, Cambridge MA, 1992.

[10] Fleming, W.H. (Chair), *Future directions in control theory : a mathematical perspective*, Report of the Panel on Future Directions in Control Theory, Society for Industrial and Applied Mathematics, Philadelphia, 1988.

[11] Gitt, W., "Information : the third fundamental quantity," *Siemens Review*, 1989, 56, pp. 36-41.

[12] Gardner, H. *The mind's new science*, Basic Books, 1987.

[13] Grubel, G., Joos, H-D., Finsterwalder, R., and Otter, M., "The ANDECS design environment for control engineering," German Aerospace Research Establishment Report TR R79-92, July 1992.

[14] Kolmogorov, A.N., "Logical basis for information theory and probability theory," *IEEE Transactions on Information Theory*, IT-16, 1968, pp.662 - 664.

[15] Lewis, C.I., "The pragmatic element in knowledge," *University of California Publications in Philosophy*, 1926, 6 , (3). Reproduced in Moser and Vandernat, *op cit.*, pp. 201-211.

[16] Macfarlane, A.G.J., *Feedback, Measurement and Control*, 9, 1976, pp. 449 - 462.

[17] Macfarlane, A.G.J., "The development of frequency-response methods in automatic control," *IEEE Transactions on Automatic Control*, AC-24, (2), 1979, pp. 250 - 265.

[18] Macfarlane, A.G.J., Grubel, G., and Ackermann, J., "Future design environments for control engineering," *Automatica* ,1989, 25 , (2), pp. 165-176.

[19] Maciejowski, J.' *The modelling of systems with small observation sets*, Lecture Notes in Control and Information Sciences, Volume 10, Springer Verlag, Berlin, 1978.

[20] Moser, P.K., and Vander Nat, A., *Human knowledge*, Oxford University Press, New York, 1987.

[21] Peitgen, H-O., Jurgens, H., and Saupe, D., *Chaos and Fractals*, Springer Verlag, New York, 1992.

[22] Pippard, A.B., *Elements of classical thermodynamics*, Cambridge University Press, 1957.

[23] Popper, K.R., *Objective knowledge*, Oxford University Press, 1972.

[24] Resnikov, H.L., *The illusion of reality*, Springer Verlag, New York, 1989.

[25] Searle, J., *Minds, brains and science*, British Broadcasting Corporation, London, 1984.

[26] Shanker, S.G. (Editor), *Godel's Theorem in focus*, Croom Helm, London, 1988.

[27] Shannon, C.E., and Weaver, W., *The mathematical theory of communication*, University of Illinois Press, Urbana, 1949.

[28] Simon, H.A., *The sciences of the artificial*, second edition, MIT Press, Cambridge MA, 1981.

[29] Solomonoff, R.J., "A formal theory of inductive inference," *Information and Control*, 7, (1), 1964, pp. 1 - 22 and pp. 224 - 254.

[30] Steward, I., "The ultimate in undecidability," *Nature*, 332, (6160), 1988, pp. 115 - 116.

[31] Winograd, T., and Flores, F., *Understanding computers and cognition*, Addison-Wesley, Reading MA, 1986.

2. Hybrid Models for Motion Control Systems

R.W. Brockett *

Abstract

This paper presents a mathematical framework for modeling motion control systems and other families of computer controlled devices that are guided by symbol strings as well as real valued functions. It discusses a particular method of organizing the lower level structure of such systems and argues that this method is widely applicable. Because the models used here capture important aspects of control and computer engineering relevant to the real-time performance of symbol/signal systems, they can be used to explore questions involving the joint optimization of instruction set and speed of execution. Finally, some comparisons are made between engineering and biological motion control systems, suggesting that the basic ideas advanced here are consistent with some aspects of motor control in animals.

1   Introduction

Because motion control embraces large and important classes of examples having industrial, medical and scientific importance and because it has a number of special features, it is worthy of study as a subject in its own right. There are a range of challenging questions that serve to distinguish it from other applications involving control but perhaps the most important center on the specification of the instruction sets and languages used to describe motion and on the specification of the capabilities that should be required of the devices that interpret the language. For the purposes of this paper, the

*Division of Applied Sciences, Pierce Hall, Harvard University, Cambridge MA 02138, USA. This work was supported in part by the National Science Foundation under Engineering Research Center Program, NSF D CRD-8803012, the US Army Research Office under grant DAAL03-92-G-0164 (Center for Intelligent Control Systems), DARPA and the Air Force under contract F49620-92-J-0466, and by the Office of Naval Research under contract N00014-92-J-1887, while the author was a visiting Professor at the Charles Stark Draper Laboratories, Cambridge, Mass.

subject of motion control is to be thought of as pertaining to engineering systems such as robots, numerically controlled machine tools and pen plotters and also to the neuroanatomy and musculature used by living systems to generate movement. It is to be understood as including both the geometry of the trajectory and the concomitant force/torque program. In most cases of interest, purposeful motion involves some interaction with physical objects having uncertain positions and characteristics. For this reason robustness is especially important. Truly, one encounters an astounding range of scientific and engineering problems in attempting to build models of, or develop designs for, systems of this type. The disciplines of control theory, computer science, mechanics, and psychology, among others, are involved. However, this diversity not withstanding, there are common patterns in the way motion control devices are designed and used. In this paper we discuss the structure of these systems and propose a class of mathematical models that capture important distinguishing features of these systems.

Conventional motion control systems usually have two types of inputs. There are the strings of symbols, coming to the system by means of a digital "communication port," and there are the forces and displacements that act on the system through the "effector port," i.e. at the interface between the motion generation device and its environment. The digital port is connected to a digital communication network and the effector port can be thought of as being connected to some type of mechanical network. Even though it is often awkward to consider them, the reactive forces on the effector are inputs and they affect the motion. For example, it is an oversimplification to imagine that the goal of a robot is simply to traverse a prescribed path in position and orientation space. Even error-free devices of this type can be programmed to successfully carry out only a rather restricted range of tasks because the position of the objects being manipulated is always uncertain. Effective motion control involves controlling not just the position of the various parts of a mechanism but also the compliance associated with that position. One must not only specify a nominal path but must also specify the incremental force-displacement relationship along the nominal path. Whitney [1] gives a carefully reasoned discussion of the need for, and problems associated with, compliance control.

In order to implement compliance control, forces and displacements must be measured and actuation provided in accordance with the desired compliance. However, there are many different ways in which one can reduce the effects of uncertainty by means of feedback. To the extent that the possibility of adjusting compliance makes a robot easier to program than it would be without this feature, compliance control supports the more general notion that the inclusion of adjustable feedback gains is useful in specifying effective motions. The effectiveness of compliance control can even be taken as an argument for the inclusion of feedback gain adjustment in the repertoire of artificially intelligent machines. This latter point is an important part of the philosophical position taken by Brooks [2,3] who argues that when building intelligent motion generation devices one should focus on "behaviors," organizing the motion program in terms of rules that couple observations to actuation. One way to turn this suggestion into a mathematical statement is to identify it with the idea that gain control should be supported in the formal scheme used to describe motion. Our papers [4,5] describe a formal method of specifying motion programs in terms of a series of feedback algorithms. However, earlier work stops short of providing a mathematical model describing how the dynamics of the symbol manipulation process interacts with the more overtly physical aspects of the motion generation process. Without such a model it is impossible to say with certainty how the so-called behavioral models behave.

2 Modeling Hybrid Systems

Our first step is to present a class of mathematical models in which symbol manipulation processes interact strongly with physical processes. We will show how to model situations in which, for example, there are communication ports whose buffers hold bits that have been received but not yet read, with the rules governing the process by which the buffers are filled and emptied involving trajectories of objects governed by Newton's laws. In a robotics context we do not necessarily want an electrical impedance model for the communication port but do need a mechanical impedance model for the end effector. In this section we introduce a series of hybrid models capturing aspects of such systems with different levels of detail.

In the present context the basic sampled data approach to computer control based on difference equations models (see e.g. [6]) is unrealistic; something closer to the scope of [7] is required. If we think of a robot as suggested above, i.e. as a device with one or more communications ports connected to a digital communication network and one or more end effectors that interact with the physical world, an appropriate model will necessarily reflect aspects of the behavior of finite automata as well as aspects of the behavior of sampled data systems. Only with this diversity can we can hope to capture the essentials of a machine that is capable of converting a symbolic description of a task into appropriate movements and to return a report relating to its accomplishments. Standard sampled data theory is predicated on a fixed sampling rate, basing the analysis on difference equations that relate the value of the state vector at successive values of the sampling times. There are multi-rate versions in which some variables are sampled at a frequency that is a multiple of a basic sampling frequency, but these are still tied to the idea of synchronous behavior. Such models are not very useful when trying to represent situations in which the initiation of events, such as the reading of a symbol or the changing of a feedback gain, is not determined by the passage of time but rather by the evolution of the state of the system. In such cases more flexibility is required. In this paper the additional flexibility is achieved through the introduction of one or more monotone increasing *triggering signals*. We denote these by the letter p; the interpretation is that certain events are blocked until p (or some component of it, if p is vector valued) reaches the next greatest integer value. We will argue that some of the most significant limitations on modeling with automata or sampled data systems can be circumvented by incorporating one of more triggering signals together with the corresponding *rate equations* describing the relationship between the evolution of time and the growth of the triggering signal.

If p is a scalar, as we shall assume until Section 4, we follow the common convention of writing $\lfloor p \rfloor$ to denote the largest integer less than or equal to p. When dealing with a function defined on the integers, we will not indicate its value at $\lfloor p \rfloor$ by $v(\lfloor p \rfloor)$, $z(\lfloor p \rfloor)$, etc. but instead, prefer to write $v\lfloor p \rfloor$ or $z\lfloor p \rfloor$. Consider the state x, the input function u and the measurement y, all vector valued functions of time, taking on values in open subsets of Cartesian spaces X, U and Y, respectively. Let p be a scalar valued function of time. The variables v and w, are string valued or, what amounts to the same thing, functions defined on the nonnegative integers, taking on values in the sets V and W, respectively. By a *type A hybrid system* we understand an input-output model of the form

$$x(k\delta + \delta) = a(k(v\lfloor p \rfloor, y(k\delta)), u(k\delta), x(k\delta)) \; ; \qquad y(k\delta) = c_e(x(k\delta), v\lfloor p \rfloor)$$
$$p(k\delta + \delta) = r(v(\lfloor p \rfloor, u(k\delta), y(k\delta)) \; ; \qquad w(\lfloor p \rfloor) = h(v(\lfloor p \rfloor, y(k_p \delta))$$

subject to the assumption that $0 < r(v(\lfloor p \rfloor, y(k\delta)) < 1$ for all values of $v(\lfloor p \rfloor)$ and $y(k\delta)$ and the understanding that k_p denotes the value of k at which p most recently reached or passed through an integer value. These are input-output systems that accept, as inputs, the symbol string v and the input function u. The variable y represents the values being measured by the sensors and is considered to be available for monitoring and control. The subscript e on c serves to call attention to the fact that the values read by the sensors depend on the environment in which the system is operating. We do not allow k or r to depend on x explicitly; the evolution of the trajectory is influenced by the state through the measurements y which enter via the v dependent feedback term $k(v\lfloor p \rfloor, y(k\delta))$. A large part of the novelty in this definition arises from the role of the rate equation p whose evolution triggers the reading of the string v. The upper bound on the size of r insures that the symbols in the string v will be read without skipping. We will refer to the set V as being the *instruction set* and call

W the *response set.*

Given initial values $x(0)$ and $p(0)$, and given inputs u and v, the difference equations that describe the type A hybrid system clearly generate a unique trajectory. The x component of the solution is casually related to u in that the first j values of x depend only on the first j values of u. Similarly, the output w is casually related to the input v in the sense that the first k elements in the w string depend only on the first k elements of the v string. However, the exact length of the substring of v that is needed to specify x on a given interval will not be known in advance because the rate at which instructions are read depend on the trajectory. Likewise, the exact number of samples of u required to determine the first k elements of w will not be known in advance. We will refer pattern as *bicausal behavior.* It is a consequence of the use of the trigger function.

Obviously the behavior of models in this class reflects aspects of automata theory and aspects of sampled data system theory. The models that arise in sampled data theory are obtained from continuous time dynamical system by means of a sampling process and the discretization of the time. The increments of k are identified with the passage of a fixed amount of time. On the other hand, in considering questions of interest in automata theory there is no reason to identify the increments in k with any thing other than the appearance of a new symbol in the string $v(0)$ followed by $v(1)$, followed by ... . If one works exclusively in the domain of dynamical systems, or exclusively in the domain of automata, this distinction is of no consequence. It is only when the automaton is part of a model whose evolution is coupled to the flow of time that this becomes an issue.

It is not uncommon to study sampled data control systems using a more detailed model consisting of a differential equation coupled with a difference equation as exemplified by

$$\dot{x}(t) = a(x(t), z(\lfloor t \rfloor)u(t))$$

$$z(\lfloor t \rfloor + 1) = f(x(\lfloor t \rfloor), z(\lfloor t \rfloor), v(\lfloor t \rfloor)).$$

Likewise, in some cases it will be preferable to use a continuous time version of the hybrid model just discussed. By a *type B hybrid system* we understand a model in which u, x, y take on values as above with their evolution being governed by the set

$$\dot{x}(t) = a(u(t), k(v(\lfloor p \rfloor, y(t)), x(t))) ; \qquad y(t) = c_e(v(\lfloor p \rfloor, x(t)))$$
$$\dot{p}(t) = r(v(\lfloor p \rfloor, u(t), y(t))) ; \qquad w(\lfloor p \rfloor) = h(v(\lfloor p \rfloor, y(\lfloor t_p \rfloor)))\,.$$

We require that r be nonnegative for all arguments but need not impose an upper bound on it; t_p denotes the value of t at which p most recently became an integer. Given $x(0)$ and $p(0)$ there will exist a unique solution

$(x(\ ), p(\ ))$ on an interval $[0, T]$ provided that we make appropriate assumptions on the right-hand side of the differential equations. For example, if for each instruction v_0 in V the function

$$f(x, u) = a(u, k(v_0, c_e(v_0, x), x))$$

satisfies a Lipschitz condition with respect to x, if it is continuous with respect to u, and if u has at most a finite number of discontinuities on any interval of finite length, we see that on any interval in which p does not take on an integral value, there exists exactly one everywhere continuous and piecewise differential pair $(x(\ ), p(\ ))$ satisfying the equation for all values of t except at the points of discontinuity of u. On any finite interval of time p passes through only a finite number of integers and so changes in $v(\lfloor p \rfloor)$ can contribute only finitely many points of discontinuity to the derivatives of x and p. At the points of discontinuity of the derivatives of x and p the variables themselves are continuous. Their derivatives satisfy the differential equation except at these isolated points. Thus under these conditions there exists a unique (x, p) with x and p continuous and differentiable almost everywhere satisfying the equation. (See Chapter 2 of [8] for a background discussion on the existence and uniqueness of solutions of ordinary differential equations with weak continuity hypotheses.)

Example 1: Consider the following simplified model for the engine-transmission system of an automobile with manual transmission. We view the system as having throttle position and gear shift lever position as inputs and engine rpm and ground speed as outputs. The throttle position and the gear shift position can be thought of as the u and v, respectively, of a hybrid system. The gear shift is necessary because the torque-speed relationship of an engine is such that little power can be developed at very low, or very high, engine rpm. To circumvent this difficulty the system is equipped with an accelerator that controls engine speed and a finite set of gear configurations determining the gear ratio between the engine and the drive wheels. The function of the variable gear ratio transmission is to let the driver get more power from the motor at a given ground speed by making it possible to operate on an advantageous portion of the speed-torque curve. A crude, but illustrative (completely nondimensionalized) hybrid model is provided by

$$\dot{x}_1(t) = x_2(t)$$
$$\dot{x}_2(t) = (-a(x_2(t)) + u(t))/(1 + v\lfloor p \rfloor)$$
$$\dot{p}(t) = k \ .$$

This can be thought of as a system that satisfies Newton's second law of motion in rotational form. The $a(\ )$ term models the decrease in the ability of the system to produce torque at high rpm. The term $(1 + v\lfloor p \rfloor)$ represents the fact that the effective inertial changes when the gear ratio changes. Because we do not model the dynamics of the clutch actuation process the rate equation is not an important component of the model.

An important practical question that arise in treating hybrid systems of type A or B is the question of expressiveness. Can one find input strings using the instruction set V that will generate a desired motion with adequate robustness. A regulation problem might involve finding a string of instructions that will maintain a desirable relationship between u and y without requiring more information than is available through c_e. A motion control problem might depend on finding a sequence of instructions $v(1), v(2), ..., v(n)$ that causes a mobile robot to negotiate a maze or an unmanned airplane to reach a given altitude with sufficient robustness.

In the definition of type A and type B hybrid systems the symbol string v is not subjected to any intersymbol constraints. We think of the set V as being the instruction set used to guide the trajectory $x(\ )$. Because the elements of v are used directly in the evolution equation, the achievable resolution in the space of x-trajectories is directly related to the cardinality of the instruction set V. Of course practical communication networks transmit with a small symbol sets, coding large symbol sets in terms of strings of symbols from a set of smaller cardinality. Decoding the substrings takes time and uses computational resources. In order to faithfully model this aspect of reality we need to introduce a somewhat more elaborate framework. As above we use the notation $\lfloor p \rfloor$ for the greatest integer less than or equal to p and, in addition, we will use $\lceil p \rceil$ to denote the smallest integer greater than p. Continuing the convention introduced previously, we write $z\lceil p \rceil$, instead of $z(\lceil p \rceil)$, etc.

Let x, u and y be functions of time taking on values in open subsets of Cartesian spaces X, U and Y, respectively. Let p a function of time taking on real values. Let z, v and w be functions defined on the integers, taking on values in the sets Z, V and W, respectively. By a *type C hybrid system* we understand an input-output model of the form

$$x(k\delta + \delta) = a(u(k\delta), x(k\delta), z\lfloor p \rfloor) ; \qquad y(k\delta) = c_e(x(k\delta), z\lfloor p \rfloor)$$
$$p(k\delta + \delta) = r(u(k\delta), y(k\delta), z\lfloor p \rfloor)$$
$$z\lceil p \rceil = f(u(k_p\delta), v\lfloor p \rfloor, y(k_p\delta), z\lfloor p \rfloor) ; \quad w(k) = h(y(k_p\delta), z\lfloor p \rfloor).$$

In this case we require $0 < r(y(k\delta), z\lfloor p \rfloor) < 1$ for all values of $u(k\delta)$, $y(k\delta)$, and $z\lfloor p \rfloor$. Given the initial values of $x(0)$, $p(0)$, and $z\lfloor p(0) \rfloor$, the model defines a unique response corresponding to each string v and function u. It is bicausal in the sense that this term was used above. The output w represents the information made available on the communications network.

Type C hybrid systems are more general than type A hybrid systems by virtue of the fact that they can interpret symbolic inputs satisfying nontrivial syntactical rules, i.e., strings coming from a language, not just a free monoids.

We complete this family of definitions with the definition of a *type D hybrid system*. By this we understand a model of the form

$$\dot{x}(t) = a(u(t), x(t), z\lfloor p \rfloor) ; \qquad y(t) = c_e(x(t), z\lfloor p \rfloor)$$
$$\dot{p}(t) = r(u(t), y(t), z\lfloor p \rfloor)$$
$$z\lceil p \rceil = f(v\lfloor p \rfloor, y(t_p), z\lfloor p \rfloor) ; \qquad w(k) = h(y(t_p), z\lfloor p \rfloor)$$

with r being nonnegative. The meaning of the variables are with minor differences, as above. The first equation, specifying the evolution of x, describes those aspects of the system for which differential equations are the appropriate basis for modeling. The variable p is to be thought of as modeling the pace of interaction between the real-time dynamics represented by x and the flow of information represented by changes in z. The last equation, specifying the way in which z changes, describes the part of the system whose evolution is triggered by events, i.e. p advancing through integral values, and represents the symbolic processing done by the system.

Clearly the type C and D models are more elaborate than the type A and D models. For example, these models allow one to estimate the length of time required to collect and process the string data and if this is an important design parameter for determining the channel capacity and buffer sizes for the symbolic input, then this additional complexity is necessary. The more elaborate models also allow one to predict the transient response of a feedback system with an automaton in the feedback loop. We will give an example below.

Example 2: Consider the following pair of first-order scalar differential equations.

$$\dot{x}_1(t) = x_2(t)$$
$$\dot{x}_2(t) = -ax_2(t) + \sum u_i(t)\sin(bx_1(t) - 2\pi i/q) .$$

If b is a positive integer and q is a positive integer larger than two then these equations can be thought of as modeling the angular rotation of a stepper motor having bq steps per revolution. The variable $x_1(t)$ represents the rotor angle in radians and $x_2(t)$ represents the angular velocity. The normal operation would be to pulse the inputs $u_i(t)$ in order. That is, one sets $u_1(t)$ equal to one for T seconds with all the other $u_i(t)$ equal to zero, then sets $u_2(t)$ equal to one for T seconds with all others off, ..., sets $u_q(t)$ equal to one for T seconds with all others off, and then repeats the cycle. T is chosen to be large enough so that each pulse advances $x_1(t)$ by $1/bq$ radians. Now consider the hybrid model

$$\dot{x}_1(t) = x_2(t)$$
$$\dot{x}_2(t) = -ax_2(t) + \sum u_i(z\lfloor p \rfloor)\sin(bx_1(t) - 2\pi i/q)$$
$$\dot{p}(t) = v\lfloor p \rfloor$$
$$z\lceil p \rceil = z\lfloor p \rfloor + 1$$

with z being an element of the set $\{0, 1, ..., q-1\}$ and the addition $z\lceil p \rceil = z\lfloor p \rfloor + 1$ being interpreted modulo q. If we define $u_i(z\lfloor p \rfloor)$ to be one if $z\lfloor p \rfloor$ equals i, and zero otherwise, and if $v\lfloor p \rfloor$ is a positive real number, then $p(t)$ will pass through an integer value every $1/v\lfloor p \rfloor$ units of time and x will advance through one complete revolution in $bq/v\lfloor p \rfloor$ units of time. Of course if $v\lfloor p \rfloor$ is too large the p equation will advance so rapidly that $x_1(t)$ cannot keep up and our qualitative description will be inaccurate.

Example 3: One of the standard instructions in a multivariable motion control environment is the instruction commanding each of several variables to move in such a way as to have the overall motion proceed along a straight line, moving at some rate f from an initial location to a new location characterized by coordinates $(a, b, ..., c)$. It often happens that the rate is taken to be the maximum of the absolute values of the individual time derivatives of the coordinates. Adopting this definition of the rate, we denote this type of instruction by moveto$[(a, b, ..., c), f]$. If the device uses stepper motors without feedback to generate the motion, then such a moveto instruction can be implemented by an enhancement of the system we discussed in Example 2. Consider the scalar case. We add to the four equations introduced above a fifth which halts the generation of pulses after a certain number of pulses have been generated. The instruction move$[(a), f]$ is then executed by

$$
\begin{aligned}
\dot{x}_1(t) &= x_2(t) \\
\dot{x}_2(t) &= -ax_2(t) + \sum u_0(z_2\lfloor p \rfloor)u_i(z_1\lfloor p \rfloor)\sin 2\pi b(x - i/p) \\
\dot{p}(t) &= v\lfloor p \rfloor \\
z_1\lceil p \rceil &= z_1\lfloor p \rfloor + 1 \ (\text{modulo } q) \\
z_2\lceil p \rceil &= z_2\lfloor p \rfloor + 1
\end{aligned}
$$

with $z_2\lfloor p \rfloor$ taking on integer values starting at zero and $u_0(z_2\lfloor p \rfloor)$ being one for all $z_2\lfloor p \rfloor < a$ and zero thereafter. Notice that the time required to execute an instruction is dependent upon the instruction and that the time interval between successive interrogations of the instruction stream is, therefore, variable. Commercially available stepper motor drivers and motion control subsystems usually provide additional functionality such as provisions for ramping up the speed and ramping down the speed, etc. Systems of this type are clearly hybrid in nature.

We will argue that the type of hybrid models just introduced are useful for posing and answering significant questions about the behavior of a variety of computer controlled devices and that the type of generality they represent is natural and appropriate in that context. The two examples just cited indicate how aspects of the performance of stepper motors and their associated electronics fit this model. Systems with z variables present can have stability problems because of the highly nonlinear feedback that couples the $x - p$ system and the z system. We will take up some of these questions below.

3   Feedback Systems

One of the reasons for introducing the definitions given above is to put in place a vocabulary that is rich enough to allow one to distinguish between the types of computer control in which the system operates as a basic sampled data, synchronous mode system and the more general type of computer control in which the basic synchronous mode compensation algorithms are themselves under computer control. Systems of this latter type are common as is discussed in reference [9]. Gain scheduled systems are used in flight control, self-tuning systems have been used in process control, and active compliance control is under investigation in robotics. In all these applications the hybrid nature of the system arises in the following way. There is a fixed "plant" such as the masses, lengths of linkages, motor constants, etc. associated with a motion generation device, a set of sensors that generate measurements, and a set of inputs that give one some control over the system. Usually these parts will not change in response to symbolic input. What can be made to change are any digitally realized feedback algorithms involving gains, coupling constants, etc. Changing these aspects of the feedback control algorithm only involves changing the contents of certain registers.

We model the uncompensated, fixed, part of the system by equations of the form

$$x(k\delta + \delta) = a(x(k\delta), u_1(k\delta), u_2(k\delta)) \; ; \;\; y(k\delta) = c_e(x(k\delta), v\lfloor p\rfloor)$$

with the understanding that the inputs labeled u_1 are available for feedback control whereas those labeled u_2 are exogenous. Upon replacing u_1 by the feedback term $K(v\lfloor p\rfloor, y(k\delta))$ we get

$$x(k\delta + \delta) = a(x(k\delta), K(v\lfloor p\rfloor, y(k\delta)), u_2(k\delta)) ; \;\; y(k\delta) = c_e(x(k\delta), v\lfloor p\rfloor)$$
$$p(k\delta + \delta) = r(u(x\delta), v\lfloor p\rfloor, y(k\delta)) ; \qquad\qquad w\lfloor p\rfloor = h(v(\lfloor p\rfloor, y(k_p\delta)) .$$

Alternatively, in those cases for which a differential equation model for the x-system is more appropriate, the model takes the form

$$\dot{x}(t) = a(x(t), K(v\lfloor p\rfloor, y(t)), u_2(t)) ; \;\;\; y(t) = c_e(x(t), v\lfloor p\rfloor)$$
$$\dot{p}(t) = r(u(t), v\lfloor p\rfloor, y(t)) .$$

Hybrid systems of this, more specific, form are of particular interest. They serve to focus attention on the fact that the feedback compensation loops may be designed to operate at high speed, even though the trigger equation may only permit infrequent changes to the compensation algorithm itself. This is of critical importance when one is attempting to capture "reflexive" response.

Motion controllers available from vendors in the form of single printed circuit boards compatible with some standard family of computer buses, often give the user the possibility to implement feedback control laws having integral control action. From the present point of view this is equivalent to creating a situation in which the controller itself is a dynamical system and suggests that these additional variables be lumped in to the z part of a type D hybrid system. However, there is a second possibility that is usually more useful and that is to model these additional variables as being real numbers and put them in the x part of the model. Saying, in effect, that the x vector can be split into two parts, $x = (x_1, x_2)$ with the second part being associated with variables that reside in memory in the controller. In so doing we arrive at a model of the form

$$\dot{x}_1(t) = a(x(t), K_1(v\lfloor p \rfloor, y(t)), u_2(t)) ; \quad y(t) = c_e(x(t), v\lfloor p \rfloor)$$
$$\dot{x}_2(t) = K_2(v\lfloor p \rfloor, y(t))$$

with the understanding that the components of the $y(t)$ vector include all of the components of x_2. The derivative of x_2 is fixed by K_2, which is, itself, selected by the feedback rule. In this case the components of x_2 can be regarded as being entirely known and at the disposal of the feedback control law. We will call such components of x *uncommitted variables*. Of course the designers of control systems make frequent use of PID control and other forms of dynamic compensation. In order to implement such schemes it is necessary to build in a number of uncommitted variables so that the results of computations based on them are available to be put to use by the instruction set. Uncommitted variables can also be used to generate temporal patterns such as might facilitate the design of an algorithm for walking or to simulate the adjoint equations if they are needed for trajectory optimization.

This situation may be specialized further, focusing on the case in which the uncontrolled dynamics of the system are linear and the compensation is constrained to be linear as well. A continuous time version of this is

$$\dot{x}(t) = Ax(t) + BK(v\lfloor p \rfloor)y(t) + B(v\lfloor p \rfloor)u ; \quad y(t) = C(v\lfloor p \rfloor)(x(t) - e)$$
$$\dot{p} = r(v\lfloor p \rfloor, y(t)) .$$

The matrices A, B, $C(\ )$ and $K(\ )$ can be thought of as defining various linear systems. The system switches from one linear system to another on the basis of the input v and the measurement y.

4 Vector Triggers and Buffering

Under suitable circumstances hybrid systems can be interconnected and sometimes a complex system is most easily understood by viewing it as the interconnection of simpler hybrid modules. Implicit in this point of view are many possibilities for analysis and design. If interconnection involves the symbol string ports of two subsystems, the alphabets of the two subsystems must agree; if it involves the physical ports of two subsystems, the physical descriptions of the subsystems must agree. The second type of interconnection is more straightforward than the first because physical time is universal whereas the triggering signals for two distinct systems will not necessarily be synchronized. In either case, the overall system that results from the interconnection of two or more hybrid systems can be expected to have more than one trigger signal. Some modes of operation will require buffering or some other special provisions to deal with problems associated with this lack of synchronization.

If p is a k-dimensional vector with components $p_1, p_2, ..., p_k$ then we use $\lfloor p \rfloor$ to denote the vector whose ith component is $\lfloor p_i \rfloor$. Likewise, by $\lceil p \rceil$ we mean the vector whose ith component is $\lceil p_i \rceil$. Suppose that p is a k-vector and that v, w, and z can be thought of as a k-tuples, $v = (v_1, v_2, ..., v_k)$, $w = (w_1, w_2, ..., w_k)$, $z = (z_1, z_2, ..., z_k)$. Let t_{pi} denote the most recent value of t at which the ith component of p passed through an integral value. With this understanding, a model of the form

$$\dot{x}(t) = a(u(t), x(t), z\lfloor p \rfloor); \qquad y(t) = c_e(x(t), z\lfloor p \rfloor)$$
$$\dot{p}_i(t) = r_i(u(t), y(t), z_i\lfloor p_i \rfloor) \qquad i = 1, 2, ..., k$$
$$z_i\lceil p_i \rceil = f_i(v_i\lfloor p_i \rfloor, y(t_{pi}), z\lfloor p \rfloor); \quad w_i\lceil p_i \rceil = h_i(y(t_{pi}), z\lfloor p \rfloor); \; i = 1, 2, ..., k$$

with each of the r_i being nonnegative, is unambiguous in the sense that an infinite input string v, an input function u defined on $[0, \infty)$, and initial conditions for x, p, and z determine a unique trajectory. Each of the "components" of z advance through integer values of their arguments. We will call such systems *hybrid systems with vector triggering* and will adopt the notation introduced earlier for the scalar trigger case,

$$\dot{x}(t) = a(u(t), x(t), z\lfloor p \rfloor); \qquad y(t) = c_e(x(t), z\lfloor p \rfloor)$$
$$\dot{p}(t) = r(y(t), z\lfloor p \rfloor)$$
$$z\lceil p \rceil = f(v\lfloor p \rfloor, y(t_p), z\lfloor p \rfloor) \qquad w(k) = h(y(t_p), z\lfloor p \rfloor).$$

Clearly the type of causality associated with such systems, while more complicated than the bicausal situation discussed previously, parallels that idea.

Buffers. Consider two distinct hybrid systems, with scalar riggers, p_1 and p_2, respectively. Suppose that the symbolic output of system one is to be used as the symbolic input of system two. Because the triggers are not necessarily synchronized, some provision for temporary storage of symbols must be provided if we are to avoid the possibility that some instructions generated by the first system will never be acted on because a subsequent instruction was generated before the earlier one was read. If system one produces symbols faster than system two can read them then any buffer of finite capacity will eventually run out of room. However, even if in some average sense the trigger rate of system two is greater than that of system one, this temporary storage must have sufficient capacity so as to allow for the fluctuations in rate of the triggers of the systems being interconnected. A simple model is provided by the system with two analog inputs, one string valued input, and one string valued output, taking the form

$$\dot{p}_1(t) = r_1(u_1(t))$$
$$\dot{p}_2(t) = r_2(u_2(t))$$
$$z\lceil p \rceil = f(z\lfloor p \rfloor, v\lfloor p_1 \rfloor) \; ; \;\; w\lfloor p_2 \rfloor = h(z\lfloor p \rfloor) \, .$$

If the function f is such that when p_1 passes through an integer value the value of z changes in such away as to make it possible to determine $v\lfloor p_1 \rfloor$, $v\lfloor p_1 - 1 \rfloor$, ..., $v\lfloor p_1 - k + 1 \rfloor$ from a knowledge of $z\lceil p \rceil$, that is $z\lceil p \rceil$ serves to store the k most recent values of v, and this system can act as a temporary storage system. If, in addition, every time p_2 passes through an integer value the output function w assumes the value of oldest previously unread value of v, then this serves the function of a first-in first-out buffer. More precisely, it is a *FIFO buffer of capacity k*. Notice that the rate functions for p_1 and p_2 are governed by a continuous time signals that are exogenous inputs to the system. Of course the interconnection of systems may require two buffers if bidirectional communication is needed.

One specific use for vector valued trigger functions is to provide a means of rapidly signaling the need to interrupt the normal behavior. If p has two or more components, several may be used for orchestrating the normal input of symbolic data with one being reserved to signal a special condition in which some normal actions are halted. Clearly vector valued triggers are useful for solving practical problems but they also raise the possibility of modeling new phenomena such as deadlock, in which all symbol strings are suppressed.

5 Hybrid Systems as Dynamical Systems

The study of hybrid systems gives rise to a whole range of new questions. The most interesting of these seem to be of two types. One group relates to modeling existing engineering devices, such as analog to digital converters, and systems, such as insect robots, whereas the other relates to a range of mathematical questions having to do with the stability and predictability of the trajectories they define. We have given some examples of the engineering aspects of these models and will discuss some biological applications in a later section. This section is devoted to the discussion of some of the dynamical systems aspects.

Multimodal Dynamical Systems. Given a type B hybrid system, one can postulate that its input string v is generated by a finite automaton driven by its output string w. Let us describe this automaton by

$$z_1(k+1) = f_1(z(k), w(k)) \; ; \; v(k) = h_1(z_1(k)) \; .$$

We do not explicitly display a trigger equation for this automaton preferring, at this point, to think of it as being classical in the sense that it responds without delay upon the appearance of a new input symbol in the w sequence. (This is realistic if, as is common, one is implementing the automaton using dedicated programmable logic.) If we interconnect such an automaton with a type B hybrid system the overall model is of the general form

$$\dot{x} = a(x(t), z\lfloor p \rfloor, u(t)) \; ; \quad y(t) = c_e(x(t), z\lfloor p \rfloor)$$
$$\dot{p} = r(y(t), z\lfloor p \rfloor)$$
$$z(\lceil p \rceil) = f(z(\lfloor p \rfloor), y(t_p)) \; .$$

In this case there is no symbolic input or output but the internal dynamics involve a finite automaton. In addition to microprocessor controlled physical systems, these models can arise through the application of singular perturbations to a family of ordinary differential equations. The paper [10] is devoted to the study of systems whose dynamical model switches among finitely many possibilities, depending on rules defined in terms of the trajectory. In [10] the system is modeled using a graph, with the nodes of the graph being associated with particular control system models and the branches describing the conditions under which one model is to be replaced by another. The models being considered here are more complete than these by virtue of the inclusion of the trigger function p.

Example 4: Recall Example 1. Now, however, we wish to model the engine-drive train system when the automobile has an automatic transmission. The gear shift is absent and the only input is the throttle position. We can model the overall system as a multimode system with the values of v being generated by a finite automaton. Thus we must replace the equations given there by ones having no symbolic input

$$\dot{x}_1(t) = x_2(t)$$
$$\dot{x}_2(t) = (-a(x_2(t)) + u(t))/(1 + z\lfloor p \rfloor)$$
$$\dot{p}(t) = k$$
$$z\lceil p \rceil = f(z\lfloor p \rfloor, x_1(t_p), x_2(t_p)) \; .$$

In this case f determines the shifting rule. Alternative choices for f can be compared on the basis of the smoothness of the resulting trajectory, the quality of the regulation of the engine speed, the fuel consumption, etc.

Problem 1: Given a multimodal dynamical system, under what circumstances is the value of x ultimately constant, regardless of the value of initial state.

Problem 2: Suppose that we are given a control system model of the form

$$\dot{x}(t) = a(x(t), u(t)) \; ; \;\; y(t) = c(x(t))$$

such that for every asymptotically constant u in some class U, the output y approaches a constant. Under what circumstances does there exist a lower dimensional vector x_1 and a hybrid system

$$\dot{x}_1(t) = a(x_1(t), z\lfloor p \rfloor, u(t)) \; ; \;\;\;\; y(t) = c(x_1(t), z(k))$$
$$\dot{p}(t) = r(x_1(t), y(t), z\lfloor p \rfloor)$$
$$z(\lceil p \rceil) = f(z\lfloor p \rfloor, x_1(t), u(t))$$

such that the response y of this hybrid system approaches a steady-state value with the steady state being the same as that of the hybrid system.

A second type of interconnection leads to a model with no analog input. Suppose that a type B hybrid system

$$\dot{x}(t) = a(u(t), k(v\lfloor p \rfloor, y(t)), x(t)) \; ; \;\;\;\; y(t) = c_e(v(\lfloor p \rfloor, x(t)))$$
$$\dot{p}(t) = r(v(\lfloor p \rfloor, y(t))) \;\;\;\;\;\;\;\;\;\;\;\;\;\;\;\; w(\lfloor p \rfloor) = h(v(\lfloor p \rfloor, y(t_p)))$$

is connected to a an ordinary continuous time model of the form

$$\dot{x}_1 = a_1(x_1(t), y(t)) \; ; \;\; u(t) = c_1(x(t)) \,.$$

In this case (u, y) are no longer exogenous and we may use the given relationship to eliminate u and y. By adopting a notation that suppresses the fact that the overall state is now the pair $(x(t), x_1(t))$, the system takes the form

$$\dot{x}(t) = \hat{a}(x(t), v(\lfloor p \rfloor))$$
$$\dot{p}(t) = r(v\lfloor p \rfloor, y(t)) \; ; \;\;\;\; w(\lfloor p \rfloor) = h(v\lfloor p \rfloor, y(t_p))$$

i.e., it is a type B hybrid system without continuous input or output. The input and output are both string valued but the internal dynamics are governed, by differential equations. The rate at which the input is read and the output returned is explicitly modeled by the equation describing p. In this sense the model is more specific than the one ordinarily used in the study of automata and we call these *dynamical automata*. In the special case when the mapping from input strings to output strings is realizable by a finite state machine this would be a digital system. See [11] for examples of this type.

The type of questions that most naturally arise in studying the engineering aspects of dynamical automata have to do with the influence of the changes in the string on changes on the state. The following stability problem provides an example.

Problem 3: Give conditions on the form of the equations of a dynamical automaton such that for all infinite input strings v that end with a specific repetitive pattern, all solutions approach a constant, regardless of the initial condition. This means, of course, that p will approach a constant and asymptotically v will be read at a constant rate.

Remark 1: The subject of computer networks (see [12] for example) seems to have gone farther than any other field in formalizing a set of ideas pertaining to hybrid systems. Borrowing from the language of that field, our hybrid systems describe the interaction between the $u - y$ variables relating to the "physical layer" of the network and the $v - w$ variables relating to the "data link" and "data link protocol" layer of the network. However, because the questions of interest in computer networks tend to center around multi-user questions the subsequent analysis done in that field is not central to our concerns.

6   Devices and Their Instruction Sets

The physical implementation of a motion control system requires actuators, amplifiers, position sensors, etc. It also requires the digital electronics necessary to interpret the instruction set and realize the feedforward, feedback and cross-coupling interconnections necessary to generate coordinated motion. Important engineering specifications include the choice of the sampling frequency for the sampled data control system, the choice of fixed or floating point representation for the digitized signals, the number of bits to be allocated for the representation of digitized signals, and the choice of the capacity of, and communication protocol for, the interface to the network. These characteristics will influence the cost and effectiveness of the device, regardless of the choice of the instruction set. However, the tradeoffs between these choices will be influenced by the choice of instruction set. To achieve the best performance for a given cost the computing hardware and the instruction set will need to be suitably matched. It is here, perhaps, that classical ideas from control theory have the most to say about the design of hybrid systems.

The sampling rate will limit the fastest response time. No "reflex" can respond in less than a sampling period. The lowest level loops, those that interact with the x variables in our model, must be implemented using dedicated digital "circuitry" that maintains a strict relationship with physical time and is not susceptible to the sort of interrupts that may be acceptable at the higher levels of a digital system. This does not mean that this circuitry need have fixed characteristics. Although it is not necessary that these lowest level circuits be capable of changing their characteristics every time the output is sampled, the various gains and set points should be un-

der program control. Thus, a second important performance characteristic
is the rate at which the feedback rule can be updated.

In order to make this more specific, we focus our discussion of instruction
sets on two types of instructions. These relate to the choice of control law
and to the choice of the information to be made available on the network.
We use the notation adopted in section 3 whereby the input vector is divided
into two parts, u_1 and u_2, with u_1 being those inputs that are under direct
control and u_2 being the disturbances. An instruction set V for a type A or
type B hybrid system will be said to be an *instruction set of the MDL Type*
if the instructions are triples (K, r, F) interpreted by the hybrid system
according to the rule that on reading such a triple the sampled data system
implements algorithms such as to force the trajectory to satisfy

$$x(k\delta + \delta) = a(K(v\lfloor p \rfloor, y(k\delta)), x(k\delta), u_2(k\delta)) ; \quad y(k\delta) = c_e(x(k\delta))$$
$$p(k\delta + \delta) = r(u(k\delta), y(k\delta)) \qquad\qquad w(k) = F(y(k_p\delta)) .$$

Or, if a differential equation model is more appropriate, the trajectory
should satisfy

$$\dot{x}(t) = a(K(v\lfloor p \rfloor, y(t)), x(t), u_2(t)) ; \quad y(t) = c_e(v(\lfloor p \rfloor, x(t)))$$
$$\dot{p}(t) = r(u_2(t), y(t)) \qquad\qquad w\lfloor p \rfloor = F(y(t_p)) .$$

This algorithm stays in force until p reaches the next integer value at which
time new (K, r, F) is read and used.

This type of instruction set includes the possibility of using feedback to
guide the motion provided that the feedback can be expressed as a function of y. It incorporates observationally dependent mode switching and
selectable reporting. The stopping rules are trajectory dependent but only
through the trigger function p. In this situation, the compliance is reflected
in the relationship between u_2 and y. Typically this can be influenced by
the choice of instruction. Because y and u_2 influence p through r, depending on the sensors available, moves can be "guarded" in the sense that this
term is used in the robotics literature [13].

Affine instruction sets are a particularly simple type of MDL instruction
set, characterized by restricting the types of K's, r's, and F's that can
appear. Specifically, they require that K be affine, $K(y) = Ky + k$, that
r be the positive part of an affine function, $r = \text{pos}(\langle d, y \rangle + d_0)$, and that
F be linear, $F(y(t_p)) = Fy(t_p)$. Although reference [4] treats a slightly
different situation, one can see from the results given there that restricting
the instruction set to be affine does not limit the trajectories one may
approximate with arbitrary accuracy but may make it necessary to use
more instructions to achieve a given accuracy.

Work on flight control systems using gain scheduling makes it clear that
gain scheduling can be an effective tool in that context, making certain
types of on line identification unnecessary. Controlling a system with an

affine instruction set is not quite the same as gain scheduling in the sense
that the term is used in the control literature [9], [14]. In the present
situation the feedback control law is not scheduled by reference to nominal
"flight conditions" but rather it is selected in advance by the task planner.
Questions that arise in the present context relate to the minimum length
of the description of a given motion program based on these primitives as
opposed to some other set.

7   Quantization and Device Independence

As remarked above, common motion control devices are programmed using symbol strings that provide an open loop description of the tool paths,
paths through the workspace, etc. making use of some pre-established con-
ventions built into the motion generating device. In principle, a significant
part of the design effort associated with developing a new system would
go into the specification of the motion description language and symbol
processing characteristics of the device that interprets it. However, in some
areas of application this is unnecessary because of the existence of standard-
s. One of the most widely used motion control languages is the industry
standard EIA code used to specify tool paths for numerically controlled machine tools. Seames [15] is a recent reference describing this language and
its use. Most pen plotters use one of a small number of, de facto standard,
instruction sets. In order to justify the investment necessary to develop
and maintain a motion control language and to promote the development
of sophisticated application programs, the language must be broadly applicable and sufficiently independent of implementation details so as to allow
it to be used with a variety of hardware setups. Its structure must not
depend on details such as the number of encoder counts per revolution, the
values of the motor constants, etc. Insofar as possible one wants device
independence so that the language can be used with a range of more or less
capable devices having the same basic function.

  For example, instruction sets that deal with coordinates of points should
be structured so that the device that interprets the instruction can round
off a given coordinate as appropriate for the degree of accuracy supported
by device at hand. A conceptual scheme for achieving this type of device
independence is to define an *ideal instruction set*, in which instructions are
parametrized using real numbers, together with an *ideal device* capable of
faithfully interpreting all instructions from the ideal instruction set. Each
actual instruction set must be a subset of the ideal instruction set. Each
actual device interprets faithfully a subset of the ideal instruction set. The
actual device treats instructions that are in the ideal instruction set, but
not in its (necessarily finite) subset, as instructions to be rounded off to the

nearest instruction in its subset.

The elements of this aspect of device independence are, then, an ideal instruction set V, the set of executable instructions V' and a rounding algorithm $f : V \rightarrow V'$. An ideal device might take the form

$$\dot{x}(t) = a(u(t), x(t), z\lfloor p\rfloor) ; \qquad y(t) = c_e(x(t), z\lfloor p\rfloor)$$
$$\dot{p}(t) = r(y(t), z\lfloor p\rfloor)$$
$$z\lceil p\rceil = f(v\lfloor p\rfloor, y(t_p), z\lfloor p\rfloor) \qquad w(k) = h(y(t_p), z\lfloor p\rfloor)$$

with v, w, and z taking on values in V, W, Z, now to be thought of as open subsets of Cartesian space. In such a model the instructions are specified and executed with complete accuracy and actual systems would approximate this system. The approximation would involve the approximation of real numbers by rational numbers having a binary representation of a certain type.

Remark 2: Of course the stability of systems of the above type is of interest. If p evolves uniformly then one could, at least conceptually, convert the equation in x into a difference equation

$$x(t+1) = F(x(t), u(\ ), z\lfloor t\rfloor)$$

and then apply standard methods to the simultaneous difference equations defining the evolution of (x, z). In the actual situation of interest, p evolves nonuniformly and one is in a situation analogous to that found in the study of asynchronous computation, see [16]. One advantage associated with passing to the representation of the instructions in terms of real numbers is that it facilitates this type of analysis. Notice that if the above system has sufficiently strong type of stability then the trajectory associated with the quantized will be close to that of the unquantized system.

Motion description languages, like other computer languages, necessarily represent an attempt to trade off a number of vaguely defined factors such as generality versus ease of use, performance versus expense of implementation, etc. Although one can achieve certain efficiencies by tailoring the syntax and semantics of a motion control language to a specific hardware implementation, overly specialized languages will be replaced by ones designed to be more broadly applicable because only popular languages attract the hardware and programming support necessary to keep them successful. Providing for device independence is, therefore, an important aspect of language design.

8 Motion Scripts

The EIA language for specifying machine tool paths was mentioned above. In using this language to describe a pocketing operation for a milling machine one might need an instruction X2Y3Z.5F4 followed by the further instruction X2Y4Z.5, etc. The first of these instructs the mill to move in a straight line at speed 4 from its present position to $X = 2$, $Y = 3$, $Z = .5$. The second commands the mill to move from there to $X = 2$, $Y = 4$, $Z = .5$. A string of such instructions can be thought of as a script. When the script is followed it yields a pocket in the work piece. In terms of this way of thinking, mechanical cad/cam software can be thought as being an aid to script writing, playing a role analogous to that played by word processors. Of course the domain of applicability EIA code is highly specialized. It only provides an open loop description of the motion because in the domain to which it is applied there are, typically, no feedback signals available.

In a more general setting, such as might be appropriate for mobile robots, the lack of provision for feedback would be a fatal flaw. At the present time there does not seem to be any general software that is designed to help the user write scripts that make use of feedback rules as well as open loop instructions. It does not seem to be known, for example, whether or not the complexity results on open loop path planning based of the solution of algebraic inequalities, see [17] for a survey, change appreciably if one allows for the use of feedback. We do not have an estimate for the minimum length of the instruction string required to express a path through obstructed space using the type of instruction set we are considering here, nor do we have an estimate on the amount of computing required to produce it. Part of any such investigation would have to be a discussion of admissible coordinate systems. The number of coordinate systems used and their relationship to the objects in the space are factors in determining the ease with which paths can be described.

9   Motion Control in Higher Animals

In the pervious sections we have presented a conceptual scheme for modeling and analyzing the type of hybrid systems commonly found in man-made motion control systems. The discussion was influenced by current engineering practice in the areas of numerically controlled machine tools and robotics, and also by ideas from other areas of control engineering practice leading to the incorporation of programmable gains and analog driven triggers. Many of the ideas we have discussed translate directly into biological terms. In this section we point out a few results from the neuroscience literature that appear to support the approach to motion control discussed above.

In very crude form, the standard view of the operation of the voluntary motor system is as follows. (Compare with "Introduction to the Motor System," by C. Ghez, Chapter 23 of [18].) At the lowest level, signals from particular areas of the spinal column command particular muscle groups to contract or relax; sensors stimulated by the resulting motion provide feedback signals to nearby areas of the spinal column. At this level of the hierarchy the feedback loops involve only a few synapses and operate on a time scale of around 10 milliseconds. Based on what is known about the speed of transmission of signals in nerves, it can be asserted that such control mechanisms respond before higher level intervention takes place. Up one level in the hierarchy, the brain stem serves as a coordinator of the activity of various muscle groups, making sure that the groups of muscles needed for an effective motion will move appropriately. The brain stem itself receives its major input from the motor cortex, located in the frontal lobe and from the ascending pathways relaying sensory information up the spinal column. At least part of the function of the motor cortex is thought to involve reformatting the overall motion plan partially formulated in the premotor cortex.

In speculating about the possibilities for mapping known aspects of this picture of voluntary motor control onto a type D hybrid model, we would say that x equations describe the dynamics of the muscles and skeleton, with the various components of the vector $K(z, y)$ being generated at appropriate places in the spinal column. The choice of z is made in the brain stem brain and sent down the spinal column so as to make available $K(z, \cdot)$ for use at the lowest level of control. In a similar way, the value of z at any moment in time is the result of a computation based on sensed data y and higher level input data v. It is tempting to speculate that the motor cortex provides v-like input to the brain stem and that the value of $f(z, v, y)$ needed to propagate z is evaluated in the brain stem.

Consider now the lowest level of this hierarchy. If x_0 is a stable equilibrium point for a system

$$\dot{x}(t) = a(u_1(t), u_2(t), x(t)) \; ; \; y(t) = b(x(t))$$

and if we apply feedback in the form of $K(y, z)$ to change the system to the form

$$\dot{x}(t) = f(K(y(t), z\lfloor p\rfloor), u_2(t), x(t)) \; ; \; y(t) = b(x(t))$$

thereby shifting the stable equilibrium point to x_1, then we can say that $K(z\lfloor p\rfloor, .)$ has adjusted the *posture* of the system. Animals may assume many different postures just as motion control system may have a large number of stable equilibrium states. Compliance is a manifestation of *muscle tone*. *Reflexes* are generated by high bandwidth, low level, feedback loops coupling stimulus and response. It is often argued that reflex motions are used as building blocks for more complicated learned motions.

Our remarks below relate to four specific points: i) animals control motion by controlling a combination of set point and gain, ii) complex motions are segmented consistent with the instruction set model of motion,

iii) pattern generators, effectively free running oscillators, are used to generate motion patterns, and iv) the premotor cortex appears to act as script writer for motion. We take up these points in the order given.

Independent control of muscle length and stiffness. Experimental evidence for this phenomenon is discussed in many places. Chapter 24 of [18] "Muscles and Muscle Receptors" by T.J. Carew and C. Ghez, surveys experimental work, e.g. Rack and Westbury [19] and Polti and Bizzi [20] and discuss phenomenological models. In mathematical terms, the experimental work is interpreted as showing that one can model the force-displacement relationship of a muscle as $f = k(x - x_0)$ with k and x_0 being independently specifiable. Neuroscientists like to interpret this as representing a spring with spring constant k and rest length x_0. Of course muscle only generates contractive force so that this overall affine relationship is achieved by means of antagonistic sets of muscles. Thus, the expression $f = \mathrm{pos}\,(k(x - x_0) - \mathrm{pos}\,(x_0 - x)$ with $\mathrm{pos}\,(a)$ being zero for $a < 0$ and being a for $a > 0$, is more suggestive of the underlying biological realization. The independence of the control of position and compliance plays a sizable role in the more recent work of Bullock and Grossberg [21] who attempt to relate it to mechanisms for learning and optimizing motions. Reflexes can, however be suppressed. This lends credibility to the idea that the motion control scheme used by animals involves gain selection.

Motions are segmented. At least one form of the idea of motion segmentation goes back to the previous century and the work of Woodworth who proposed that motion could be divided into a ballistic phase and a controlled phase. The former being thought of as an open loop unguided phase and the latter being slower and subject to active control. Experimental data supporting a subdivision of motion into a ballistic and controlled phase is given by Desmedt in "Patterns of motor commands during various types of voluntary movement in man," in [22]. Bullock and Grossberg [21] identify the optimization of motion with a process whereby the low compliance, tightly controlled, modality is gradually replaced by a ballistic motion. In a different direction, the work of Sternberg *et al.* on typing [23] contains a number of suggestions as to how finger motions may be segmented.

Motions generation uses internal dynamics. A number of experimentalists have worked with cats that have had their brain stems cut in such a way as to eliminate inputs to the spinal column from the higher brain centers. (See the paper by A.R. Arshavsky *et al.*, "The cerebellum and the control of rhythmical movements," in [22] and the work of Grilliner [24].) Basic findings include the observation that cutting the descending motor pathways just below the cerebellum does not destroy the normal rhythmic walking pattern of a cat on a moving treadmill. From this one can infer that the oscillatory pattern of stimuli necessary for walking is generated by neural circuits in the spinal column, with the help of sensory feedback from the environment. Grilliner presents evidence that the sensory feedback is, in fact, not necessary. It has also been observed that the loss of descending input changes the muscle tone with the muscles that normally act to oppose the force of gravity becoming less compliant.

Motions are scripted using gain control. Experiments with Macaque monkeys have shown that approximately 40 milliseconds before the initiation of arm movement there is activity in the premotor cortex. This is followed by activity in the motor cortex, followed by the actual arm motion. It is generally thought that the premotor cortex generates a plan for motor behavior and that this plan is transmitted through the motor cortex for execution. It is not known with certainty what resources are used by the premotor cortex in this planning process. It is to be expected that the synthesis occurring in the premotor cortex would draw on a data base of previously used motion segments together with a motion sequence editor which would provide the various types of geometric transformations, scaling and time warping necessary to adapt these motion segments to the purpose at hand.

The literature on motion control is very large. Whole books have been written about particular subsystems [25]. One should not take these brief remarks as anything other than suggestions about where to begin a serious study. Moreover, we should point out that there are those who believe that any attempt to organize motion control along hierarchical lines as we have suggested here is doomed to failure [26].

10 Conclusions and Future Work

We have presented a new class of models and have attempted to show that they are appropriate for modeling a range of phenomena that cut across the usual boundary between control engineering and computer engineering. These models address issues having to do with the way signal processing interacts with symbol processing, one of the most poorly understood aspects of information processing machines. Our models reflect current engineering practice in the area of motion control. We have argued for greater emphasis on feedback in the instruction sets designed for motion control and raised the question of motion planning with feedback.

For the last fifty years biologists, economists and psychologists have adopted the classic feedback loop diagram and used it to try to explain a range of phenomena including personality disorders and the growth of the GNP. In more recent times it has become clear to engineers that the notation of driving the input with the output in a direct way, powerful as it is in explaining both exponential growth and stabilization, must be supplemented with additional ideas if we are to develop effective ways of thinking about anything but the most classical types of control. An enlarged framework, with a definite place for higher level abstraction and information processing seems to be required. The models presented here can play a role in establishing such a framework.

11 References

1. D.E. Whitney, "Historical Perspective and State-of-the-Art in Robot Force Control," *IEEE Conference on Robotics and Automation*, IEEE, 1985, pp. 262–268.

2. R.A. Brooks, "Elephants Don't Play Chess," in *Designing Autonomous Agents*, (Pattie Maes, Ed.), MIT Press, Cambridge, MA, 1990.

3. R.A. Brooks, "A Robust Layered Control System for a Mobile Robot," *IEEE Conference on Robotics and Automation*, RA-2, 1986, pp. 14–23.

4. R.W. Brockett, "On the Computer Control of Movement," *IEEE Conference on Robotics and Automation*, RA-4, 1988.

5. R.W. Brockett, "Formal Languages for Motion Description and Map Making," in *Robotics*, (R.W. Brockett, Ed.) American Math. Soc., Providence RI, 1990.

6. G.F. Franklin and J.D. Powell, *Digital Control of Dynamic Systems*, Addison Wesley, Reading, MA, 1980.

7. S. Bennett and D.A. Linkens, Eds. *Computer Control of Industrial Processes*, IEE, London, 1982.

8. E.A. Coddington and N. Levinson, *Theory of Ordinary Differential Equations*, McGraw-Hill, New York, 1955.

9. K.J. Astrom and B. Wittenmark, *Adaptive Control*, Addison-Wesley, Reading, MA, 1989.

10. R.W. Brockett, *Smooth Multimode Control Systems*, Proceedings of the 1983 Berkeley-Ames Conference on Nonlinear Problems in Control and Fluid Mechanics, (L. Hunt and C. Martin, Eds.) Math. Sci. Press, 1984, pp. 103–110.

11. R.W. Brockett, "Smooth Dynamical Systems Which Realize Arithmetical and Logical Operations," in Lecture Notes in Control and Information Sciences. *Three Decades of Mathematical System Theory*. (H. Nijmeijer and J.M. Schumacher, Eds.) Springer Verlag, 1989, pp.19–30.

12. Mischa Schwartz, *Telecommunication Networks*, Addison Wesley, Reading, MA, 1987.

13. J.J. Craig, *Introduction to Robotics*, Addison Wesley, Reading, MA, 1986.

14. Gunter Stein, "Adaptive Flight Control: A Pragmatic View," in *Applications of Adaptive Control*, (K. Narendra and R. Monopoli, Eds.) Academic Press, 1980.

15. Warren S. Seames, *Computer Numerical Control*, Delmar Publishers, Albany, New York, 1990.

16. D.P. Bertsekas and J.N. Tsitsiklis, *Parallel and Distributed Computation*, Prentice Hall, Englewood Cliffs, NJ, 1989.

17. J. Schwartz, M. Sharir, and J. Hopcroft, Eds., *Planning, Geometry and Complexity of Robot Motion*, Ablex Publishing, Norwood, NJ, 1987.

18. E.R. Kendal and J.H. Schwartz, *Principles of Neural Science*, Elsevier, New York, 1985.

19. P.M.H. Rack and D.R. Westbury, "The effect of length and stimulus rate on the tension in the isometric cat soleus muscle," *J. of Physiology* **204**, pp. 443–460.

20. A. Polti and E. Bizzi, "Characteristics of motor programs underlying arm movements in monkeys", *J. of Neurophysiology* **42**, 1979, pp. 183–194.

21. D. Bullock and S. Grossberg, "Adaptive neural networks for control of movement trajectories invariant under speed and force rescaling," *Human Movement Science* **10**, 1991, pp. 3–53.

22. E.V. Evarts, S.P. Wise, and D. Bousfield, Eds., *The Motor System in Neurobiology*, Elsevier Biomedical Press, Amsterdam, 1985.

23. S. Sternberg, S. Monsell, R.L. Knoll, and C.E. Wright, "The Latency and Duration of Rapid Movement Sequences: Comparisons of Speech and Typewriting," in *Information Processing in Motor Control and Learning*, (G.E. Stelmach, Ed.) Academic Press, New York, 1978.

24. S. Grilliner, "Locomotion in vertebrates: central mechanisms and reflex interaction," *Physiol. Rev.* **55**, 1975, pp. 247–304.

25. M. Ito, *The Cerebellum and Neural Control*, Raven Press, New York, 1984.

26. J.F. Kalaska and D.J. Crammond, "Cerebral cortical mechanisms of reaching movements," *Science* **225**, 1992, pp. 1517–1523.

3. Some Control Problems in the Process Industries

M. Morari*

Abstract

The two problems discussed in this paper, the control of systems with constraints and statistical quality control, have in common that they have attracted much of the academic and industrial interest in the last decade. The problem formulations are motivated, progress toward a solution is summarized, and outstanding issues are indicated.

1 Introduction

The chemical process industries (CPI) are generally defined to include the following major industrial sectors: food, textiles, paper, chemicals, petroleum, rubber and plastics; stone, clay and glass; and nonferrous metals. The CPI play a key role in the economies of most countries. In the U.S., they generated over \$217 billion in value added in 1985, or 21.7 percent of all manufacturing value added. Various forms of process control are applied throughout these industries. Naturally the problems encountered in each one of the segments of this diverse group are quite different from each other, and so are the techniques which are applied. An excellent survey indicating what control design techniques have been applied successfully in which industry sectors in Japan was reported by Yamamoto and Hashimoto [55]. The survey shows not only the expected dominance of PID variants throughout the industry, but also that just about any academic idea has been tried by somebody on some process, or has at least been considered for application. The techniques range from the well publicized fuzzy logic and neural net controllers, to various types of linear and nonlinear optimal control techniques for which the theoretical foundations were laid in the 1960s.

*Department of Chemical Engineering, 210-41, California Institute of Technology, Pasadena CA 91125, USA, Tel.: (818)356-4186. Fax: (818) 568-8743. Email: MM@IMC.CALTECH.EDU

It is not possible to discuss the broad range of problems and techniques in depth in an overview paper of this type. It is necessary to concentrate on a few problems. The choice made here is biased by the personal experience and involvement of the author and is not meant to imply that the discussed problems and techniques are the only important ones. The two problem areas discussed here stand out in that they have received much interest from industry and the research community during the last decade. They span a wide spectrum: the control of large (many inputs and outputs) systems subject to constraints, and modern versions of statistical process control. In both cases there is a large number of manipulated variables or degrees of freedom. In both cases it can be safely assumed that the underlying process is linear, or at least that the nonlinearity does not dominate the problem. In the former case we have a reasonably good description of the cause/effect relationships. In the latter our understanding of the input/output relationships is rather incomplete, and one *control* objective is to determine the major effects. Usually, in the former case the throughput is high, and little value is added in the production process; in the latter we are dealing with low volume production lines with high product value. In the former case dynamics are important; in the latter often not or very simple assumptions about the dynamics suffice.

2   Linear Systems With Constraints

All real world control systems must deal with constraints. For example, the control system must avoid unsafe operating regimes. In process control, these constraints typically appear in the form of pressure, temperature and concentration limits. Physical limitations impose constraints–pumps and compressors have finite throughput capacity, surge tanks can only hold a certain volume, motors have a limited range of speed. Often targets and both soft and hard constraints for the controlled variables are dictated by an economic optimization carried out at a supervisory control level. This optimization takes into account changes in the feed, the environment (e.g. ambient temperature), and sometimes also the system itself (e.g. catalyst deactivation). The constraints may change in real time as equipment malfunctions are recognized by the process monitoring and diagnostics system.

All these problems require nonlinear controllers. Linear control systems designed to deal with constraints [6] provide no answer here. The traditional approach has been to implement simple single loop controllers and to provide nonlinear elements like split ranges, overrides, and selectors at a higher level. These types of schemes are described in more practice oriented textbooks and monographs like those by Luyben [34] and Åström and Hägglund [1]. In simple situations and high-speed applications (e.g., compressor surge), their popularity has remained unchanged.

During the last two decades because of economic pressures and environmental concerns, many systems in the petroleum and chemical industries have become much more integrated, involving many manipulated inputs and controlled outputs which are highly interactive. A number of industrial examples of this type are referenced in the review by García et al. [22]. Nowadays a typical problem studied in industry may have twenty inputs and forty outputs [16]. A detailed model of a fluid catalytic cracker (FCC) published by Amoco [40] is in many ways representative of a real industrial problem and seems an excellent test bed for any algorithm which addresses large multivariable constrained problems. In these situations the selector and override schemes tend to become very complex [7] and can be hard to design by anybody but the experts.

Thus, in practice, two approaches may be needed to deal with the control problems arising from constraints: A computationally fast modification of linear control systems for SISO situations, or situations with only a few inputs and outputs and simple objectives; and a computationally more involved approach which allows the designer to tackle problems of almost arbitrary complexity. These two approaches, referred to as Anti-Windup Bumpless Transfer (AWBT) and Model Predictive Control (MPC) will be discussed in the following. [1]

2.1 Anti-Windup Bumpless Transfer

All physical systems are subject to actuator saturation. We will refer to such a constraint as an *input limitation*. In addition, commonly encountered control schemes must satisfy multiple objectives and hence need to operate in different control modes. Each mode has a linear controller designed to satisfy the performance objective corresponding to that mode. If the operating conditions demand a change of mode, an override or selection scheme chooses the appropriate mode and executes a mode switch. The switch between operating modes is achieved by a selection of the plant input from among the outputs of a number of parallel controllers, each corresponding to a particular mode. We will refer to such a mode switch as a *plant input substitution* since the output of one controller is replaced by that of another.

As a result of substitutions and limitations, the actual plant input will be different from the output of the controller. When this happens, the controller output does not drive the plant and as a result, the states of the controller are wrongly updated. This effect is called *controller windup*. The adverse effect of windup is in the form of significant performance deterioration, large overshoots in the output and sometimes even instability [11].

---

[1] An excellent oral review of these topics was presented by E. Gilbert at the 1992 American Control Conference.

Performance degradation is especially pronounced if the controller is stable with very slow dynamics and gets even worse if the controller is unstable. In addition to windup, when mode switches occur, the difference between the outputs of different controllers results in a bump discontinuity in the plant input. This in turn causes undesirable bumps in the controlled variables. What is required is a smooth transition or *bumpless transfer* between the different operating modes.

We refer to the problem of control system analysis and controller synthesis for the general class of linear time invariant (LTI) systems subject to plant input limitations and substitutions as the *anti-windup bumpless transfer* (AWBT) problem. Presently, there are few results applicable to the general AWBT problem. Several somewhat ad-hoc schemes are available which work successfully for specific problems. Some attempts towards generalizing these schemes have been reported [2, 53, 52].

In the following we will first review some of the conventional approaches. For more details the reader is referred to two surveys [2, 53]. All approaches to solving the AWBT problem can be summarized as follows:
Design the linear controller ignoring limitations and substitutions and then add AWBT compensation to minimize the adverse effects of limitations and substitutions on closed loop performance.
While many of these schemes have been successful (at least in specific SISO situations), they are by and large intuition based and have little theoretical foundation. Specifically:

- no attempt has been made to formalize these techniques and advance a general AWBT analysis and synthesis theory.

- with the exception of a few [25, 24, 17, 9], no rigorous stability analyses have been reported for anti-windup schemes in a general setting.

- the issue of robustness has been largely ignored (notable exceptions are [11, 10]).

- extension to MIMO cases has not been attempted in its entirety. As pointed out by Doyle *et al.* [17], for MIMO controllers, the saturation may cause a change in the direction of the plant input resulting in disastrous consequences.

- a major void in the existing AWBT literature is a clear exposition of the objectives (and associated engineering trade-offs) which lead to a graceful performance degradation in any reasonably general setting.

A recent thesis [9] has addressed the AWBT problem in a more general framework which is "novel" insofar as the AWBT problem is concerned but is standard in linear theory. After the review we will introduce this general framework, because we feel that it has significant implications for both the AWBT analysis and synthesis problems.

2.1.1 Anti-Reset Windup

Anti-reset windup [11, 8] has also been referred to as *back-calculation and tracking* [2, 18] and *integrator resetting* [53]. Windup was originally observed in PI and PID controllers designed for SISO control systems with a saturating actuator.

Consider the output of a PI controller as shown in Figure 2.1:

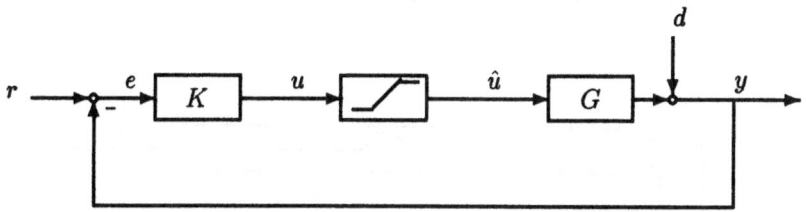

Figure 2.1: Interconnection for PI control of plant $G(s)$

$$K(s) = k(1 + \frac{1}{\tau_I s}) \tag{1}$$

$$= \left[\begin{array}{c|c} 0 & \frac{k}{\tau_I} \\ \hline 1 & k \end{array} \right] \tag{2}$$

$$u = k(e + \frac{1}{\tau_I} \int_0^t e \, dt) \tag{3}$$

$$\hat{u} = sat(u) \tag{4}$$

$$= \begin{cases} u_{min} & \text{if } u < u_{min} \\ u & \text{if } u_{min} \le u \le u_{max} \\ u_{max} & \text{if } u > u_{max} \end{cases} \tag{5}$$

$$e = r - y. \tag{6}$$

Here we used the following notation for a transfer matrix in terms of its state-space data:

$$G(s) = C(sI - A)^{-1}B + D = \left[\begin{array}{c|c} A & B \\ \hline C & D \end{array} \right] \tag{7}$$

If the error e is positive for a substantial time, the control signal gets saturated at the high limit u_{max}. If the error remains positive for some time subsequent to saturation, the integrator continues to accumulate the error causing the control signal to become "more" saturated. The control signal remains saturated at this point because of the large value of the integral. It does not leave the saturation limit until the error becomes negative and remains negative for a sufficiently long time to allow the integral part to come down to a small value. The adverse effect of this integral windup is in the form of large overshoots in the output y and sometimes even instability.

To avoid windup, an extra feedback path is provided in the controller by measuring the actuator output $\hat{u}$ and forming an error signal as the difference between the output u of the controller and the actuator output $\hat{u}$. This error signal is fed to the input of the integrator through the gain $\frac{1}{\tau_r}$. The controller equations thus modified are (refer to Figure 2.2)

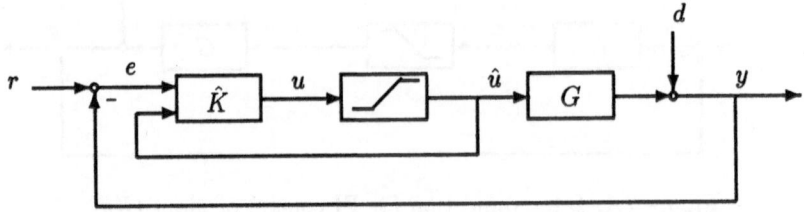

Figure 2.2: Anti-reset windup setup for PI control of plant $G(s)$

$$\hat{K}(s) = \left[\begin{array}{c|cc} -\frac{1}{\tau_r} & k(\frac{1}{\tau_I} - \frac{1}{\tau_r}) & \frac{1}{\tau_r} \\ \hline 1 & k & 0 \end{array}\right] \tag{8}$$

$$u = k\left[e + \frac{1}{\tau_I}\int_0^t \left(e - \frac{\tau_I}{k\tau_r}(u - \hat{u})\right) dt\right] \tag{9}$$

$$\hat{u} = sat(u) \tag{10}$$

$$e = r - y \tag{11}$$

When the actuator saturates, the feedback signal $u - \hat{u}$ attempts to drive the error $u - \hat{u}$ to zero by recomputing the integral such that the controller output is exactly at the saturation limit. This prevents the integrator from winding up. Several incremental forms of this strategy have been outlined in [53]. An extension to a general class of controllers has been reported and is commonly referred to as high gain conventional anti-windup (CAW). Unfortunately, guidelines for choosing this gain rely on simulations.

2.1.2 Conditional Integration

The essential idea behind this approach is summarized as: *Stop integration at saturation!* Inherently a strategy for dealing with integrator windup, it prescribes that when the control signal saturates, the integration should be stopped [53].

Åström *et al.* [2] describe how the limits on the actuator can be translated into limits on the output when using a PID controller for the error feedback case. These limits on the output constitute the "proportional band" and are such that if the instantaneous output value of the process is within these limits, then the actuator does not saturate. Conditional integration then takes the form: *Update the integral only when the process output is in the proportional band.*

One obvious disadvantage is that like the conventional anti-reset windup technique, this scheme is limited to integrator windup. Secondly, the control signal may be held in saturation indefinitely because the integrator is "locked" without being updated. This may cause severe overshoots in the process output. This problem of a "locked integrator" can be resolved by stopping the integrator update only when the update causes the control signal to become more saturated and to allow the update when it causes the control signal to "desaturate" [53]. Krikelis [32] has suggested the use of a pure integrator with a dead-zone nonlinearity as a feedback around it to automate the process of conditional integration in an "intelligent" way.

2.1.3 Internal Model Control (IMC)

The IMC structure [44] was never intended to be an anti-windup scheme. Nonetheless, as pointed out in [11, 44, 17, 52], it has potential for application to the anti-windup problem, for the case where the system is open loop stable.

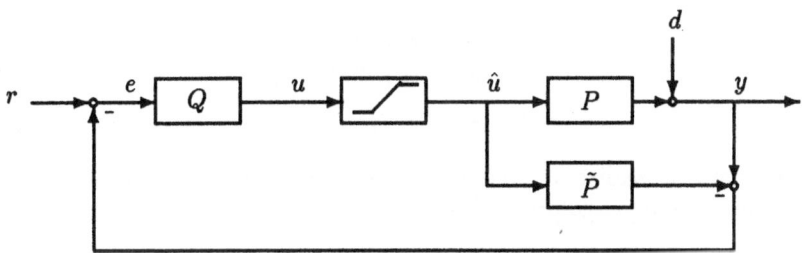

Figure 2.3: The IMC structure

Figure 2.3 shows the IMC structure. If the controller is implemented in the IMC configuration, actuator constraints do not cause any stability problems provided the constrained control signal is sent to both the plant and the model. Under the assumption that there is no plant-model mismatch ($P = \tilde{P}$), it is easy to show [11] that the IMC structure remains effectively open loop and stability is guaranteed by the stability of the plant (P) and the IMC controller (Q). Stability of P and Q is in any case imposed by linear design and hence stability of the nonlinear system is assured.

Thus the IMC structure offers the opportunity of implementing complex (possibly nonlinear) control algorithms without generating complex stability issues, provided there is no plant-model mismatch. Unfortunately, the cost to be paid is in the form of somewhat "sluggish" performance. This is because the controller output is independent of the plant output in both the linear and nonlinear regimes. While this does not matter in the linear regime, its implication in the nonlinear regime is that the controller is unaware of the effect of its actions on the output, resulting in some sluggishness. This effect is most pronounced when the IMC controller has fast dynamics which are "chopped off" by the saturation. Moreover, unless the IMC controller is designed to optimize nonlinear performance, it will not give satisfactory performance for the saturating system.

2.1.4 The Conditioning Technique

The conditioning technique [28, 27] was originally formulated as an extension of the back calculation method proposed by Fertik and Ross [18]. In this technique, windup is interpreted as a lack of consistency between the internal states of the controller and the controller output when the control signal saturates. Consistency is restored by modifying the inputs to the controller such that if these modified inputs (the so-called "realizable references") had been applied to the controller, its output would not have saturated.

Consider a simple error feedback controller with a saturating actuator:

$$
\begin{aligned}
\dot{x} &= Ax + B(r - y) \qquad\qquad (12)\\
u &= Cx + D(r - y) \qquad\qquad (13)\\
\hat{u} &= sat(u) \qquad\qquad\qquad\quad\ (14)
\end{aligned}
$$

The modified "realizable" reference r^r to the controller based on the conditioning technique [28] is:

$$r^r = r + D^{-1}(\hat{u} - u); \qquad\qquad (15)$$

The modified "conditioned" controller is:

$$
\begin{aligned}
\dot{x} &= (A - BD^{-1}C)x + BD^{-1}\hat{u} \qquad (16)\\
u &= Cx + D(r - y) \qquad\qquad\qquad\ (17)\\
\hat{u} &= sat(u) \qquad\qquad\qquad\qquad\qquad\ (18)
\end{aligned}
$$

The same controller was also derived by Campo and Morari [11] as a special case of the observer-based approach of Åström and Wittenmark [3].

Several drawbacks of the conditioning scheme are obvious and some not-so-obvious drawbacks have recently been reported [52, 29]. Firstly, the strategy fails for controllers having rank deficient D matrices [29]. Secondly, in terms of design, the strategy is "inflexible" since it modifies the linear controller without using any additional "tuning" parameters for optimizing nonlinear performance. Thirdly, it suffers from the so-called "inherent short-sightedness" problem [52] because the technique can only handle one saturation level (either the upper limit or the lower limit). Walgama *et al.* [52] have attempted to resolve these deficiencies by proposing that conditioning be performed on a filtered set-point signal instead of the direct setpoint. However, the choice of the filter remains an open question with no sound guidelines for its design. Extension to MIMO cases is also not clear at this time. In particular, stability analysis has not been attempted except in its most rudimentary form (checking that all closed-loop poles are stable).

2.1.5 The Observer-Based Approach

As has been pointed out before, an interpretation of the windup problem under saturation is that the states of the controller do not correspond to the control signal being fed to the plant. This inaccuracy in the state vector of the controller is due to a lack of correct estimates of the controller states. To obtain correct state estimates and avoid windup, an observer is introduced [3].

For a controller defined by (11), (12) and (13), the nonlinear observer (assuming (C,A) detectable) is defined by

$$\dot{\hat{x}} = A\hat{x} + B(r - y) + L(\hat{u} - C\hat{x} - D(r - y)) \tag{19}$$

$$u = C\hat{x} + D(r - y) \tag{20}$$

$$\hat{u} = sat(u) \tag{21}$$

Instead of having a separate controller and a separate observer, both are integrated into one scheme to form the anti-windup compensator. Thus, the observer comes into the controller structure only when the saturation is reached and does not affect the unsaturated controller. The unstable modes of the controller can be stabilized by a suitable choice of L so that $(A - LC)$ is stable.

For specific choices of the observer gain L, this approach reduces to one of several existing anti-windup schemes:

- Hanus conditioned controller: $L = BD^{-1}$.

- CAW compensator (assuming a strictly proper controller): $L = \alpha B$, where α is the gain in the conventional anti-windup scheme.

- Anti-reset windup: $L = \frac{1}{\tau_r}$.

In [53], several other schemes have been generalized based on this "inherent observer" property of anti-windup compensators. Despite the significant generalization offered by this approach, no theoretically rigorous guidelines exist for the design of L which would optimize nonlinear performance.

2.1.6 A New General AWBT Scheme

The standard interconnection for an idealized linear design is the linear fractional transformation (LFT) of an LTI plant P and an LTI controller K as shown in Figure 4.

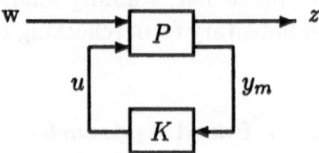

Figure 2.4: Idealized linear design

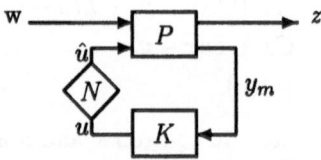

Figure 2.5: Linear design with nonlinearity N

Due to limitations and/or substitutions, a nonlinearity N is introduced into the structure as shown in Figure 5.

The general AWBT problem is based on the interconnection shown in Figure 2.6. $\hat{P}$ is obtained from P by providing an additional output $\hat{u}$. $\hat{K}$ is the AWBT compensated controller to be designed.

The design criteria to be satisfied by $\hat{K}$ are as follows:

- $\hat{K}$ is LTI

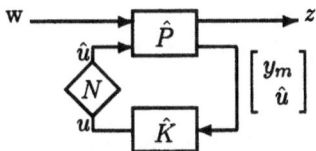

Figure 2.6: The AWBT problem

- ensure stability of the nonlinear closed loop system.

- recover linear performance when $N = I$.

- show "graceful performance degradation" (anti-windup) when $N \neq I$.

Under some other fairly general assumptions one can show [9] that $\hat{K}$ must have the following form:

$$\hat{K}(s) = [\ U(s) \quad I - V(s)\ ] \tag{22}$$

where

$$V(s) \;\;=\;\; \left[\begin{array}{c|c} A - H_1 C & -H_1 \\ \hline H_2 C & H_2 \end{array} \right] \tag{23}$$

$$U(s) \;\;=\;\; \left[\begin{array}{c|c} A - H_1 C & B - H_1 D \\ \hline H_2 C & H_2 D \end{array} \right] \tag{24}$$

$$\tag{25}$$

It is easy to verify that

$$K(s) = V^{-1}(s)U(s) \tag{26}$$

for any H_1, H_2 [10], with the restriction that H_2 be invertible and that $A - H_1 C$ be stable. Any interconnection of linear system elements with an actuator nonlinearity can be brought into the general form as in Figure 4. For specific choices of H_1 and H_2, this framework reduces to one of several known AWBT techniques:

1. Anti-reset windup: $H_1 = \frac{1}{\tau_r}$, $H_2 = 1$.

2. Hanus conditioned controller: $H_1 = BD^{-1}$, $H_2 = I$.

3. Observer-based anti-windup: $H_1 = L$, $H_2 = I$.

4. Conventional anti-windup: $H_1 = \alpha B(I + \alpha D)^{-1}$, $H_2 = (I + \alpha D)^{-1}$.

5. IMC: $H_1 = \begin{bmatrix} 0 \\ B_p \end{bmatrix}$, $H_2 = I$, where subscript p refers to the plant model.

Clearly, the degrees of freedom available in this AWBT framework (in the form of H_1 and H_2) allow the consideration of a wide variety of AWBT approaches. Also, the assumptions made impose few restrictions on AWBT design. In his thesis Campo [9] indicated various approaches to address stability and performance – affected by controller memory and "directionality" in multivariable systems – in this novel framework, but fell short of providing a solution to the synthesis problem.

2.2   Model Predictive Control

Model predictive control (MPC) became known in the late 1970s and early 1980s under a variety of names including Dynamic Matrix Control, Model Algorithmic Control, Model Predictive Heuristic Control, etc. While the details of these various algorithms differ, the main idea is very much the same. Generalized predictive control (GPC) was popularized by Clarke and coworkers [13, 14, 15] and is viewed by some to be in the class of model predictive controllers, or at least to offer an alternative to MPC. Clearly the main idea is similar: the *moving* or *receding horizon*.

The origin of the two techniques, however, is quite different. From its beginning, MPC was conceived to deal with multivariable constrained problems, while GPC was proposed as an alternate approach for the adaptive control of unconstrained single input/single output systems. All expositions of MPC employ a state space description, while the presentations of GPC employ exclusively a transfer function description. Soeterboek [51] analyzed the similarities and differences between the various techniques in the single input/single output case employing a transfer function approach. He also discusses optimization algorithms to deal with input and rate constraints. For stable SISO systems, however, we are not aware of any example where an on-line optimization effort is justified versus a simple anti-windup scheme as discussed above.

Model predictive control was reviewed by García et al. [22] recently. The reader is referred to this paper for an overview and examples of industrial applications. Several excellent theoretical insights have been put forward by Rawlings [49], Mayne [37, 38, 39] and Polak [56] since publication of this review and will be discussed here in their simplest form.

2.2.1   Finite-Horizon Model Predictive Control

We will assume for the moment that all the system states are known and that the discrete time system is described by the following standard state space model

$$x(k) = Ax(k-1) + Bu(k-1) + w(k-1) \qquad (27)$$

where x is the system state, u the manipulated input, and w, in the deterministic framework employed here, a vector of impulses. At time k the objective function

$$J_k = \sum_{\ell=0}^{m-1} x(k+\ell)^T Q x(k+\ell) + u(k+\ell)^T R u(k+\ell) + \sum_{\ell=m}^{p} x(k+\ell)^T Q x(k+\ell) \qquad (28)$$

is to be minimized through the appropriate choice of

$$\mathcal{U}(k) = [u(k), u(k+1), \cdots u(k+m-1)] \qquad (29)$$

subject to the constraints

$$\mathcal{C}^u \mathcal{U}(k) \geq \mathcal{C} \qquad (30)$$

Note that this form of constraints is rather general and allows constraints to be enforced on any linear combination of inputs and outputs [48]. We assume $u(k+\ell) = 0, \ell \geq m$. Thus, a finite number m of future u's is sought to minimize the objective specified over a finite horizon p $(p \geq m)$. The main idea is to implement $u(k)$ obtained from the solution of this optimization problem, to obtain the next measurement $x(k+1)$, and then to resolve the optimization problem stated above, shifting the time index by 1, thus, leading to a moving or receding horizon strategy.

Without constraints and with $m = p = \infty$ the stated optimization problem leads to the standard linear quadratic optimal regulator which has many nice well studied and well known properties. As we make m and p finite, no such nice general properties have been found, and it is doubtful that there are any. For finite horizon problems, various specific results related to stability have been derived [5, 13, 14, 15, 19, 20, 21]. In all cases either specific parameter choices, specific types of systems or other system properties (e.g. monotonicity of the solution of the Riccati equation [5]) have been assumed. These results are not particularly useful for the designer whose system will most likely not fit into any of the studied categories and who has to resort to trial and error to obtain a controller which provides closed loop stability.

The finite horizon strategies have been claimed to exhibit certain robustness properties [5, 12]. The robustness is achieved indirectly by trial and error adjustment of various tuning parameters. It is doubtful that these receding horizon techniques have any robustness advantage over loop shaping techniques à la LQG/LTR or internal model control [44]. They are certainly inferior to techniques specifically conceived to produce robust designs like H_∞ or μ Synthesis. Thus, after having worked with these finite horizon techniques for quite some time, we had to conclude that at least for unconstrained problems, they only have clearly identifiable disadvantages when compared to infinite horizon strategies.

For a long time we adhered to the finite horizon strategies, however, because we believed that they were essential for dealing with system constraints. Needless to say, the stability and even more so the robustness analysis of receding horizon control strategies with constraints is very difficult and has been tackled by only a few researchers [57, 58, 23]. However, some simple ideas due to Rawlings and Mayne have shown that even for the constrained case a finite horizon formulation is not necessary.

2.2.2   Infinite-Horizon Model Predictive Control

Consider first an open-loop stable system and the typical case (30) where the constraints do not change with time. It is intuitively clear that when the *unconstrained* problem is solved with an *infinite* horizon p, then all the constraints will be automatically satisfied after some point in the future, say $\tilde{p}$, because the state will tend to 0. Thus, if $p \geq \tilde{p}$, we can extend the horizon to infinity and easily solve the constrained problem with constraints at all times to infinity. This would be accomplished by solving a quadratic program with the finite horizon objective

$$J_k = \sum_{\ell=0}^{m-1} x(k+\ell)^T Q x(k+\ell) + u(k+\ell)^T R u(k+\ell)$$

$$+ \sum_{\ell=m}^{\infty} x(k+\ell)^T Q x(k+\ell) = \sum_{\ell=0}^{m-1} x(k+\ell)^T Q x(k+\ell)$$

$$+ u(k+\ell)^T R u(k+\ell) + x(k+m)^T \bar{Q} x(k+m) \tag{31}$$

and with the constraints (30). Here $\bar{Q}$ is obtained from the Lyapunov equation

$$A^T \bar{Q} A - \bar{Q} = -Q \tag{32}$$

For an unstable system, we clearly cannot turn off the control after some finite time, unless the unstable modes have been zeroed by that time [49]. Alternatively, if for the *unconstrained* solution the constraints are automatically satisfied beyond $m = \tilde{p}$, we can set m and p to infinity and implement state feedback, beyond $\tilde{p}$ [37]. Again we can tackle the infinite horizon constrained problem by using the same finite horizon objective and constraints as above but where $\bar{Q}$ would now be obtained from the Riccati equation

$$\bar{Q} - Q + A^T \bar{Q} A - A^T \bar{Q} B (R + B^T \bar{Q} B)^{-1} B^T \bar{Q} A = 0 \qquad (33)$$

It should be noted that by solving the constrained infinite horizon problem with the final term of the objective function modified on the basis of the solution of a Lyapunov or Riccati equation, the on-line computation effort is not affected. The advantage of this strategy is enormous, however: Lyapunov arguments [49, 37] can be used to show that as long as there is a feasible solution to the optimization problem, this receding horizon strategy leads to a *closed loop stable system in the presence of general actuator/output constraints.*

If the states are not measurable, the algorithm can be modified in an obvious manner by employing a Kalman filter for state estimation and using the standard optimal prediction techniques [26]. However, no separation principle exists because of the constraints. Thus, closed loop stability does not necessarily result. We expect, however, that stability can be proven in the presence of constraints, at least for systems which are open loop stable. There are other details like sustained disturbance inputs and nonzero setpoints which will have to be resolved in this promising new formulation before a practical algorithm results.

The noise and disturbance parameters in the Kalman filter can be used not only to improve the disturbance rejection properties of the control scheme, but also to affect the speed of response and the robustness of the closed loop system as is standard in LQG designs and illustrated in [43, 33].

The model predictive control algorithm of the form described above has not been implemented in industry, but is expected to find wide acceptance because of the stability guarantees. Most currently employed industrial algorithms have the following characteristics [33], not all of them specified explicitly: the filter design assumes Wiener process (integrated white noise) disturbances on all outputs (to obtain integral control action) and no measurement noise; a state space formulation with an FIR or a step response model is employed; and the horizon is chosen by default so large that stability problems are rarely encountered. The modifications necessary to convert current industrial algorithms to take advantage of the new theoretical results are very minor [42].

3 Quality Control

The problem discussed in this section is not as widely studied by the academic community, but is not any less important than the issues addressed in the previous section. The reason for the neglect may be the dominant interest in problems with interesting dynamics. In the situations considered here, dynamics are generally not important or can be addressed in a straight forward and ad-hoc manner.

Let us start with a few motivating examples.

Making batches of photographic emulsions[41]. Making photographic emulsions requires a mixture of art and science. It involves a series of complicated steps executed by skilled operating personnel according to recipes developed over many years. There are several measures of quality of the final product, the most important of which is grain size. Time histories of about 40 variables are tracked at one-second intervals over the course of production process. The cause and effect relationships between the measured process variables and the final product quality are not well understood. In this particular situation a history of over 200 batches was developed and the objective is to detect the causes of problem batches as well as to detect trends or clusters or other problems that may lead to process improvements.

Fluidized bed reactors [31]. Hydrogenolysis of N-butane on 10% nickel on silica catalyst is carried out in a fluid bed reactor. Due to the small vapor hold up and the large thermal hold up, the dynamics of the concentrations can be considered at steady state with respect to temperature. Under normal operating conditions, the process is roughly linear. Five concentrations related to product quality (methane, ethane, propane, butane and hydrogen) are measured and so are seven "input variables" which may affect the observed product quality. A monitoring scheme is to be developed which detects deviations in operating conditions which significantly affect product quality.

Solution polymerization of butadiene[35]. The process consists of a train of reactors in which all the reagents and a solvent are fed to the first reactor in the train. The rubber is coagulated and the solvent and unreacted monomer are recovered and recycled. The most important control problem in the process is that of controlling the final properties of the rubber. These quality variables cannot be measured on-line, but rather samples leaving the final reactor are taken approximately every two hours and analyzed in the quality control laboratory. Property data are contaminated with substantial analytical error, and since the measurements are available only with two hour delays, it is not possible to filter out this measurement noise in a conventional manner.

The described problems are not particular to the process industries, but are typical for many other branches of manufacturing, for example the manufacturing of semiconductor materials. What is needed is a quality control procedure which should fulfill at least the following four conditions [41]:

1. The procedure should be multivariate and take into account the relationships among the variables.

2. A single answer should be available to the question "Is the process in control?"

3. An overall Type 1 error should be specified.

4. Procedures should be available to answer the question "If the process is out of control, what is the problem?"

Condition 4 is usually more difficult to address than the other three, particularly as the number of variables increases. The classic approach to tackle problems similar to those described in the examples has been via the techniques of statistical quality control, the origins of which lie back at least 60 years and which are used extensively in the process industries. The basic idea of quality control charts to monitor and control processes are due to Shewhart [50]. On a Shewhart chart data are plotted sequentially in time. The chart contains target value and upper and lower action limits.

The cumulative sum (CUSUM) procedure was developed by Page [45, 46] and Barnard [4] for testing the hypothesis, that the process mean is equal to the target value against the alternative hypothesis that it is not. Statistical quality control charts are used extensively in industry. DuPont, for example, has more than 10,000 CUSUM charts being actively used [36].

Unfortunately, these techniques are woefully inadequate to deal with any of the situations described above because they are inherently univariate in nature. MacGregor [35] has shown that for pure gain or steady state processes in which there are nonzero costs associated with taking control actions, statistical quality control charts can be equivalent to stochastic optimal controllers. However, if there are no costs associated with taking control action or if there exist process dynamics, then standard charting methods can be greatly inferior.

Techniques which have gained acceptance to deal with multivariate problems are principal component analysis (PCA) and partial least squares or projection to latent structures (PLS). The underlying idea of these techniques is that although a large number of variables may be measured in most process situations, those variables are often highly correlated. The true dimension of the space in which the process moves is almost always much lower than the number of measurements [31]. PCA and PLS are multivariate statistical methods which consider all the noisy correlated measurements and project this information down onto lower dimensional subspaces which (hopefully) contain all the relevant information. Both techniques are examples of "factor analysis" commonly used in statistics. PCA was introduced by Pearson [47] and is now a standard tool in regression analysis. PLS was invented by Wold [54] and has become a standard in chemometrics. An alternative derivation of PLS based on the concept of "restriction regressors" was developed recently by [30]. In the three example cases described above, one- or two-dimensional subspaces captured all the relevant variations in the input and output spaces. This makes it possible to devise simple two-dimensional operating charts which signal to the personnel when deviations occur in combinations of possibly numerous operating conditions which have a detrimental effect on final product quality.

4 Conclusions

These days, in large segments of the CPI, control techniques which deal with large multivariable constrained problems are of primary interest and MPC is clearly the method of choice. Much progress has been made in our understanding of MPC. It is to be expected that the finite-horizon horizon formulation will be abandoned in favor of the infinite-horizon formulation with its stability guarantees. The robustness properties in the presence of constraints remain poorly understood and should be the focus of future work in this area.

New general frameworks for classifying AWBT schemes have been developed which provide much insight into the similarities and differences between the numerous proposed techniques. For stability analysis various small-gain arguments have been extended so that robust stability can also be addressed, albeit in a conservative manner. Short of simulation, there is no tool for quantitative performance evaluation and very few guidelines for synthesis have emerged.

The developments of the last decade have extended the old proven control chart methods to the complex multivariate situations which arise routinely in today's demanding production environments. In the future we will see modifications to deal with process dynamics and nonlinearities.

Acknowledgment: Partial support from the National Science Foundation and the Department of Energy is gratefully acknowledged. The author wishes to acknowledge the help of M. Kothare in preparing the review on AWBT.

References

[1] K. J. Åström and T. Hägglund. *Automatic Tuning of PID Controllers.* Instrument Society of America, Research Triangle Park, N.C., 1988.

[2] K. J. Åström and L. Rundqwist. Integrator windup and how to avoid it. In *Proceedings of the 1989 American Control Conference*, pages 1693–1698, 1989.

[3] K. J. Åström and B. Wittenmark. *Computer Controlled Systems Theory and Design.* Prentice-Hall, Inc., Englewood Cliffs, N.J., 1984.

[4] G.A. Barnard. Control charts and stochastic processes. *Journal Roy. Statis. Soc.*, B21:239–271, 1959.

[5] R. R. Bitmead, M. Gevers, and V. Wertz. *Adaptive Optimal Control.* Prentice Hall, 1990.

[6] S. Boyd, V. Balakrishnan, C. Barratt, N. Khraishi, X. Li, D. Meyer, and S. Norman. A new CAD method and associated architectures for linear controllers. *IEEE Transactions on Automatic Control*, AC-33(3):268–283, 1988.

[7] E.H. Bristol. After DDC idiomatic control. *Chemical Engineering Progress*, 76(11):84–89, 1980.

[8] P. S. Buckley. Designing overide and feedforward controls. *Control Engineering*, 18(8):48–51, 1971.

[9] P. J. Campo. *Studies in Robust Control of Systems Subject to Constraints.* PhD thesis, California Institute of Technology, 1990.

[10] P. J. Campo, M. Morari, and C. N. Nett. Multivariable anti-windup and bumpless transfer: A general theory. In *Proceedings of the 1989 American Control Conference*, 1989.

[11] P.J. Campo and M. Morari. Robust control of processes subject to saturation nonlincarities. *Computers & Chemical Engineering*, 14(4/5):343–358, 1990.

[12] D.W. Clarke. Adaptive generalized predictive control. In Y. Arkun and W.H. Ray, editors, *Proceedings of Fourth Internation Conference on Chemical Process Control CPCIV*, pages 395–417, South Padre Island, Texas, 1991.

[13] D.W. Clarke and C. Mohtadi. Properties of generalized predictive control. In *Proc. 10th IFAC World Congress*, volume 10, pages 63–74, Munich, Germany, 1987.

[14] D.W. Clarke, C. Mohtadi, and P.S. Tuffs. Generalized predictive control-I. The basic algorithm. *Automatica*, 23:137–148, 1987a.

[15] D.W. Clarke, C. Mohtadi, and P.S. Tuffs. Generalized predictive control-II. Extensions and interpretations. *Automatica*, 23:149–160, 1987b.

[16] C.R. Cutler and F.H. Yocum. Experience with the DMC inverse for identification. In Y. Arkun and W.H. Ray, editors, *Proceedings of Fourth Internation Conference on Chemical Process Control CPCIV*, pages 297–318, South Padre Island, Texas, 1991.

[17] J.C. Doyle, R.S. Smith, and D.F. Enns. Control of plants with input saturation nonlinearities. In *Proceedings of the 1987 American Control Conference*, pages 1034–1039, 1987.

[18] H.A. Fertik and C.W. Ross. Direct digital control algorithm with anti-windup feature. *ISA Transactions*, 6(4):317–328, 1967.

[19] C.E. García and M. Morari. Internal model control 1. A unifying review and some new results. *Ind. Eng. Chem. Process Des. & Dev.*, 21:308–232, 1982.

[20] C.E. García and M. Morari. Internal model control 2. Design procedure for multivariable systems. *Ind. Eng. Chem. Process Des. & Dev.*, 24:472–484, 1985.

[21] C.E. García and M. Morari. Internal model control 3. Multivariable control law computation and tuning guidelines. *Ind. Eng. Chem. Process Des. & Dev.*, 24:484–494, 1985.

[22] C.E. García, D.M. Prett, and M. Morari. Model predictive control: Theory and practice–a survey. *Automatica*, 25:335–348, 1989.

[23] H. Genceli and M. Nikolaou. Stability of predictive control systems with hard and soft constraints on process outputs. In *Proceedings AIChE 1992 Annual Meeting*, Miami Beach, Florida, 1992.

[24] A.H. Glattfelder and W. Schaufelberger. Stability of discrete override and cascade-limiter single-loop control systems. *IEEE Trans. Automat. Contr.*, AC-33(6):532–540, June 1988.

[25] A.H. Glattfelder, W. Schaufelberger, and H.P. Fassler. Stability of override control systems. *Int. J. Contr.*, 37(5):1023–1037, 1983.

[26] G. C. Goodwin and K. S. Sin. *Adaptive Filtering Prediction and Control*. Prentice-Hall, 1984.

[27] R. Hanus and M. Kinnaert. Control of constrained multivariable systems using the conditioning technique. In *Proceedings of the 1989 American Control Conference*, pages 1711–1718, 1989.

[28] R. Hanus, M. Kinnaert, and J. L. Henrotte. Conditioning technique, a general anti-windup and bumpless transfer method. *Automatica*, 23(6):729–739, 1987.

[29] R. Hanus and Y. Peng. Conditioning technique for controllers with time delays. *IEEE Transactions on Automatic Control*, 37(5):689–692, May 1992.

[30] T. R. Holcomb, H. Hjalmarsson, and M. Morari. Significance regression: A statistical approach to biased regression and partial least squares. CDS Technical Memo CIT-CDS 93-002, California Institute of Technology, Pasadena, CA 91125, February 1993.

[31] J. Kresta, J.F. MacGregor, and T.E. Marlin. Multivariate statistical monitoring of process operating performance. In *AIChE Annual Meeting*, San Francisco, California, 1989.

[32] N. J. Krikelis. State feedback integral control with intelligent integrators. *Int. J. Contr.*, 32(3):465–473, 1980.

[33] J.H. Lee, M. Morari, and C.E. García. State space interpretation of model predictive control. *Automatica*, in press, 1993.

[34] W. L. Luyben. *Process Modeling, Simulation, and Control for Chemical Engineers*. McGraw-Hill Publishing Company, 2nd edition edition, 1990.

[35] J.F. MacGregor. Interfaces between process control and on-line statistical process control. *AIChE Computing and Systems Technology (CAST) Division Communications*, 10(2):9–20, September 1987.

[36] D.W. Marquardt. New technical and educational directions for managing product quality. *Amer. Statistician*, 38:8–14, 1984.

[37] D. Mayne and H. Michalska. An implementable receding horizon controller for stabilization of nonlinear systems. In *Proc. 29th CDC*, Honolulu, Hawaii, 1990.

[38] D. Mayne and H. Michalska. Receding horizon control of nonlinear systems. *IEEE Transactions on Automatic Control*, pages 814–824, 1990a.

[39] D. Mayne and H. Michalska. Model predictive control of nonlinear systems. In *Proceedings American Control Conference*, pages 2343–2348, 1991.

[40] R.C. McFarlane, R.C. Reineman, J.F. Bartee, and C. Georgakis. Dynamic simulator for a model IV fluid catalytic cracking unit. *Computers & Chemical Engineering*, 17(3):273–300, 1993.

[41] P. Miller. Batch process diagnostics. Seminar, California Institute of Technology, February 10 1993.

[42] M. Morari, C.E. García, J.H. Lee, and D.M. Prett. *Model Predictive Control*. Prentice-Hall, 1993. in preparation.

[43] M. Morari and J.H. Lee. Model predictive control: The good, the bad and the ugly. In Y. Arkun and W.H. Ray, editors, *Proceedings of Fourth Internation Conference on Chemical Process Control CPCIV*, South Padre Island, Texas, 1991.

[44] M. Morari and E. Zafiriou. *Robust Process Control*. Prentice-Hall, Inc., Englewood Cliffs, N.J., 1989.

[45] E.S. Page. Continuous inspection schemes. *Biometrika*, 41:100–114, 1954.

[46] E.S. Page. Cummulative sum charts. *Technometrics*, 3:1–9, 1961.

[47] K. Pearson. On lines and planes of closest fit to systems of points in space. *Pholosophical Maganize*, 2:559–572, 1901.

[48] D. M. Prett and C. E. García. *Fundamental Process Control*. Butterworths, 1988.

[49] J. Rawlings and K. Muske. The stability of constrained receding horizon control. *IEEE Transactions Automatic Control*, in press, 1993.

[50] W.A. Shewhart. *Economic Control of Quality*. Van Nostrand, 1931.

[51] R. Soeterboek. *Predictive Control - A Unified Approach*. Prentice Hall, 1991.

[52] K. S. Walgama, S. Rönnbäck, and J. Sternby. Generalization of conditioning technique for anti-windup compensators. *IEE Proceedings-D*, 139(2):109–118, March 1992.

[53] K. S. Walgama and J. Sternby. Inherent observer property in a class of anti-windup compensators. *International Journal of Control*, 52(3):705–724, 1990.

[54] S. Wold, A. Ruhe, H. Wold, and W. Dunn. The collinearity problem in linear regression: The partial least squares approach to generalized inverses. *SIAM J. Sci. Stat. Comput.*, 5(3):753–743, September 1984.

[55] S. Yamamoto and I. Hashimoto. Present status and future needs: The view from japanese industry. In Y. Arkun and W.H. Ray, editors, *Proceedings of Fourth Internation Conference on Chemical Process Control CPCIV*, pages 1–28, South Padre Island, Texas, 1991.

[56] T. Yang and E. Polak. Moving horizon control of linear systems with input saturation, disturbances, and plant uncertainty: Part 1. Technical report, Electronics Research Laboratory, University of California, Berkeley, 1991.

[57] E. Zafiriou. Robust model predictive control of processes with hard constraints. *Computers Chemical Engineering*, 14(4/5):359–371, 1990.

[58] E. Zafiriou and A. Marchal. Stability of SISO Quadratic Dynamic Matrix Control. *AIChE Journal*, 37(10):1550–1560, 1991.

4. Engineering Aspects of Industrial Applications of Model-Based Control Techniques and System Theory

A.C.P.M. Backx *

Abstract

This paper describes some main engineering aspects of the design of robust, high performance industrial control systems. The importance of accurate modelling of relevant process dynamics is outlined. An identification procedure is discussed that robustly results in compact accurate models of all relevant process dynamics. The models obtained from process identification are embedded in internal model based control systems. The techniques have been successfully applied to many different processes. Two examples of applications are discussed: the control of a binary distillation column and the control of a low pressure gasphase polymerization reactor.

1 Introduction

Industries are nowadays confronted with a very dynamic and hardly predictable marketplace. Due to large production capacities in almost all industry segments the market has been turned into a full customer-oriented market. Process industries have to operate their production processes in ever closer agreement with market opportunities and market demands to remain competitive and profitable [Garcia, 1986,1988; Morari, 1988].

Evolution of the past decade, showing world wide competition in many segments of industry and high social attention for environmental problems especially related to production in processing industries, has resulted in a need for reliable methods, techniques and tools to operate processes at maximum efficiency, total quality controlled and with maximum flexibility. A trend is observed towards increasing demands with respect to tailor-made products that have to be produ ced with existing installations and process

---
*IPCOS b.v., Gebouw Waelstate, Breeven, De Waal 32, 5682 PH Best, The Netherlands

equipment. At the same time products often have to meet tighter specifications. This innovation requires flexibility in the use of existing production
facilities with minor adjustments in the process hardware only for the production of existing and new products. Plants have to be conside red to be
general purpose where the operation and control of these plants has to enable production of the various products in accordance with actual demands.

Industries are frequently forced to operate with low margins and n-
evertheless remain profitable in order to survive. Reliable and profitable
operation of production environments with low margins and under mentioned quality and flexibility constraints implies a need for operation of
processes in a very predictable way. Detailed knowledge of dynamic behaviour of critical units depicted in the form of a mathematical model and
extensive use of this knowledge for operation and control is crucial to meet
these demands.

The classical approach for modelling of relevant unit process dynamics
is rigorous modelling. In general, these techniques appear not to be very
well suited for accurate modelling of all relevant process dynamics.

By the end of the fifties people started investigating techniques for the
modelling of process dynamics on the basis of observed system input output
behaviour [Zadeh, 1962]. In the early seventies these process identification
techni ques for single input single output system identification on the basis of a stochastic approach were rather well understood [Åström, 1970;
Eykhoff, 1974]. In the past decade, these process identification techniques
have matured to a level that starts suiting the requirements for accurate
modelling of relevant dynamics of indus trial processes [Ljung, 1983, 1987;
Swaanenburg, 1985; Willems, 1986, 1987; Backx, 1987, 1992; Söderström,
1989; Wahlberg, 1989, 1991; Schrama, 1992]. Some of these techniques enable true multi-input multi-output system identificati on.

Simultaneously, techniques have been developed for the design of model
based control systems for robust multivariable control of industrial unit pro-
ces ses [Richalet, 1978; Cutler, 1980; Doyle, 1982, 1984, 1989, 1991, 1992;
Garcia, 1988; Maciejowski, 1989; Morari, 1989; Soeterboek, 1990; Boyd,
1991]. These techniques together with in-line model based optimization
techniques enable operation of unit processes so that undesired dynamic
properties of the system are compensated for and that just desired behaviour is observed at the process outputs. Of course the compen sation of
non-desired dynamic process properties is restricted by the internal mechanisms of the process as compensation has to be done through appropriate
variati ons of manipulated variables within operating constraints. These

techniques give flexibility in the operation of unit processes and allow for adjustment of process characteris tics in accordance with current demands.

Operation of a plant in accordance with above outlined demands requires fully integrated optimal control of all interacting unit processes, which significantly contribute to the overall performance of the plant. Decisions taken at management level with respect to product types, product volume, processed raw materi als and intermediate products have to result in the most economic producti on of the desired products in time, in the right quantities, at the demanded quality.

An outline of the relation between process control and industrial needs is given in section 2. Modelling of relevant process dynamics appears to be of crucial importance for the design of robust high performance multivariable unit process control systems. The engineering aspects related to modelling of relevant process dynamics are discussed in section 3. Use of mathematical models inside control systems for prediction of process behaviour has become a standard approach for multivariable control system design. Internal model based control strategies and model predictive control have become defacto standards in industry. Realization of high performance and high robustness simultaneously in the control of critical unit processes requires the use of new modelling and control system design techniques. Section 4 gives an outline of these techniques and their application in industry. Some industrial applications are discussed in section 5. The applications selected are a high performance control system for simultaneous control of both top and bottom product quality of a distillation column and a control system for a low pressure gasphase polymerization reactor.

2   Industrial needs and the relation with process control

Operation of industrial processes has already drastically changed over the past decade. This change can be considered to be the first result of the evoluti on of computer technology of the past decades. Automation of processes on the basis of latest computer and control technology is enabling operation of these processes in a very flexible way.

Today's technology is starting to allow for operation of processes at low margins in a very predictable way. Production scheduling can already be adjusted on the basis of current market and production priorities and tech-

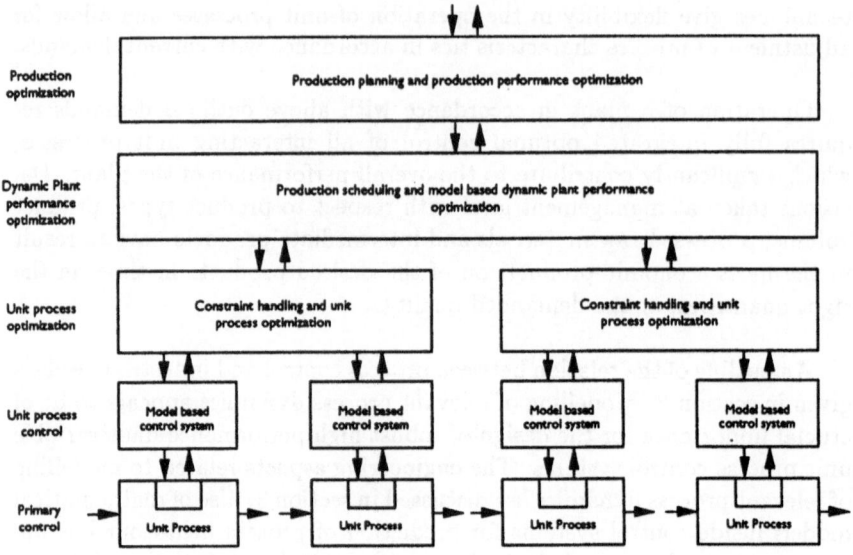

Figure 2.1: Hierarchical process control

nical difficulties. This can be compared with operating a plant in the same way as flying an airplane at extremely low altitude at high speed through a valley with hills. Processes can, with today's technology, be operated efficiently, flexibly at the demanded quality level enab ling operation in a predictable way. This indeed allows for production at low margins without endangering profit. Combination of process technology and control technology even makes the design of processes with completely new capabilities possible, where control systems compensate for undesired process properties and allow less robust process designs. The main advantage of a combined design of a process and an additio nal model based control system is its flexibility in operation. Processes get a General Purpose character whereas the control systems determine the momentary properties, qualities and behaviour.

Process operation close to the process constraints requires a hierarchical approach (cf. Fig. 1). The first levels of control, which can be discerned in this approach, are:

- level 0 → field instrumentation and primary signal processing
- level 1 → primary process control

- level 2 → unit process control

- level 3 → constraint handling and optimization of unit process dynamics

- level 4 → dynamic plant performance optimization

- level 5 → production optimization

The **primary process control** level covers the control of primary, mostly physical, process inputs (e.g. pressure, flow, force, velocity, speed, ...). Control of these variables is mainly done by SISO PID control systems. Depen ding upon the specific process input configuration, physical properties, applied sensors/actuators and the operating range of the variable the applied control system tuning can be fixed or automatically adjusted (autotuning, gain scheduling, ...). The primary control systems in general result in a smoothly responding linear or weakly non-linear response of the controlled primary process input variables. The setpoints of the primary controllers are the manipulated varia bles for the unit process control systems.

The **unit process control** level, in general, covers the model based control of a MIMO unit process. A unit process is a part of a plant consisting of a set of process input variables, which can be manipulated within a well-defined operating range, a set of process disturbances, which cannot be manipulated and a set of process output variables, which have to be controlled in accordance with given specifications (cf. Fig. 2). Control at this level mostly requires model based MIMO control as the unit processes are true multi input multi output systems with strong dynamic interactions between the various manipulated variables and controlled variables. Especially for critical unit processes — processes that predominantly determine overall plant performance — adequate manipulation of the MIMO dynamics, in general, significantly improves overall plant performance.

As the transfers between manipulated variables and controlled variables of these unit processes are usually smoothly nonlinear and while they are often operated in specific operating points or operating ranges, their dynamic behaviour can well be approximated by a set of linear models $\mathcal{F}$, which describes the process dynamics in the selected operating points.

$$y(t) = \mathcal{F}(Y_0, U_0, u(t)) \tag{1}$$

with

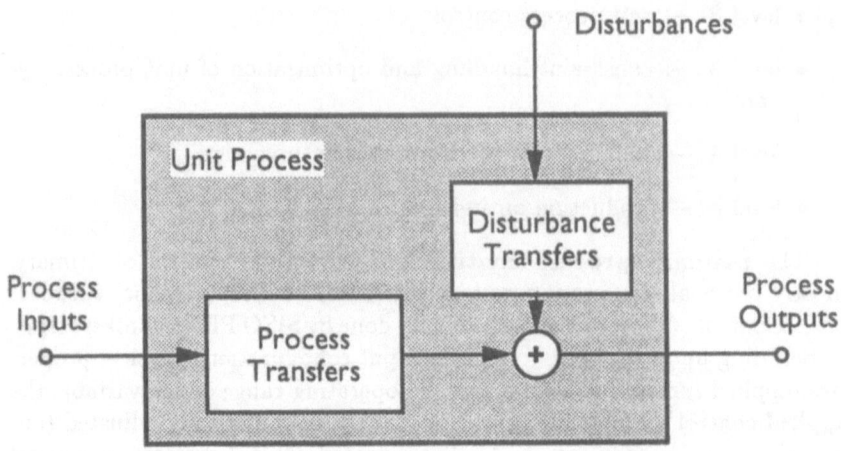

Figure 2.2: Unit process

$y(t)$	process outputs	$\dim[y] = q$
$u(t)$	process inputs	$\dim[u] = p$
Y_0	nominal process output	
U_0	nominal process input	
$\mathcal{F}$	linear approximation in operating point	

The dynamic properties of these unit processes are mainly determined by the dynamic properties of the internal mechanical, physical, chemical, biological processes and of the applied process equipment. The unit processes quite often show non-desired dynamic behaviour like overshoots, badly damped oscillations, combined fast and slow dynamics resulting in slowly drifting output responses on input manipulations, as this behaviour is an inherent part of the original process and its corresponding equipment design. Furthermore, the disturbances acting upon the unit processes often cause a too large variance of the process outputs.

Model based control systems enable manipulation of system dynamics up to physical process constraints and can compensate a significant part of the disturbances at the controlled process outputs. They may not only compensate for the non-desired system properties by appropriate system input manipulations on the basis of known system behaviour, they also enable decoupling, smooth setpoint tracking and adequate disturbance rejection. Application of state–of–the–art System Theory results in extensive degrees of freedom for manipulation of process dynamics in accordance with desired system characteristics defined in control system design criteria. The difficulty encountered, however, is the translation of desired process response criteria into control system design criteria. Response criteria like overshoot, rise time, response time, crosstalk, etc. for the various input/output transfers have to be translated into appropriate weighting functions in the mathematical criteria available for controller design and optimization or have to be satisfied by using numerical optimization techniques. Design criteria available from system theory are the H_2-optimal controller, which minimizes the 2-norm ("average value") of the sensitivity operator (transfer of output disturbance to the process outputs) with input weight W_1 and output weight W_2:

$$\min_C \| W_2EW_1 \|_2^2 = \min_C \frac{1}{2\pi} \int_{-\infty}^{\infty} \text{trace}[(W_2EW_1)^T (W_2EW_1)]d\omega \qquad (2)$$

with

E sensitivity operator $E = (I + PC)^{-1}$
P process transfer function
C controller

and the controller, which minimizes the ∞-norm ("maximum value") of the weighted sensitivity function:

$$\min_C \| W_2EW_1 \|_\infty = \min_C \sup_\omega \bar{\sigma}(W_2EW_1(i\omega)) \qquad (3)$$

with $\bar{\sigma}$ the maximum singular value of the sensitivity function.

A useful model predictive control system structure is the Internal Model based Control (IMC) system. The most general form of the internal model based control system [Backx, 1987; Morari, 1989] is depicted in Fig. 3. This scheme contains separate controllers for feedback disturbance rejection, feedforward setpoint tracking and feedforward disturbance compensation. Each of these controllers can be designed on the basis of the modelled process and disturbance dynamics. Design of the feedback controller can be done with H_∞ control system design techniques. The feedforward setpoint compensator is designed in such a way that changes in the operating point of the process will, in general, not violate opera ting constraints of the process. The feedforward disturbance compensator compensates measu red disturbances by using known disturbance transfer characteristics. As the model does not simulate disturbances, the disturbances reconstructed by the model in the disturbance compensator need not only be subtracted from the error signal, but must be added to the model outputs as well (cf. Fig. 3).

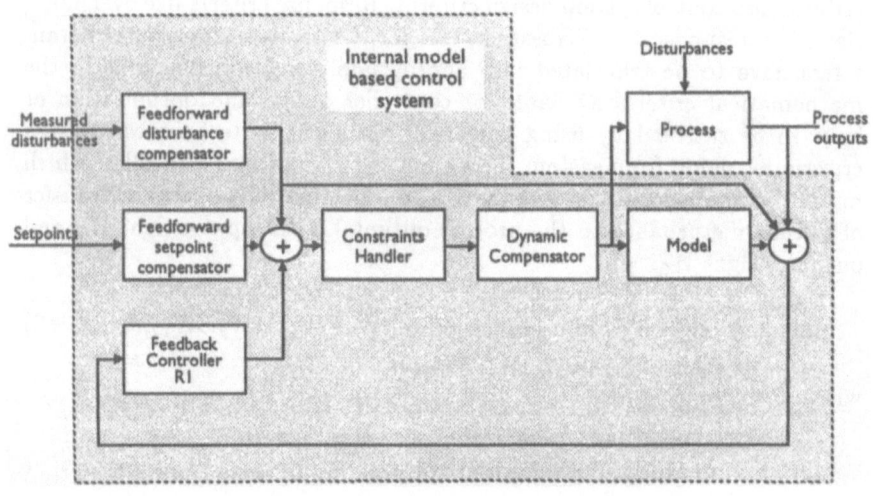

Figure 2.3: Internal Model based Control scheme

A special module in the control scheme is the constraint handler. It is a module that minimizes a quadratic criterion function J, which describes the difference between outputs desired by the controllers and outputs that can actually be realized without violating both amplitude and speed constraints on inputs, outputs and states of the process.

$$\min_C J = \min_C \| y_p - y_e \|_2 \qquad (4)$$

with

y_p expected process outputs after manipulation

y_e desired process outputs generated by the controllers

C set of constraints (amplitudes and speeds, on inputs outputs and related variables or states)

This constraint handler has to prevent operation of the process outside its operating constraints. The constraints handler uses the process model to simulate process responses over some horizon and to detect whether constraints at inputs, outputs or states are violated. The constraint handler will adjust the signals transferred so that no violation of constraints will occur in future. The constraint handler interfaces the unit process control level and the unit process optimization level in the control hierarchy (cf. Fig. 1).

The **unit process optimization** level takes care of optimal operation of the supervised unit processes. This implies operation of the unit processes up to physical or operational constraints in accordance with requirements determined by overall dynamic performance optimization for the whole plant. The unit processes usually have more manipulated variables than controlled variables. The corresponding degrees of freedom can be used for further optimization of unit process opera tion. Options for using these additional degrees of freedom are:

- introduction of additional, so called, soft constraints

- operation of (some) input variables at preferred values, the so-called ideal resting values

- minimization of an additional criterion that will mostly be related to operation economics

The soft constraint values and the ideal resting values depend upon instantaneous operating conditions of the plant. These values have to be determined from plant wide optimization.

The **dynamic plant performance optimization** level covers the optimization of overall plant performance. Production schedules are determined on the basis of known characteristics of the production facilities. Up–to–date information on process capabilities and production planning information are used to calculate the best schedules with corresponding process operation conditions. All costs, including changeover costs are included for the calculation of the best production schedule.

In general linear programming techniques on the basis of *steady state plant simulation* are used nowadays for determination of the best operating conditions for each of the unit processes. Optimization criteria mainly reflect product quality properties and operation economics.

Further optimization can be achieved by using *rigorous dynamic model based simulators* for the critical unit processes and by execution of the optimization including all relevant process dynamics. Special data reconciliation techni ques are required in this case for updating the data generated by the rigorous process models so that simulation results are reliable reflections of real process behaviour.

As a result of the optimization the best operating conditions for each of the unit processes are determined. These conditions include the setpoint values, soft constraints, ideal resting values and additional optimization criteria for the model based unit process control systems.

The **Production optimization** level covers overall optimization of production. At this level optimal production planning is determined considering the products requested by the market, required maintenance of installations and equipment and total production costs related to suggested production schedules. The optimization is based on *economic models* of operation of the production facilities, product and material (e.g. feedstock, raw materials, intermediate products, ...) prices, maintenance costs etc. As a result of the optimization a target planning is obtained, which is used to determine a best corresponding production schedule.

3   Engineering aspects of modelling of relevant process dynamics

Optimum operation of processes requires knowledge of all relevant dynamics of the process transfers represented in the form of a compact mathematical model. Relevant process dynamics are the dynamics, which can be used for significantly influencing the behaviour of the process output variables (cf. Fig. 2). The relevance of dynamics is in practice determined by:

- the operating range and operating constraints of process inputs;

- the range and operating constraints of process outputs;

- the resolution of the measurements of the output variables;

- operating constraints of some process intermediate or state variables;

- the range and dynamic properties of output disturbances.

In order to determine the relevance of process dynamics, the following observation can be made. The selected p process inputs can be manipulated simultaneously within their operating constraints to control a process. The constraints are both amplitude constraints and rate of change as a function of time constraints. The process input manipulations can be considered to span a subspace $\mathbf{R}^{*,p-,t}$ of the complete input space $\mathbf{R}^{p,t}$ which is the space spanned by all possible input manipulations without considering constraints. In this description * refers to the subspace, p refers to the number of inputs and t refers to time. Each of the inputs is considered to be a function of time. Manipulation of the input variables within their constraints invokes responses at the q outputs of the process. The output responses will reflect the whole process dynamics corresponding to the applied input manipulations. If the input manipulations span the whole allowed input subspace, the output responses will represent a subspace $\mathbf{R}^{*,q,t}$ of the output space. Whereas the output space $\mathbf{R}^{q,t}$ is the space spanned by the output vectors without accounting for constraints, only those dynamics are considered to be relevant, which significantly contribute to the behaviour observed at the process outputs. A significant contribution being a contribution that exceeds the output noise level and that exceeds the resolution of the measurements at the process outputs. As a result the input subspace $\mathbf{R}^{*,p,t}$ is mapped to the output subspace $\mathbf{R}^{*,q,t}$:

$$\mathbf{R}^{*,p,t} \longrightarrow \mathbf{R}^{*,q,t}. \tag{5}$$

In general, the restriction to relevant dynamics implies that all process input output dynamics, which can be used for manipulation of the behaviour of the process outputs have to be part of the model. All the dynamics that do not meet this condition should not be included in the model to prevent unnecessary increase of model complexity as model complexity is directly related to ultimate control system complexity.

It is clear that the selection of appropriate process inputs, which will be used for process manipulation, is important. The inputs have to be selected so that the complete output subspace also spanned by the output disturbances can be spanned with permitted input manipulations. Furthermore, those inputs have to be selected that enable well-conditioned output manipulations. Preferably, the inputs should not invoke process responses in preferred directions. The output space has to be equally spanned. Due to the restricted dynamic ranges of both input manipulations via actuators and of output measurements via sensors, the condition number of the transfers has to be better than the available minimum dynamic range of the transfers over the whole frequency range used for process operation.

$$C(j\omega) = \frac{\sigma_{max}(H(j\omega))}{\sigma_{min}(H(j\omega))} \tag{6}$$

In practice, this condition restricts the range of condition numbers of transfer functions that can be used for control system design to less than 10. This is a consequence of limited signal to noise ratios of sensors and of restricted resolution of actuators. Furthermore, the selection of appropriate process inputs and outputs requires an understanding of the main process mechanisms. As the inputs are to be used for manipulation of the process outputs in a predictable way, the manipulators have to be based upon using the main mechanisms of the process. As a consequence, the model looked for has to reflect these mechanisms. Insight into the main mechanisms governing process behaviour can be obtained from rough mechanistic modelling of the process. The mathematical model looked for has to compactly describe these mechanisms and their corresponding input output process transfer characteristics. Detailed modelling of all relevant process dynamics is done by means of process identification techniques. Assumptions have to be made on process behaviour to enable process identification. The assumptions made on the process input output transfer behaviour are:

- the behaviour can be well approximated by linear models

- the behaviour is assumed to be time invariant compared to the system dynamics

- the process is showing causal transfer characteristics

- disturbances may be considered to be stationary noise

- process transfers are bandwidth restricted

- process dynamics can be well approximated by finite dimensional sets of difference equations

Process identification techniques with the above-mentioned assumptions enable modelling of all relevant process dynamics on the basis of observed input output behaviour of the process. However, due to lack of knowledge of detailed process transfer properties a multi-step identification procedure is required to ultimately end up with a compact model describing all relevant dynamics. This is related to the additional choices that have to be made for enabling process identification and the a-priori knowledge required for making a choice on a well funded basis:

- choice of modelset

- choice of parameter estimation criterion

- choice of parameter estimation method

These choices respectively depend on the knowledge of

- the complexity of the process transfer characteristics respectively expressed in the order of the difference equations and in the structure of the transfer characteristics

- type of application of the model (e.g. output simulation, input reconstruction, one or a specific number of samples ahead prediction of the process outputs)

- specific characteristics of the noise present at the outputs of the process

A property related to process identification that is quite often forgotten, but that is important for understanding the qualities of the models obtained by applying process identification techniques, is the restricted frequency range over which the models are valid. This frequency range is restricted by the duration of the testrun used for identification at the lower frequency side and by half the selected sampling frequency at the high frequency side. The models obtained also cover parts outside this frequency range but will, in general, not be valid there. In particular, the steady state gain of an identified model is based on extrapolation of the model properties to frequency $\omega = 0$. Special precautions have to be taken during parameter estimation to get valid DC-gain values. In general, the a-priori knowledge available at the start of the process identification is very limited. This implies that mostly no a-priori knowledge on model order nor on model structure is available. Furthermore, at the start of the process identification procedure, quite often no knowledge on the stochastics of the output noise is available. The identification procedure, therefore, usually has to consist of a sequence of identification steps with each subsequent step being directed to further extension of the level or detail of the knowledge and/or to further compaction of the models. A procedure developed and successfully applied over the past years consists of the following four main identification steps [Backx 1987]:

- non-parametric identification of input/output behaviour using finite impulse response (FIR) finite step response (FSR) or high order AR-MA models

- initial semi-parametric (parametric with respect to order, non-parametric with respect to structure identification using Minimal Polynomial Start Sequence of Markov Parameter (MPSSM) models:

$$y_k = \sum_{j=1}^{\infty} F_j(a_i, M_i, M_0 \mid i = 1, 2, \ldots, r)u_{k-j} \qquad (7)$$

with

$$F_j = \begin{cases} M_0 & j = 0 \\ M_j & j = 1, 2, \ldots r \\ \sum_{j=1}^{r} a_i F_{j-i} & j > r \end{cases}$$

- final semi-parametric identification using quadratic optimization techniques for finetuning of the MPSSM model parameters

- parametric identification using full parametric (both order and structure) state space models

The first step is required due to the fact that it is very difficult to
get information on the complexity of a model required for describing all
relevant process input output dynamics. A non-parametric model does
not require such information, it simply describes the observed dynamics
in the form of a matrix valued weighting function giving the weights to
be applied to past input signal vectors to include their contribution to the
presently observed vector value. If no additional a-priori knowledge on
output noise characteristics is available ordinary least squares parameter
estimation methods can be applied. Output error minimization then results
in a model well suited for simulation of future process outputs. The criterion
function to be minimized for FIR estimation is

$$\min_{M} \arg V = \min_{M} \arg \| W(\hat{Y}(M) - Y) \|_2 \qquad (8)$$

with

$$Y = [y_k, y_{k-1}, \ldots, y_{k-m}]$$
$$\hat{Y}(M) = [\hat{y}_k, \hat{y}_{k-1}, \ldots, \hat{y}_{k-m}]$$

Y are the recorded process output responses. $Y(M)$ are the simulated
outputs and M are the model parameters of the non-parametric model in
this expression. W is a weighting function that may be applied. This function is convex in case the selected modelset is the FIR or the FSR modelset.
As a consequence, no initial value is needed for the model parameters to
guarantee the minimum being global. The model obtained from (8) will actually be the best linear approximation of the observed process behaviour.
If this behaviour properly represents the actual process dynamics, so will
the model. The conditions to be satisfied for the model to ultimately converge to the best linear approximation of true process behaviour are

- the applied input test signal may not be correlated with the process
output noise

- the applied input test signal must be uncorrelated amongst the various
input entries

- the applied input test signal must have almost equal power over the
whole relevant frequency range

- the length of the testdata sequence has to exceed at least 10 times
the largest relevant process time constant

On the basis of the initial non-parametric model the degree of the minimal polynomial of the semi-parametric MPSSM model can be determined. This degree is given by the number of independent Markov parameters. A good estimate for the degree of the minimal polynomial corresponding with all relevant process dynamics may be obtained by looking for this number of significantly independent Markov parameters from the available non-parametric model. Significance in this case is determined by the noise level on the initial model parameters. The minimal polynomial degree in this case can be determined from the singular values of the finite block Hankel matrix filled with Markov parameters transformed to vectors using the Vec operator

$$H = U \Sigma V^t \tag{9}$$

with

$$H = \begin{bmatrix} \text{Vec}(M_1) & \text{Vec}(M_2) & \ldots & \text{Vec}(M_i) \\ \text{Vec}(M_2) & \text{Vec}(M_3) & \ldots & \text{Vec}(M_{i+1}) \\ \vdots & \vdots & \vdots & \vdots \\ \text{Vec}(M_{N-i+1}) & \text{Vec}(M_{N-i}) & \ldots & \text{Vec}(M_N) \end{bmatrix}$$

$$M_i = [m_1, m_2, \ldots, m_p]$$

$$\text{Vec}(M_i) = [m_1^t, m_2^t, \ldots, m_p^t]^t$$

An appropriate value for the degree of the minimal polynomial corresponds to the number of singular values above the noise level, which can be recognised to be the steady level the singular values assume in cases where the noise on the Markov parameter entries is not correlated. In practice, it is advisable to test a couple of degrees of the minimal polynomial with respect to the qualities of the finally obtained model for control system design. These qualities are determined by various model properties

- non minimum phase zeros

- condition of the model related to its Hankel singular values

- condition of the transfer function (cf. (6))

Once a model order has been chosen, the initial MPSSM model can be determined. In practice, two different algorithms are applied

- the Gerth algorithm [Gerth, 1972]

- the finite block Hankel matrix realization algorithm of Zeiger and McEwen [Zeiger, 1974]

The algorithm of Gerth is a two step algorithm that immediately results in an initial MPSSM model. The first step gives a set of minimal polynomial coefficients by solving a set of linear equations in least squares sense. The second step results in the set of start sequence of Markov parameters using the earlier estimated minimal polynomial coefficients. Although the algorithm does not minimize the difference between the initial non-parametric model and the MPSSM model looked for in some well-defined norm, the model, in practice, mostly appears to be good enough to start the final optimization of the MPSSM model parameters. The algorithm of Zeiger and McEwen gives a State Space model on the basis of the approach developed by Ho and Kalman [Ho 1966]. A state space realization is calculated from the finite block Hankel matrix. To get an initial MPSSM model, the state space model has to be one to one translated to the corresponding MPSSM model. As the initial MPSSM model is fitted to the earlier estimated non-parametric model, this model will be close to the best linear approximation of true process behaviour recorded in the process data used for identification. The subsequent identification step involves the fine tuning of the MPSSM model parameters by numerical minimization of a (weighted) output error criterion in least squares sense

$$\min_{a,M} \ \text{arg} \ V = \min_{a,M} \ \text{arg} \ \| \ W(\hat{Y}(a_i, M_i, M_0 \mid i = 1, 2, \ldots, r) - Y) \ \|_2 \quad (10)$$

As a result, a semi-parametric model is obtained with the same properties as the non- parametric model. This implies that the model will be the best linear approximation of true process behaviour if the conditions mentioned above for the convergence of the non-parametric model also hold for this model. The final identification step to be executed is a model reduction step to get rid of final redundancies related to non-parametric treatment of model structure. The model reduction technique applied is based on the calculation of the balanced realization of the state space representation [Moore 1981] of the MPSSM model obtained from (10) [Pernebo, 1982]. Both the controllability gramian P and the observability gramian Q of the balanced realization are equal to a diagonal matrix Σ. The diagonal elements of this matrix are the Hankel singular values of the system ordered to decreasing value, i.e. the largest Hankel singular value is upper left and the smallest is lower right on the diagonal.

$$\begin{bmatrix} A_{11} & A_{12} \\ A_{21} & A_{22} \end{bmatrix} \begin{bmatrix} \Sigma_{11} & 0 \\ 0 & \Sigma_{22} \end{bmatrix} \begin{bmatrix} A_{11}^t & A_{12}^t \\ A_{21}^t & A_{22}^t \end{bmatrix} - \begin{bmatrix} \Sigma_{11} & 0 \\ 0 & \Sigma_{22} \end{bmatrix}$$
$$+ \begin{bmatrix} B_1 \\ B_2 \end{bmatrix} [B_1^t \ B_2^t] = 0 \quad (11)$$

$$\begin{bmatrix} A_{11}^t & A_{12}^t \\ A_{21}^t & A_{22}^t \end{bmatrix} \begin{bmatrix} \Sigma_{11} & 0 \\ 0 & \Sigma_{22} \end{bmatrix} \begin{bmatrix} A_{11} & A_{12} \\ A_{21} & A_{22} \end{bmatrix} - \begin{bmatrix} \Sigma_{11} & 0 \\ 0 & \Sigma_{22} \end{bmatrix}$$

$$+ \begin{bmatrix} C_1^t \\ C_2^t \end{bmatrix} [C_1 \ C_2] = 0 \tag{12}$$

The reduced order model is obtained from these equations by making Σ_2 equal to zero. The corresponding reduced order state space model is

$$\begin{aligned} \tilde{A} &= A_{11} + A_{12}(I - A_{22})^{-1}A_{21} & (13) \\ \tilde{B} &= B_1 + A_{12}(I - A_{22})^{-1}B_2 \\ \tilde{C} &= C_1 + C_2(I - A_{22})^{-1}A_{21} \\ \tilde{D} &= D + C_2(I - A_{22})^{-1}B_2 \end{aligned}$$

As has been indicated in the beginning of this section, precautions have to be taken to guarantee that the steady state gains of the model will be right. The process identification procedure described above, does not take care of this. To fulfil this additional requirement the optimizations given in (8) and (10) have to be done under constraints with respect to the steady state gains. Also the model approximations based on the Gerth algorithm or the algorithm of Zeiger and McEwen require adjustments. To circumvent this problem the initial MPSSM model parameters can be adjusted by adjusting the steady state gains to the a-priori known gains.

4 Internal model-based process control

The best way to control a process is to continuously drive it within the operating constraints as fast and as smoothly as possible to the desired conditions on the basis of its known dynamic behaviour. This requires the use of detailed knowledge of the process dynamics to compute the input manipulations needed for doing so. Control of the process then actually becomes feedforward control where a mathematical model that accurately describes the process dynamics is used for calculating the required future input manipulation needed for bringing the process outputs to the desired values.

Recorded deviations from the desired values are to be compensated for in a feedforward way. Model predictive control strategies do this job. The internal model based process control scheme is an example of such a model predictive control system (cf. Fig. 3). The model obtained with the identification procedure described in the previous section is the heart of this model based control system. Three properties of the ultimate control system are strongly interrelated:

- model accuracy

- control system performance

- control system robustness

The better the model accuracy, the better both robustness and performance will be. For given model qualities, robustness and performance can be exchanged. The ultimately reachable performance in a predefined control system configuration will be dictated by model inaccuracies. Model inaccuracies are determined by three factors in general:

- too low order approximation of high order process dynamics

- process non-linearities which are not covered by the (mostly) linear process models

- time-varying processes which are approximated by a time invariant process model

Presently, it is only the first class of model inaccuracies that can be properly considered in control system design whereas the final two classes of model inaccuracies, in practice, appear to be far dominant. Further research into the effects of these model inaccuracies and their consideration in control system design is required. Design of the internal model-based control system implies the design of the various modules of the control system

- the dynamic precompensator

- the feedback control system

- the feedforward setpoint compensator

- the feedforward disturbance compensator

- the constraints handler

Depending on the demands on the ultimate control system, various design strategies can be applied. A poor man's version is the design of a dynamic precompensator using the implicit model following techniques based on classical LQ design methods [Kreindler, 1976; Backx, 1987] to make the system diagonally dominant in combination with single loop feedback control filters. This approach has two advantages

- it is a straightforward design technique that does not require specific knowledge of controller design techniques

- implementation of the controller can be done in almost every environment due to the fact that the system is made diagonally dominant first and feedback control can often be realized with single loop controllers

The disadvantage with this approach is the conservative design resulting in good robustness but restricted performance of the controller. The control system in general will be rather sluggish due to the decoupling which restricts the bandwidth of the controller to minimum of all possible input output transfers. More sophisticated versions of the internal model based control system are obtained by using H_∞ design techniques. The H_∞ techniques are used to design the control system as much as possible in line with specifications for the controlled process and up to constraints given by the process dynamics. The capabilities offered by the weighting filters (cf. (3)) to shape performance and robustness characteristics of the control system enable maximum flexibility in designing in the right system characteristics. The maximum obtainable performance and robustness is, of course, restricted by the process properties and by the model inaccuracies. The control system is designed to best approximate the inverse of that part of the process dynamics that can be inverted. In the design, all wishes with respect to performance for the various outputs and with respect to transient properties can be considered and used for the design. To enable optimization of process behaviour considering operating constraints for the process, a constraints handler is used. The constraints handler is designed to handle both hard and soft constraints on both amplitudes and rates of change of respectively inputs, outputs and states. Hard constraints in this respect are constraints given by physical restrictions in the process. These constraints may never be violated. If operation of the process within these constraints does not appear to be feasible, the process has to be brought back to safe operating conditions without bothering about performance. The soft constraints are related to preferred operating conditions for the process such as, e.g., operation with minimum energy consumption. If these constraints appear not to be feasible, they may be skipped at the price of a somewhat worse performance. The additional degrees of freedom available in non-square systems are used for minimization of an additional criterion and for operating process inputs as close as possible to preferred values.

5   Some applications

The techniques discussed have been successfully applied for the control of a broad range of processes. Two examples of these applications will be discussed. The first application concerns the control of both top and bottom product impurities of a distillation column. The C6/C7 splitter has to be operated over a broad operating range amongst which the operation at high purity levels for both top and bottom with feed disturbances. An outline of the column and of the control system configuration is given in Fig. 7.1. The multivariable control system has two manipulated variables, reboiler stream flow and reflux flow. The controlled variables are respectively top and bottom impurities. To enable operation of the process over a broad range, the process dynamics are linearized by using log functions to the measured impurities. This allows for robust control of the process over a very broad operating range with one control system. The step responses are given in Fig. 7.2. The control system also uses measured feed disturbances for feedforward compensation. Fig. 7.3 gives the variation of the feed flow. The control system has been designed using the H_∞ design techniques. Some response characteristics of the controlled process are given in Fig. 7.4, while Fig. 7.5 shows the manipulated variables.

The second example concerns the control of a low pressure gasphase polymerization reactor. This type of reactor is used for the production of polyethylene, which is about the most important plastic nowadays. The reactors are generally used for the production of a broad range of different polymers (grades). A grade of polyethylene is mainly determined by the partial pressure of the monomer in the reactor, the mole ratio of co-monomer in the gasphase reactor temperature and production rate. Control of a reactor implies simultaneous control of these four variables. A changeover from one grade to another implies loss of production capacity and mostly results in so-called wide spec product of low value Control of the reactor in such a way that both the performance of the reactor with respect to polymer quality and reduction of changeover time and the volume of wide spec product can be reduced as required. An outline of the reactor is given in Fig. 7.6.

The process is unstable due to the strong exothermic character of the process. Reaction capacity is therefore limited by the capacity of the applied heat exchanger. The control system covers three controllers (cf. Fig. 7.7)

- primary temperature control

- a cascaded pressure control system

- a multivariable controller controlling the partial pressure C_2, the C_x/C_2 ratio the temperature of the bed and the production rate

The manipulated variables of the multivariable controller are the total pressure setpoint, the C_x/C_2 setpoint, the catalyst feed and the temperature setpoint. Fig. 7.8 shows the pressure simulated by the model in comparison with recorded process pressure. Fig. 7.9 shows the pressure response of the IMC controlled pressure in comparison with the pressure controlled with a

PID controller. Fig. 7.10 shows the model validation results for the outer loop model. An overview of a simulated grade change is given in Fig. 7.11. The controller is tested against a rigorous model. With the controller a grade change can be executed in two hours compared to at least 6 hours for a conventionally controlled grade change. Fig. 7.12 shows the grade change with conventional control. Fig. 7.13 shows the grade change with IMC.

6 Conclusions

An overview has been given of the main engineering aspects of the modelling of relevant process dynamics using process identification techniques and of the design of control systems based on the obtained model. Currently applied techniques for the design of robust high performance industrial control systems have been outlined. The techniques have been successfully applied to many different processes. Two examples of application of the model based control system design techniques have been discussed: the application to a distillation column and the application to a gasphase polymerization reactor.

The techniques discussed still require further development, however, to include a maximum a-priori knowledge on model inaccuracies to get the best possible performance of the control systems in combination with guaranteed robustness. In particular, in combination with constraints handling, detailed knowledge of system properties in cases of a reduction of the number of available manipulated inputs or controlled outputs has to be included in the adjusted control strategy to still guarantee best possible performance and robustness.

7 Figures

8 References

[1] Åström, K.J. and P. Eykhoff, (1970) System identification - A survey, Automatica, Vol. 7, pp. 123-162

[2] Backx, A.C.P.M., (1987), Identification of an industrial process: A Markov parameter approach, PhD Thesis, Eindhoven University of Technology, Eindhoven, The Netherlands

[3] Backx, A.C.P.M. and A.H. Damen, (1992), Identification for the control of MIMO processes, IEEE Trans. AC, Vol. AC-37, No. 7, pp.

980-986

[4] Boyd, S.P. and C.H. Barratt, (1991), Linear Controller Design, Limits of performance, Prentice Hall, Information and System Sciences Series, Englewood Cliffs, New Jersey

[5] Cutler, C.R. and B.L. Ramaker, (1980), Dynamic Matrix Control - A computer control algorithm, Proc. Joint American Control Conf., San Francisco, California, USA

[6] Doyle, J.C., (1982), Analysis of feedback systems with structured uncertainties, IEE Proceedings, Vol. 129, Part D, No. 6, pp. 242-250

[7] Doyle, J.C., (1984), Lecture Notes. ONR/Honeywell workshop on advances in multivariable control

[8] Doyle, J.C., K. Glover, P.P. Khargonekar and B.A. Francis, (1989), State space solutions to standard H_2 and H_∞ control problems, IEEE Trans. AC, AC-34, pp. 831-847

[9] Doyle, J.C., A. Packard and K. Zhou, (1991), Review on LFT's, LMI's and μ, Proc. of the 30th Conference on Decision and Control, Brighton, UK, pp. 1227-1232

[10] Doyle, J.C., B.A. Francis and A.R. Tannenbaum, (1992), Feedback control systems, MacMillan Publishing Company, New York, USA

[11] Eykhoff, P., (1974), System Identification; Parameter and State Estimation, John Wiley & Sons

[12] Garcia, C.E. and D.M. Prett, (1986), Design methodology based on the fundamental control problem formulation, The Shell Process Control Workshop, December 15-18, 1986, Chapter 1, Workshop papers, pp. 3-25, Butterworths, Stoneham, MA

[13] Garcia, C.E., D.M. Prett and B.L. Ramaker,(1988), Fundamental Process Control, Butterworths, Stoneham, MA.

[14] Garcia, C.E., D.M. Prett and M. Morari, (1988), Model Predictive Control: Theory and Practice, Automatica

[15] Ljung, L. and T. Söderström, (1983), Theory and practice of recursive identification, The MIT press.

[16 Ljung, L., (1987), System Identification: Theory for the user, Prentice-Hall Inc., Englewood Cliffs, N.J.

[17] Maciejowski, J.M., (1989), Multivariable Feedback Control Design, Addison-Wesley Publishers, Wokingham, UK

[18] Morari, M., (1988), Process Control Theory: Reflections on the past and goals for the next decade, The Second Shell Process Control Workshop, December 12-16, 1988, Butterworths, Stoneham, MA

[19] Morari, M. and E. Zafiriou, (1989), Robust Process Control, Prentice Hall, Englewood Cliffs, New Jersey

[20] Richalet, J., A. Rault, J.L. Testud and J. Papon, (1978), Model predictive heuristic control: Application to industrial processes, Automatica, Vol. 14, pp. 413-428

[21] Schrama, R., (1992), Approximate Identification and Control Design, PhD Thesis, Delft University of Technology, Delft, The Netherlands

[22] Söderström, T. and P. Stoica, (1989), System Identification, Prentice Hall, UK

[23] Soeterboek, R., (1990), Predictive Control - A Unified Approach, PhD Thesis, Delft University of Technology, Delft, The Netherlands

[24] Swaanenburg, H.A.C., W.M.M. Schinkel, G.A. van Zee and O.H. Bosgra, (1985), Practical aspects of industrial multivariable process identification, In: Barker H.A. and P.C. Young (Eds.) Identification and System Parameter Estimation, IFAC Proc. Series 1985, York, U.K., pp. 201-206

[25] Wahlberg, B., (1989), System identification using high order models, revisited, Proc. 28th Conf. on Decision and Control, Tampa, Florida, pp. 634-639

[26] Wahlberg, B. and L. Ljung, (1991), On estimation of transfer function error bounds, Proc. of the European Control Conference, Grenoble, France

[27] Willems, J.C., (1986), From time series to linear systems; Part I: Finite dimensional linear time invariant systems, Automatica, Vol. 22, pp. 561-580

[28] Willems, J.C., (1986), From time series to linear systems; Part II: Exact modelling, Automatica, Vol. 22, pp. 675-694

[29] Willems, J.C., (1987), From time series to linear systems; Part III: Approximate modelling, Automatica, Vol. 23, pp. 87-115

[30] Zadeh, L.A., (1962), From circuit theory to system theory, Proc. IRE, Vol. 50, pp. 856-865

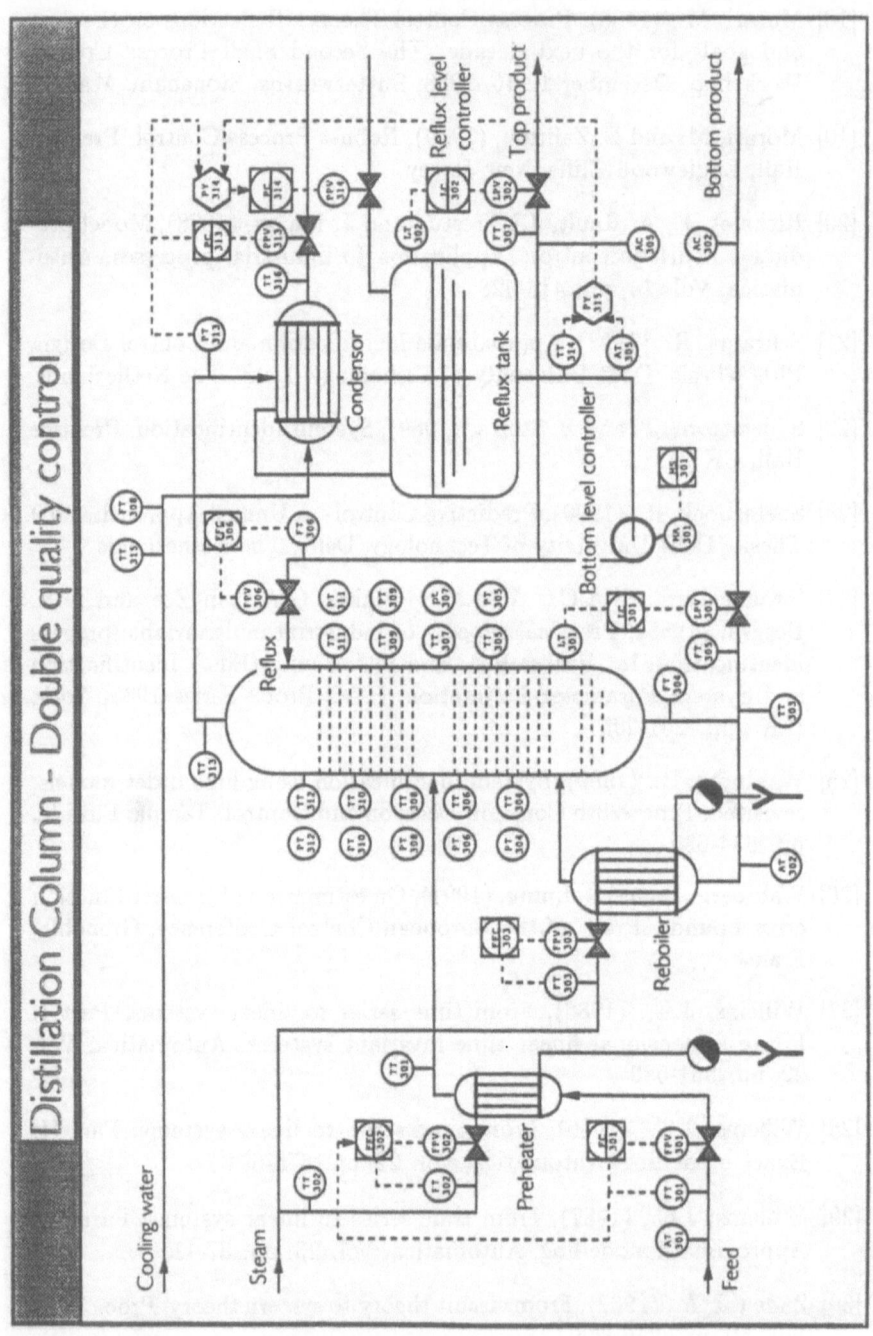

Figure 7.1: Distillation column double quality control

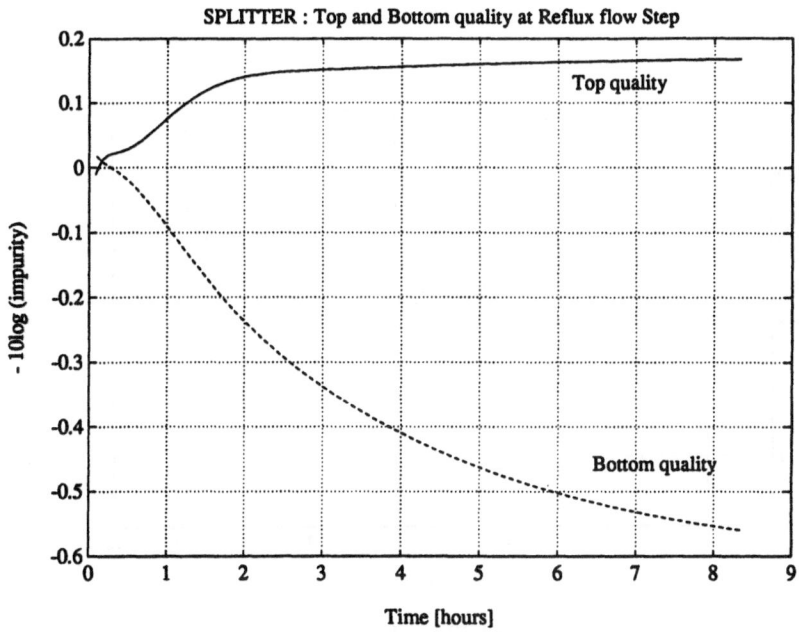

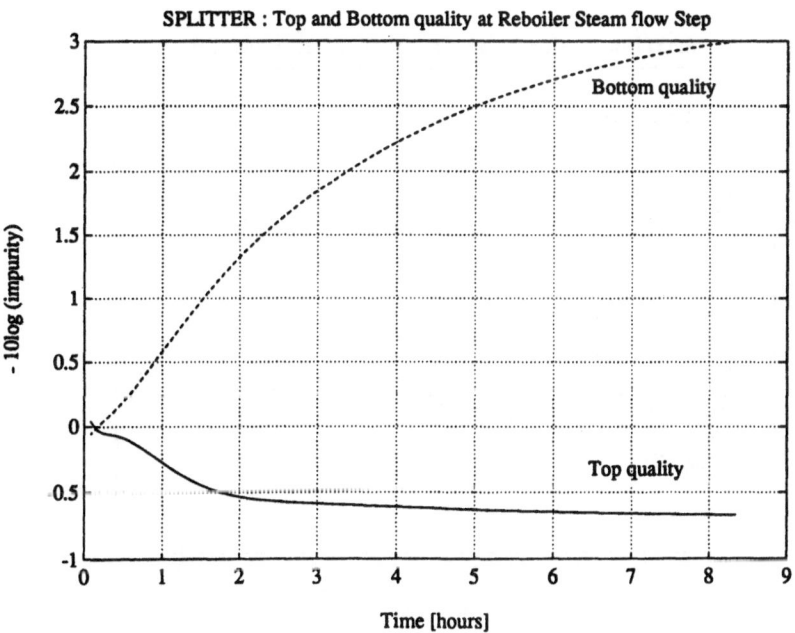

Figure 7.2:

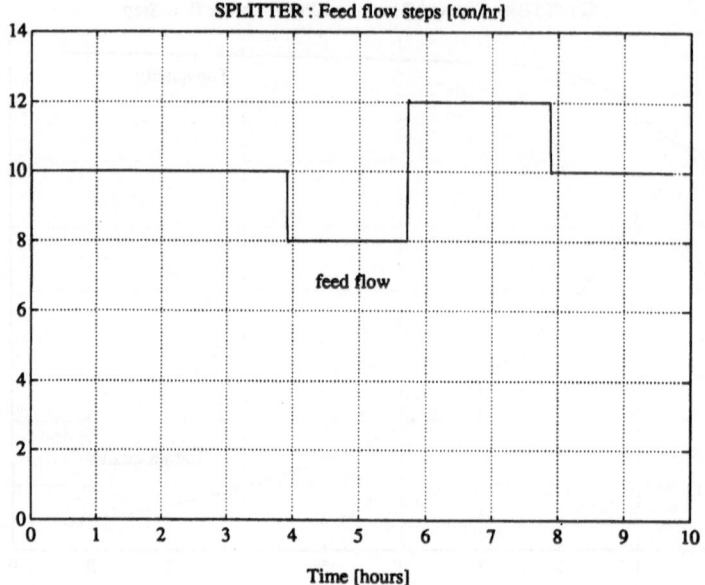

Figure 7.3:

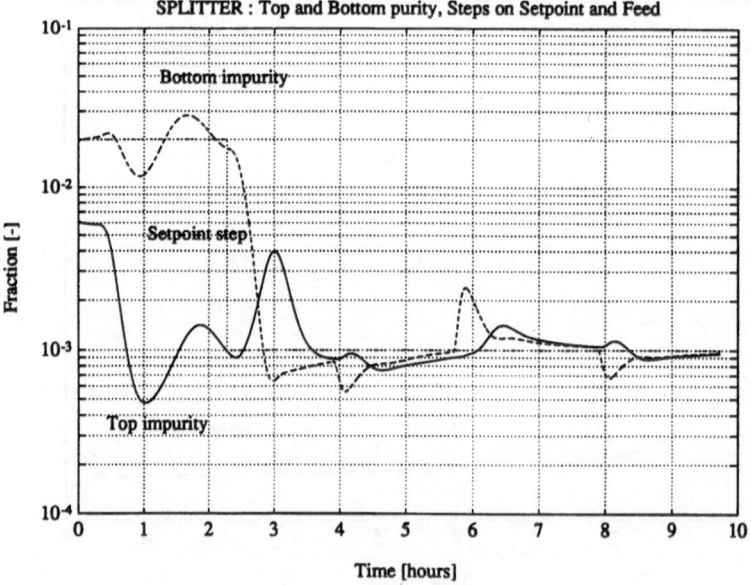

Figure 7.4:

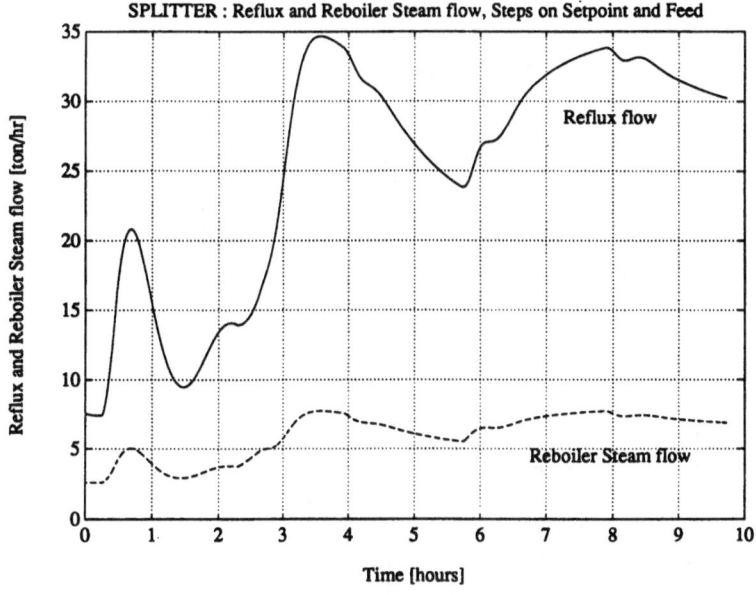

Figure 7.5:

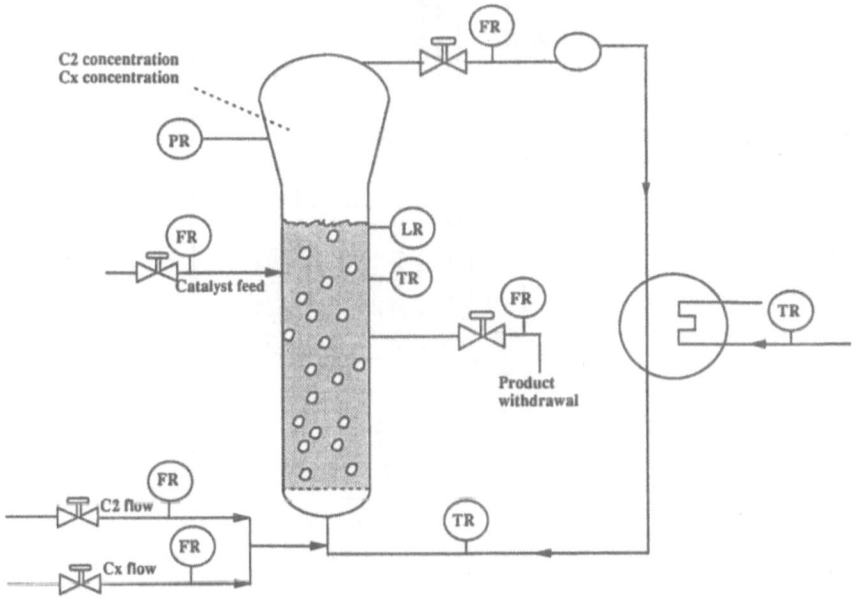

Figure 7.6:

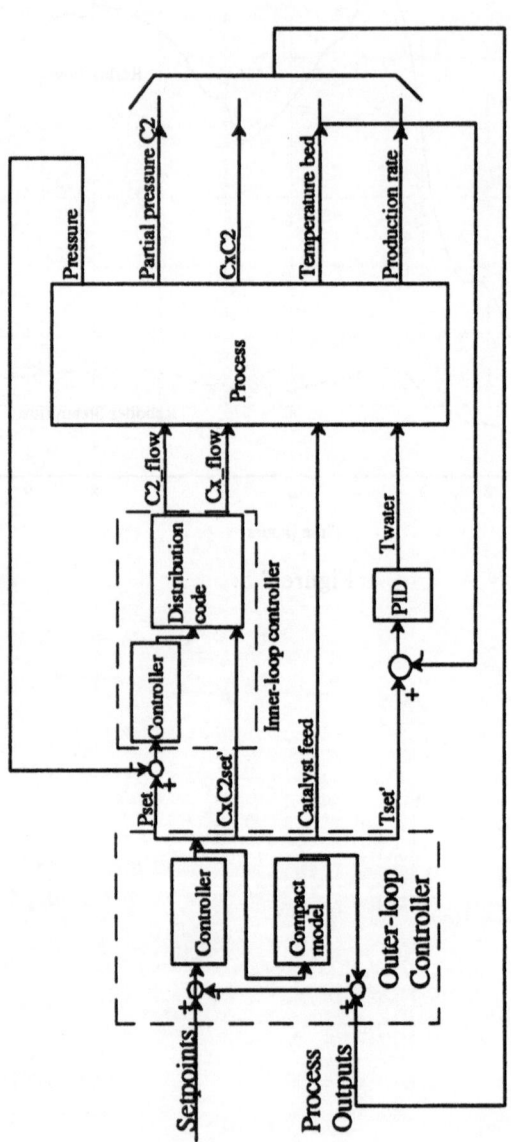

Figure 7.7:

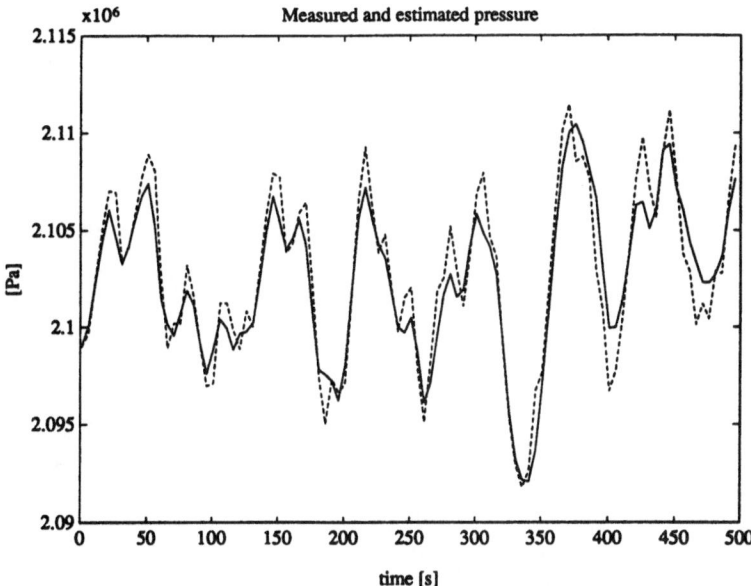

Figure 7.8:

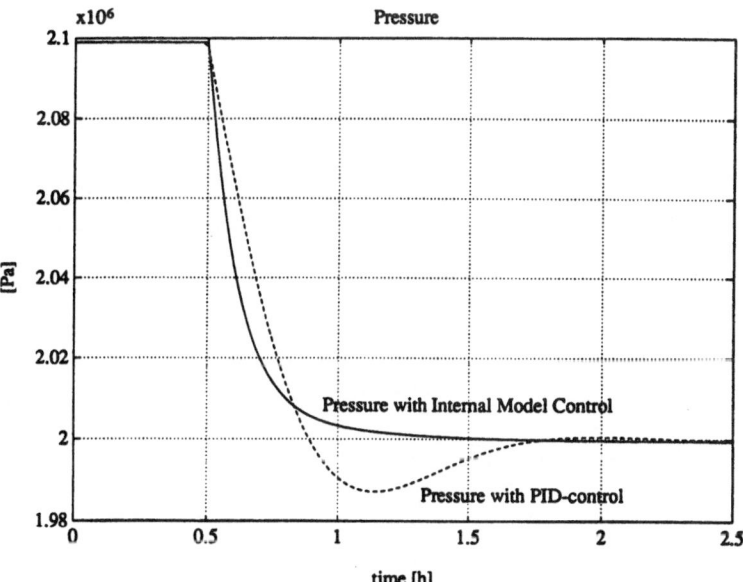

Figure 7.9:

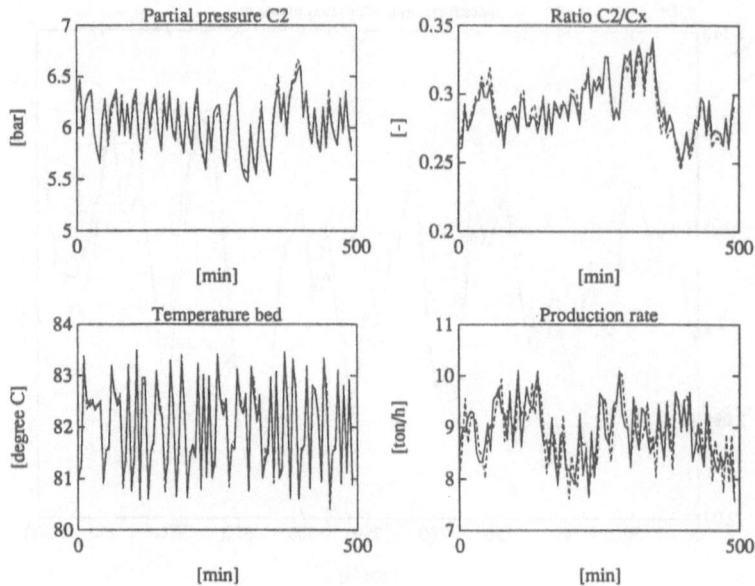

Figure 7.10:

Setpoint	From	To
Partial Pressure monomer	6 bar	5 bar
Ratio Cx/C2	0.3	0.6
Temperature	82°C	82°C
Production rate	8.6 ton/h	5 ton/h

Figure 7.11:

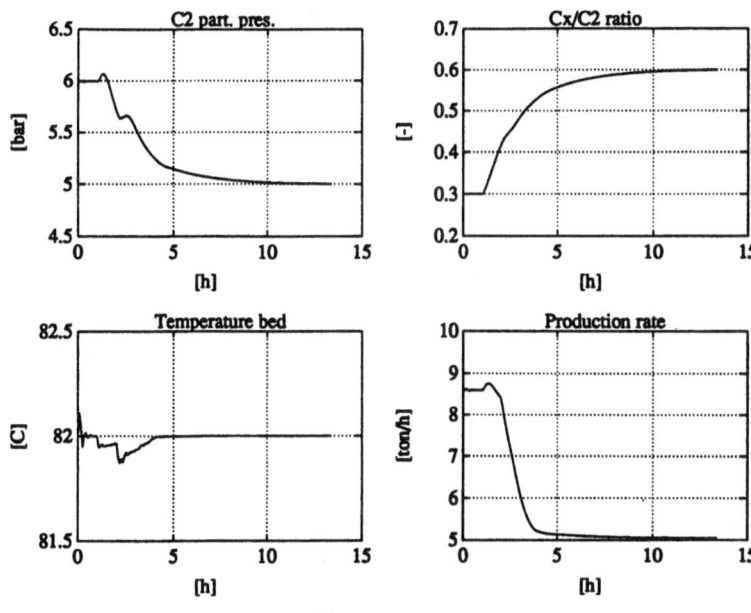

Figure 7.12:

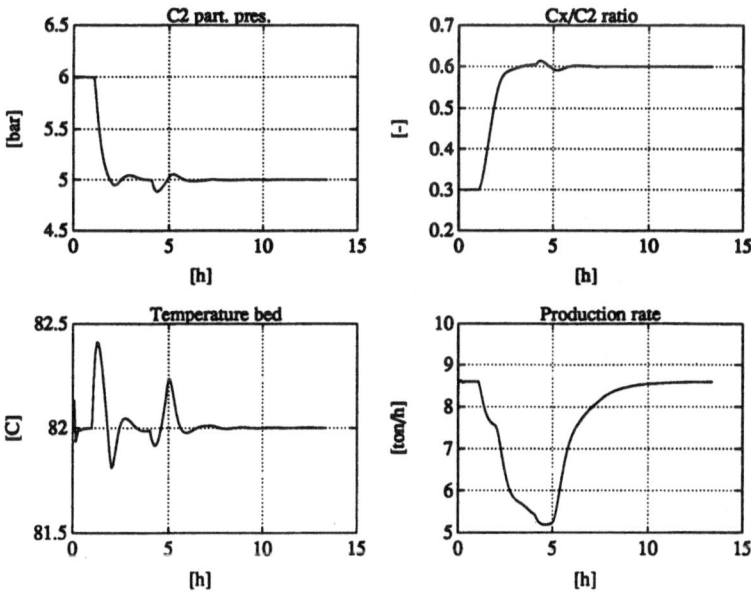

Figure 7.13:

5. Towards a Joint Design of Identification and Control ?

M. Gevers *

Abstract

This paper aims at introducing the reader to the various issues that arise in the development of a coherent methodology for the development of robust control design on the basis of models identified from data. When a reduced complexity model is identified with the purpose of designing a robust controller, the model is just a vehicle for the computation of a controller. The design of the identification and of the controller must be seen as two parts of a joint design problem. The central message of this paper is to show that the global control performance criterion must determine the identification criterion. This leads to non standard identification criteria, which can be minimized by appropriate experimental set-ups.

1 Introduction

The intensive work that is presently going on in the general area of identification in connection with robust control design finds its origin in the awareness, among people from both the identification and the robust control community, that a wide gap exists between the premises on which robust control design is built and the tools and results that 'classical' identification theory is able to deliver. (By 'classical' I mean the theory as it existed at the end of the nineteen eighties, and which can be found in classical textbooks such as [Lju87] or [SS89].) To understand how such gap has materialized, a historical retrospect is perhaps appropriate.

---

*Center for Engineering Systems and Applied Mechanics, Louvain University, Bâtiment Maxwell, Place du Levant 3, 1348 Louvain la Neuve, Belgium. Fax : +32 10 478667. Tel. : +32 10 472590. Email: gevers@auto.ucl.ac.be.
This paper presents research results of the Belgian Programme on Interuniversity Poles of attraction initiated by the Belgian State, Prime Minister's Office, Science Policy Programming. The scientific responsibility rests with its author.

A historical perspective

Until 1960 and the publication of Kalman's celebrated work on state space models [Kal60], control design was for the most part based on graphical techniques using Bode plots and Ziegler-Nichols charts. The controllers that were built were PI and PID controllers; they were not model based. Robustness concepts were incorporated in the design techniques in the form of gain and phase margins.

The major impact of Kalman's work was perhaps not so much the introduction of state space models, but the replacement of graphical design techniques by *model based certainty equivalence* control design. Linear Quadratic Gaussian (LQG) control and model reference control became major new design techniques. Parametric models became the central focus of attention, and it is therefore natural that the development of parametric identification techniques followed that of model based control design on its heels. One of the important early papers on parametric identification is that of Åström and Bohlin [ÅB65] which introduced many of the formal concepts that dominated identification theory for about 20 years.

The Achilles' heel of the model based control era of the sixties and seventies was the certainty equivalence principle. Except for the gain and phase margins naturally inherited by LQG controllers, the model based control design methods did not lend themselves easily to tractable design methods that would incorporate plant model uncertainty descriptions. Perhaps for that reason (or would it be sheer laziness?), for a long time identification theorists focused on questions of convergence and efficiency of parameter and transfer function estimates in the case when the true system was contained in the model set, rather than on the effect on controller performance of plant/model errors due to undermodeling.

It is interesting to observe that the first analyses of the *interplay between identification and control design* were produced during this certainty equivalence era. Motivated by problems in communications, Fel'dbaum introduced the concept of *dual control*, in which he showed that, when controlling a system whose parameters are unknown, the control effort must pursue the dual goal of "investigating" and "directing" [Fel60]. Fel'dbaum then called "investigation risk" the loss in the achieved performance due to the fact that the control is not optimal in view of obtaining information about the system (resulting in a subsequent suboptimal control action), and he called "action risk" the loss in the achieved performance due to the fact that the control causes a deviation from the best achievable state.

The ideas of Fel'dbaum inspired Åström and Wittenmark [ÅW71], who addressed the problem of *combined identification and control* in the context of exact modeling (i.e. the true system is in the model set) using a linear regression (ARX) model structure. They considered the direct minimization of both a one-step and a multi-step minimum variance regulation criterion. In the case of an exact model structure, the unknown parameters and their covariance matrix can be added to the state, the problem can be re-framed as an optimal control problem for an augmented system, and solved (at least in principle) using dynamic programming. Even though the model errors in [ÅW71] are only noise-induced errors - and not bias errors due to unmodeled dynamics - the contribution of Åström and Wittenmark is significant for the following reasons:

- it showed, probably for the first time and albeit in the context of exact model structure, that the combined identification and control problem can be formulated as the minimization of a global control performance criterion, leading in simple cases to a simple and computable solution;

- it produced an optimal solution to the minimum variance regulator problem with unknown parameters in the form of a *cautious controller* with reduced gains;

The publication of Zames' paper [Zam81] marked the start of the model-based H_∞ robust control design period that dominated most of the eighties. The key technical result that made it all possible was the availability of the Youla-Kucera parametrization of all stabilizing controllers for a given plant model. The structured singular value (or μ-synthesis) approach later extended worst case control design from the realm of unstructured uncertainties (i.e. error bounds on the transfer function model) to that of structured uncertainties (i.e. error bounds on parameters). It is interesting to observe that these robust control design methods have been developed entirely in the context of models that are not data-based.[1] As the robust control design methods invaded the literature (if not the world of process control applications), the pressure grew to apply these methods on models identified from real data. This required the development of identification techniques able to deliver error bounds in the case of undermodeling and, more importantly, it required an understanding of the interactions between control design and identification design in this restricted complexity modeling situation.

In the identification community - and for the reasons explained above - very few attempts had been made to quantify errors on estimated transfer functions due to the use of restricted complexity model sets. In fact, at the end of the 'classical era' of the development of identification theory (say, 1987), the only useful result available on the error due to undermodeling (also called the 'bias error') was an implicit characterization of the convergence point of the parameter estimation algorithm [WL86]. While very

---

[1] Having just presented a sketch of the robust control design steps, John Doyle, one of the key contributors to robust control theory, recently asked a bunch of identifiers: "Can you guys tell me how data can improve our designs in any way?" [Doy92]

useful for design, this integral formula did not give a clue as to a bound on the error between true and estimated transfer functions.

As for the interaction between identification and control in the case of restricted complexity models, a small step had been made in [GL86], where, for certain control performance criteria, it was shown that the performance degradation due to errors in the identified model can be minimized by performing the identification in closed loop and with an appropriate data filter. Since the 'appropriate data filter' is itself a function of the optimal controller, and hence of the unknown model, the result can only be applied in practice by replacing the optimal filter by an approximation, thus leading to the presently popular iterative control and identification design methods. The result of [GL86] was the first instance in which closed loop identification was shown to be helpful, rather than something to be avoided at all cost. The necessity of performing closed loop identification when the model is to be used for control design has since been recognized as a key element in the successful application of identification for control, as we shall see in Section 5.

State of the art at the end of the eighties

At the end of the eighties, the state of affairs concerning the connection between identification and control can thus be characterized as follows:

- robust control control design tools were being developed at a rapid pace, under the assumption that prior hard bounds were available on transfer function errors or parameter errors, or a combination of these;

- with the exception of a small school of thought that had developed "bounded error identification methods", mainstream identification theory had almost come to a halt. It was able to deliver sophisticated models and techniques, but was unable to quantify the errors on identified transfer function models in the frequency domain, as required by the H_∞ robust control design theory;

- some preliminary but scant results were available on optimal design of the identification when the model is of restricted complexity and is to be used for control design [GL86]; these early results were based on H_2 performance measures, and made no connection yet with the new theories of H_∞ robust control;

- except for the case of exact model structure (see [ÅW71]), no results were available on the interplay between identification and control, or a fortiori on their combined design.

It is thus clear that a rather huge gap had developed at the end of the eighties between the tools and assumptions of robust control design and the techniques that identification theory had produced. The most obvious manifestation of this gap, and the one that has triggered most of the present research activity, was the realization that robust control theory requires a priori *hard bounds* on the model error, whereas classical identification theory delivers at best *soft bounds*[2] in the case where the system is in the model set and no bounds at all in the case of undermodeling.

Nowadays, one has come to realize that the great 'hard-versus-soft' bound debate is not the real issue. An identification and control design method that leads to a closed loop system that is stable with probability .99 is of course just as acceptable as an H_∞-based design that leads to a 'guaranteed stable' closed loop, but that is based on prior error bounds that cannot always be verified. However, the main focus of research - and by far - is still on trying to produce identification methods that allow for the computation of uncertainty bounds, whether hard or soft. This is of course a most pertinent scientific pursuit: whatever the eventual objective, it is unreasonable to deliver a model to a user without a statement about its quality. However, if the objective of the identification is to design a robust controller, then the most important issue is probably not the estimation of uncertainty bounds on the identified model, but the design of a control-oriented identification or, even better, the synergistic design of identification and control.

This leads me to suggest that the new and fashionable research area that deals with the interconnection of identification and control can be subdivided into three areas, that correspond to three aspects of the identification and robust control design problem:

1. Estimation of uncertainty bounds on identified models;

2. Identification *for* robust control design;

3. The combined (synergistic) design of the identification and control.

The point of this classification is to stress that the goals pursued in these three areas of research are quite different.

A brief review of the present research efforts.

For the moment, the mainstream approach seems to be 'Perform the identification with a method that allows the computation of error bounds on the estimated model, then design a robust controller using that model and its bounds'. The main focus of research is therefore on the estimation of

---

[2]By soft bounds we mean confidence intervals or ellipsoids in the probabilistic sense.

error bounds, whether hard or soft, and novel identification techniques are being produced for the sake of delivering such bounds. The problem is that identification methods whose sole merit is to deliver accurate error bounds on restricted complexity models may well produce nominal models that are ill-suited for robust control design: that is, the frequency distribution of the model error may be such that they lead to poor closed loop performance. Examples have been given in [Sch92a,b].

The idea of the second line of research above is to develop identification methods that will produce models whose uncertainty distribution, over the frequency range, allows for high performance robust control design. Thus these models should have low uncertainty where closed loop control specifications require this, but they may have large uncertainty in frequency bands where this does not imperil closed loop stability or penalize closed loop control performance. Results are now available for the tuning of the identification method towards such objectives: the identification must be performed in closed loop with an appropriate data filter. These results rely on an understanding of the interactions between identification, robust stability and robust performance. As already mentioned, an early result leading to the idea of closed loop identification was [GL86]; however, that result was based on performance degradation ideas and the authors failed to consider the connections with the newly emerging robust control theory. It is probably fair to say that the book by Bitmead et al [BGW90] gave a major impetus to this second line of research (as well as the third one). There, the robust stability and robust performance criteria of H_∞ control design were used as the key ingredient for an understanding of the identification/control interactions, in an adaptive control framework. Using H_∞-based prescriptions as a guide (or an excuse), a design was proposed using a combination of Least Squares identification and LQG/LTR control design. In the scheme of [BGW90], the identification design takes account of the robust control requirements through the data filters and the prescription of closed loop identification. Thus, this scheme fits in the framework of identification *for* control. It was followed by other results, using different identification techniques and/or different control design schemes, which all came to similar conclusions. The Delft school played and continues to play a major role in the progress of control-oriented identification and approximation: [Hak90], [HSV92], [Sch91], [Sch92a], [Sch92b].

The third line of research is to combine the identification and the control design in a mutually supportive way, from the point of view of robust stability and/or robust performance. Even though the objective might seem overly ambitious and elusive, some results are now emerging. They all take the form of iterative schemes in which a succession of identification and control design steps are performed, leading to more and more performant control systems. The identification steps are performed in closed loop using data obtained with the last controller operating on the actual plant. The control design steps use the most recently identified plant model. The different schemes vary in the identification criteria and techniques, the model structures that are used, the control design criteria, the way in which the model uncertainty is or is not used in the control design step, the way in which the performance requirements are increased or not between successive design steps: see [BYM92], [LAKM92], [LS90], [PB93], [Sch92a], [Sch92b], [ZBG91], [ZBG92]. Even though the specific identification and control design techniques vary between these schemes, they all have in common a succession of performance enhancement designs. The idea of redesigning controllers using closed loop data collected on the plant in order to improve performance is what process control engineers naturally tend to do. The merit of the recent research is to develop systematic and theoretically justified procedures to achieve this performance enhancement.

Conclusion for an introduction

To summarize, most of the present focus of research is on the estimation of error bounds. This problem is not only of independent interest, but it is also an important step towards the design of robust controllers based on identified models. However, the key ingredient for the successful application of robust control design methods to identified models is not so much the computation of error bounds, but it is to *let the global control performance criterion dictate what the identification criterion should be, and to design the controller in a way that takes account of data-based information about the plant/model mismatch.* This idea will be the focal point of this paper.

To achieve this objective requires a better understanding of the interconnections between closed loop identification and control design. Thus, this paper will focus on these interconnections. In Section 2, we introduce the concepts of *optimal loop*, *design loop* and *actual loop*, and we present the robust stability and robust performance constraints that are at the heart of the identification and control interplay. This interplay leads to three questions that correspond to the three research areas delineated above. These are briefly discussed in Section 3. In Section 4, we give a brief review of prediction error identification theory in open and closed loop, with the aim of displaying the role of the experimental set-up and the design variables in the properties of the identified model. These results are used in Section 5 to show how the identification criterion can be shaped to become a performance robustness criterion, which is itself tuned by the global control design criterion. This result serves as an inspirational source for the more ad-hoc iterative design schemes that are described in Section 6. These iterative schemes are the presently available alternatives to the combined design of identifier and controller in the form of a global - but for the moment elusive

- optimization problem. The formulation of this combined problem raises many deep and challenging open questions that will undoubtedly occupy numerous researchers in the years to come.

2   The identification/control interplay

Our first task is to demonstrate how identification and robust control interact in a model based control design procedure in which the model is constructed from data collected on the process.

To make things perfectly clear[3] about the set-up to which our ensuing developments apply, we shall make the following assumptions.

Assumptions

1. The true plant will be assumed to be representable as follows,

$$y_t = P(z)u_t + v_t, \qquad (1)$$

   where $P(z)$ is a scalar strictly proper rational transfer function, u_t is the input, v_t is an unmeasurable disturbance acting on the output y_t.

2. Prior knowledge about the system may have helped the designer to select a parametric model structure or may have given him insight about the achievable bandwidth, but the information about the dynamics of the process is assumed to be derived from data collected on the process. Hence no prior model or approximation of $P(z)$ is available.

3. The exact model structure is assumed to be unknown, but we shall consider that - possibly after some initial analyses including plant data information - the designer has set his or her eyes on a certain parametrized model set,

$$\mathcal{M} \triangleq \{\widehat{P}(z,\theta), \qquad \theta \in D_\theta \subset R^d\}, \qquad (2)$$

   together (possibly) with a noise model v'_t, where $\widehat{P}$ is a strictly proper transfer function. Thus, the model structure estimation will not be part of our discussion.

4. The true plant is not contained in the model structure, i.e. there exists no value of θ for which $\widehat{P}(z,\theta) = P(z)$ for almost all z.

---

[3] As Richard Nixon used to say.

There are, of course, control design procedures that are based not on data, but on a prior model of the system, perhaps with some knowledge or estimate about its quality. However, the situation described by our assumptions is typical of process control applications, in which such prior knowledge is usually not available and in which the model is necessarily data-based.

Thus, the task that we consider here is to design a feedback controller on the basis of a model that is to be identified from data taken on the process, using the available techniques of robust control design, given that whatever the identification exploits that are accomplished by the designer, the model will not be able to represent the true system accurately.

A typical control design type situation is that the designer has a global control performance criterion, $J_{global} \triangleq J(P, C)$, in mind. For example, in an H_∞ design, one might like to minimize the following control performance criterion,

$$J_{H_\infty} \triangleq J(P, C) \triangleq \|W(z)\frac{1}{1 + P(z)C(z)}\|_\infty, \qquad (3)$$

over a class of admissible controllers, where $W(z)$ is a weighting that reflects performance specifications and $\frac{1}{1+P(z)C(z)}$ is the sensitivity of the actual feedback system. In an LQG framework, J_{global} could take the form

$$J_{LQG} \triangleq J(P, C) \triangleq \lim_{N \to \infty} \frac{1}{N} \sum_{t=1}^{N}[(y_t - r_t)^2 + \lambda u_t^2], \qquad (4)$$

where the signals y_t, r_t and u_t are, respectively, the output signal, the tracking signal, and the to be designed control signal, and where λ is a positive weighting factor that reflects the respective importance given to the tracking error and the control effort.

These criteria cannot be minimized because the first one depends explicitly on the unknown $P(z)$, while the second depends on $P(z)$ through the dynamic relationship that links r_t, u_t and y_t. Instead, one designs a controller on the basis of an estimate $\hat{P}(z, \hat{\theta})$ of $P(z)$, which we shall in this paper consider to have been obtained from plant data by identification. Thus, one has to design both an identification method (taking into account that the model set is of restricted complexity) and a model-based control design procedure, possibly taking account of plant/model error information. In analysing the interplay between the identification and the control parts of the design, it will prove useful to consider the three feedback loops represented in Figures 1, 2 and 3.

The optimal loop of Figure 2.1 contains the true system in feedback with the optimal controller $C^{opt}(z)$. This optimal controller depends on

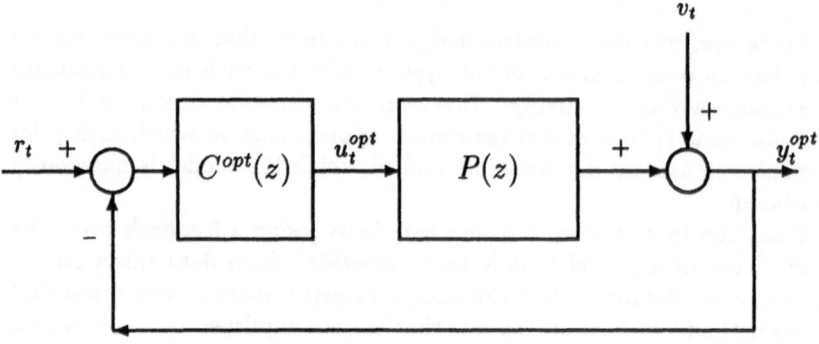

Figure 2.1: Optimal feedback loop

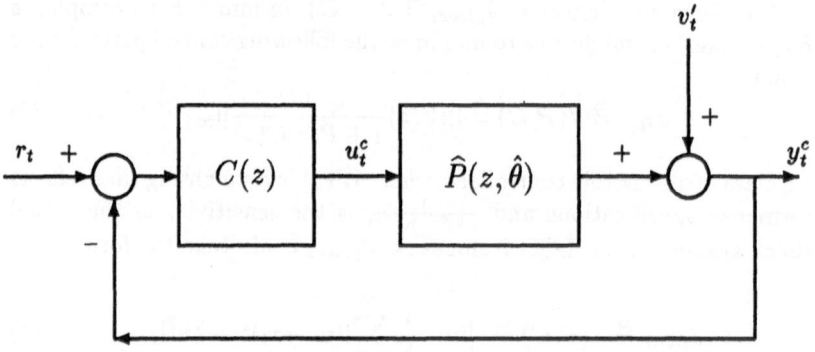

Figure 2.2: Nominal (or design) feedback loop

the unknown true system $P(z)$, and can therefore not be computed. The design of the controller $C(z)$ is conceptually performed on the basis of the nominal (or design) loop of Figure 2.2, in which the true plant is replaced by an identified model $\widehat{P}(z, \hat{\theta})$. The actual feedback loop of Figure 2.3 contains the true system $P(z)$ and the designed controller $C(z)$.

The reasons for drawing attention to these three figures is that much of the discussion about the interplay between identification and robust control is based on a comparison between these loops.

- The three loops are all driven by the same reference signal, r_t, while the noise signal, v'_t, in Figure 2.2 is an estimate of the actual noise source v_t.

- Ideally, one would like the identification and control design to be such that the performance achieved by the designed controller on the actual system is as close as possible to that achieved by the optimal controller. That is, one would like the loops of Figures 2.1 and 2.3 to

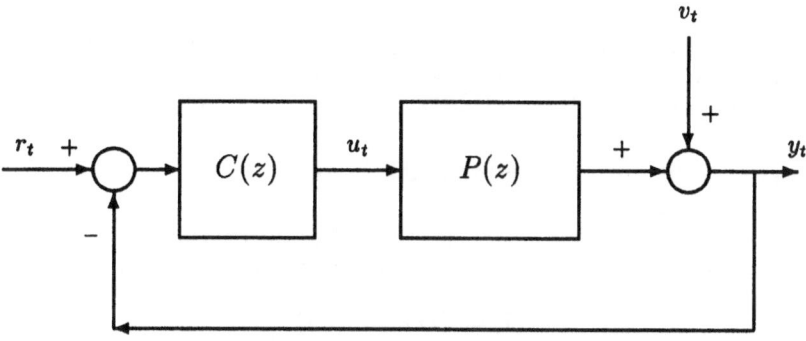

Figure 2.3: Actual (or achieved) feedback loop

be 'close to one another' in some sense. Since C^{opt} is unknown, it is usually impossible to use the closeness of these two loops as a design criterion. Instead, one compares the loops of Figures 2.2 and 2.3.

- One has to make sure that the controller $C(z)$ designed for the loop of Figure 2.2 stabilizes the actual loop of Figure 2.3: this is the concept of *robust stability.*

- The controller designed for the design loop must also produce on the actual loop an achieved performance that is not too different from the designed performance: this is the concept of *robust performance.*

- Thus, in robust analysis one wants the loops of Figures 2.2 and 2.3 to be 'close to one another' in some sense.

- Note that a comparison between the signals u_t^{opt}, u_t^c and u_t, as well as between y_t^{opt}, y_t^c and y_t, gives information about the mismatch between the corresponding closed loop systems. This is exploited in the scheme of Zang et al. [ZBG91] that we describe in Section 6.

In the classical robust stability and robust performance analyses, it is assumed that the 'nominal model' $\widehat{P}$ is obtained from prior information and/or modeling techniques, and that the plant/model error is either experimentally based or god-given. The key point is that the choice of $\widehat{P}$ is typically not a part of the control design procedure. Here, with a model $\widehat{P}(z,\hat{\theta})$ that is estimated from data within a set of $\mathcal{M} = \{\widehat{P}(z,\theta)\}$ of candidate models, the robust stability and robust performance requirements hinge both on the identification design and on the control design. *The model is only used as a vehicle to compute a high performance controller, and therefore it need not necessarily be a good open loop model of the plant.*

Without rederiving any of the theory, we now briefly summarize some fundamental formulae about robust stability and robust performance insofar as they clearly exhibit the identification/control interplay.

Robust stability of a unity feedback loop

We first introduce some notations. With $P(z)$ the true plant and $\widehat{P}(z,\hat{\theta})$ an estimated model, we define the additive plant/model error as

$$L(z,\hat{\theta}) = P(z) - \widehat{P}(z,\hat{\theta}). \tag{5}$$

We denote the designed mixed sensitivity function as,

$$\widehat{M} \triangleq C(z)[1 + C(z)\widehat{P}(z,\hat{\theta})]^{-1}, \tag{6}$$

the designed sensitivity function as,

$$\widehat{S}(z) \triangleq [1 + C(z)\widehat{P}(z,\hat{\theta})]^{-1}, \tag{7}$$

and the achieved sensitivity function as,

$$S(z) \triangleq [1 + C(z)P(z)]^{-1}. \tag{8}$$

We now consider the two feedback loops of Figures 2.2 and 2.3, with some fixed estimated model $\widehat{P}(z,\hat{\theta})$ in the nominal loop. One simple version of a robust stability result for such unity feedback loops is as follows.

Assume that $P(z)$ and $\widehat{P}(z,\hat{\theta})$ have the same number of unstable poles, and that the designed loop of Figure 2.2 is internally stable. Then the controller $C(z)$ will stabilize all plants $P(z)$ for which the following inequality holds:

$$\|L(z)\widehat{M}(z)\|_\infty < 1. \tag{9}$$

This inequality can be more explicitly restated as follows.

$$\left| [P(e^{j\omega}) - \widehat{P}(e^{j\omega},\hat{\theta})] \times \frac{C(e^{j\omega})}{1 + \widehat{P}(e^{j\omega},\hat{\theta})C(e^{j\omega})} \right| < 1 \quad \forall\omega. \tag{10}$$

We make the following observations concerning this inequality.

- The term $P - \widehat{P}$ represents the plant/model error and is essentially determined by the identification part of the design. Notice that the distribution of the model error in an identified model is strongly influenced, among other things, by the input spectrum.[4] Therefore, in closed loop identification (and, in particular, in adaptive control) the controller also influences the frequency distribution of this plant/model error.

---

[4] For those readers not too familiar with identification theory, this point will be made amazingly obvious in Section 4.

- The right hand fraction in (10) is the mixed sensitivity function of the nominal loop. Thus, for an estimated plant model and with a designed controller, this frequency dependent quantity is entirely known. We note that this quantity is therefore influenced by both the identification and the control design.

This inequality exhibits the interplay between identification and control design as far as robust stability is concerned. In classical robust control design, it is interpreted as a constraint on the controller to be designed for a given plant/model error bound: the controller must provide for a small value of the mixed sensitivity function where the plant/model error is large. In our joint identification and control design, it can alternatively be interpreted as putting constraints on the identification: the plant/model error must be small in frequency bands where the mixed sensitivity of the designed closed loop system is large.

Robust performance of a unity feedback loop

In terms of performance, there are various ways of defining the robustness of a controller design. Let us assume first that the control design is based on the minimization of some global performance criterion, $J(P, C)$, as illustrated above by two examples. In order to discuss the performance achieved by a controller designed on the basis of a reduced complexity model, it is useful to introduce the following concepts.

If the plant were known exactly, then the minimization of $J(P, C)$ over the class of admissible full order controllers would result in an optimal controller, C^{opt}, to which there corresponds an *optimal cost*, denoted $J^{opt} \triangleq J(P, C^{opt})$. This is the cost obtained for the loop of Figure 2.1.

The controller is effectively designed on the basis of a nominal model, $\widehat{P}(z, \hat{\theta})$, which we shall denote $\widehat{P}$ for short, and possibly on the basis of information about a bound on the model error $L(e^{j\omega}) = P(e^{j\omega}) - \widehat{P}(e^{j\omega})$. We denote the corresponding controller $\widehat{C}(z)$. The *designed cost* is then defined as $J^{des} \triangleq J(\widehat{P}, \widehat{C})$. It is the cost obtained on the design loop of Figure 2.2.

The quantity that really matters is not the optimal cost nor the designed cost, but the *achieved cost*, i.e. the cost achieved by the designed controller on the actual plant: $J^{ach} \triangleq J(P, \widehat{C})$. Thus, one measure of the performance robustness of an identification/control design is the comparison between J^{des} and J^{ach}. This comparison expresses how 'close' the loops of Figures 2.2 and 2.3 are, *as measured in terms of the global control performance criterion.* Notice that J^{ach} can be either larger or smaller than J^{des}.

To be more specific, we consider the two performance criteria suggested above. Consider first the H_∞ design criterion (3), and let $\widehat{C}$ be a controller designed on the basis of $\widehat{P}$ and (possibly) some known or assumed bound on the error, $L(e^{j\omega})$. We can then write:

$$W\frac{1}{1+P\widehat{C}} = W\frac{1}{1+\widehat{P}\widehat{C}} + [W(\frac{1}{1+P\widehat{C}} - \frac{1}{1+\widehat{P}\widehat{C}})].$$

By expressing successively each one of the three terms above as the sum (or difference) of the other two, and by applying the triangle inequality to each of these three expressions, Schrama showed that one can squeeze $J^{ach} \triangleq \|W\frac{1}{1+P\widehat{C}}\|_\infty$ between the following lower and upper bounds [Sch92a]:

$$\left| \|W\frac{1}{1+\widehat{P}\widehat{C}}\| - \|W(\frac{1}{1+P\widehat{C}} - \frac{1}{1+\widehat{P}\widehat{C}})\| \right|$$

$$\leq \|W\frac{1}{1+P\widehat{C}}\|$$

$$\leq \|W\frac{1}{1+\widehat{P}\widehat{C}}\| + \|W(\frac{1}{1+P\widehat{C}} - \frac{1}{1+\widehat{P}\widehat{C}})\|. \qquad (11)$$

Thus, the achieved cost is bounded above by the sum of the designed cost, $J^{des} = \|W\frac{1}{1+\widehat{P}\widehat{C}}\|_\infty$, and the H_∞ norm of the weighted difference between the sensitivity of the actual loop (Figure 2.3) and that of the design loop (Figure 2.2). We call this second term $J^{pr}_{H_\infty}$, because it is a performance robustness measure:

$$J^{pr}_{H_\infty} \triangleq \|W(\frac{1}{1+PC} - \frac{1}{1+\widehat{P}C})\|_\infty. \qquad (12)$$

$J^{pr}_{H_\infty}$ expresses the performance error that results from applying the controller $\widehat{C}$, designed for $\widehat{P}$, to the true plant P. With these notations, we can rewrite the inequalities (11) in a more suggestive way:

$$|J^{des} - J^{pr}| \leq J^{ach} \leq J^{des} + J^{pr}. \qquad (13)$$

The inequalities show that, if $J^{pr}_{H_\infty}$ is very small, the controller designed for $\widehat{P}$ achieves almost the same performance on the true plant.

Consider now the LQG criterion (4), and let $\widehat{C}$ again denote a controller computed on the basis of some nominal model together (possibly) with plant/model error information. Using the triangle inequality again, we show that the square root of J^{ach} can be squeezed between a lower and an upper bound. To do this, we first rewrite the LQG criterion as the square of a vector norm. We denote

$$J^N_{LQG} = J^N(P,C) = \frac{1}{N}\sum_{t=1}^{N}[(y_t - r_t)^2 + \lambda u_t^2]. \qquad (14)$$

We now introduce the following vector 2-norm for a vector process $\begin{pmatrix} x_t \\ y_t \end{pmatrix}$:

$$\left\| \begin{array}{c} x_t \\ y_t \end{array} \right\|_2 \triangleq (\sum_{t=1}^{N} [(\frac{1}{\sqrt{N}} x_t)^2 + (\frac{1}{\sqrt{N}} y_t)^2])^{1/2}. \tag{15}$$

Redefining α as the positive square root of λ, $\alpha \triangleq (\lambda)^{1/2}$, we can then rewrite the criterion as,

$$J_{LQG}^N = \left\| \begin{array}{c} y_t - r_t \\ \alpha u_t \end{array} \right\|_2^2 \tag{16}$$

Now, consider the signals defined in the loops of Figures 2.2 and 2.3, and observe that

$$\begin{pmatrix} y_t - r_t \\ \alpha u_t \end{pmatrix} = \begin{pmatrix} y_t^c - r_t \\ \alpha u_t^c \end{pmatrix} + \begin{pmatrix} y_t - y_t^c \\ \alpha(u_t - u_t^c) \end{pmatrix}$$

Therefore, by repeated use of the triangle inequality again, we have

$$\left| \left\| \begin{array}{c} y_t^c - r_t \\ \alpha u_t^c \end{array} \right\|_2 - \left\| \begin{array}{c} y_t - y_t^c \\ \alpha(u_t - u_t^c) \end{array} \right\|_2 \right|$$

$$\leq \left\| \begin{array}{c} y_t - r_t \\ \alpha u_t \end{array} \right\|_2$$

$$\leq \left\| \begin{array}{c} y_t^c - r_t \\ \alpha u_t^c \end{array} \right\|_2 + \left\| \begin{array}{c} y_t - y_t^c \\ \alpha(u_t - u_t^c) \end{array} \right\|_2 \tag{17}$$

For the same reasons as above, we denote

$$J^{pr,N} \triangleq \left\| \begin{array}{c} y_t - y_t^c \\ \alpha(u_t - u_t^c) \end{array} \right\|_2^2$$

$$= \frac{1}{N} \sum_{t=1}^{N} [(y_t - y_t^c)^2 + \lambda(u_t - u_t^c)^2]. \tag{18}$$

$J^{pr,N}$ expresses the performance error that results from applying the LQG controller $\widehat{C}$, designed for $\widehat{P}$, on the actual plant P. Taking the limit for $N \to \infty$, the inequalities (17) can then be rewritten as

$$|(J^{des})^{1/2} - (J^{pr})^{1/2}| \leq (J^{ach})^{1/2} \leq (J^{des})^{1/2} + (J^{pr})^{1/2}, \tag{19}$$

with an obvious definition for $J^{pr} = \lim_{N \to \infty} J^{pr,N}$. The upper bound in (19) had been obtained in [ZBG91] by a more complicated argument using the Hölder inequality, and served as the basis for the iterative design scheme to be described in Section 6.

We note that $J^{ach} = \lim_{N \to \infty} \frac{1}{N} \sum_{t=1}^{N} [(y_t - r_t)^2 + \lambda u_t^2]$, where y_t and u_t are the signals in the actual loop when the controller $\widehat{C}$ is applied to that loop, while y_t^c and u_t^c are the signals in the design loop when the same controller $\widehat{C}$ is used.

The inequalities (11) and (17) call for the following observations.

Comments

1. In both cases, the inequalities show that the achieved performance is bounded above by the sum of the designed performance, and a term expressing the performance error between the two closed loops, in a measure that is determined by the global control performance criterion. The achieved cost will be close to the designed cost, provided J^{pr} is much smaller than the designed cost. Thus, J^{pr} is a *robust performance criterion*, hence the notation. In the H_∞ case, $J_{H_\infty}^{pr}$ is indeed the classical robust performance criterion. In the LQG case, minimizing $J^{pr,N}$ corresponds to making the errors between the corresponding signals in the two loops small in the sense defined by the LQG control performance criterion. In each case, J^{pr} will be small if the actual loop and the design loop are "close to one another in the appropriate sense".

2. Both inequalities show that in order to make the achieved cost small, one should minimize the designed cost (this is what the control design classically does), and at the same time one should keep the difference between the two closed loops small, again in the norm determined by the global control performance criterion $J_{glob} = J(P, C)$.

3. We note that the estimated plant model, $\widehat{P}$, and the controller, $\widehat{C}$, both influence the two terms J^{des} and J^{pr}. Thus, ideally, one should minimize the two terms jointly over the class of admissible plant models and admissible controllers. This is an impossible task in the case of restricted complexity models.[5] On the other hand, minimizing J^{des} with respect to the controller for a given model $\widehat{P}$ is in both cases a classical control design task, whereas J^{pr} expresses in both cases a distance between two closed loop transfer functions, in the appropriate measure. Therefore, an obvious suboptimal strategy is to make J^{des} small by controller design for a given plant model, and to keep J^{pr} small by identification design for a given controller. Since J^{des} depends on the estimated plant model, and J^{pr} depends on the designed

---

[5] We recall that in [ÅW71] the achieved criterion is minimized jointly over the parametrized set of plant models and corresponding controllers, but the model set is assumed to contain the true system, and the minimization leads to a tractable solution only for the very simple minimum variance control criterion.

controller, this strategy can only be applied in an iterative manner, using a succession of *local controller designs* and *local identification designs*:[6]

$$\min_{C} J(\widehat{P}_i, C) \quad \longrightarrow \quad \widehat{C}_{i+1}$$

$$\min_{P(\theta) \in \mathcal{M}} J^{pr}(P(\theta), \widehat{C}_i) \quad \longrightarrow \quad \widehat{P}_{i+1}. \tag{20}$$

This idea is at the heart of the iterative identification/controller design methods that we discuss in Section 6.

4. In classical robust control design, a unique nominal model is given a priori together with error bounds, and the problem is restricted to designing a controller. When the model is obtained from data, as discussed in this paper, we observe that it is natural to design the controller such as to minimize the designed cost, and to design the model such as to minimize the performance robustness criterion: the term J^{pr} becomes our local identification criterion. This is certainly a non standard identification criterion: in (11) it is the H_∞ norm of the frequency weighted difference between the sensitivities of the two loops; in (17) it is a weighted Least Squares criterion of the difference between the corresponding signals in the two loops. We make the important observation that **these identification criteria are entirely determined by - and hence consistent with - the global control performance criteria**. This is what we meant in the introduction by stating that in a combined identification/control design *the control performance criterion should dictate what the identification criterion should be*.

5. It remains to be seen whether these nonstandard identification criteria can actually be minimized over a class of admissible models: this is the object of Section 5. It also remains to examine the properties of the iterative schemes that have only been sketched above. This is the object of Section 6.

To get a better understanding of the constraints imposed on the estimated plant model by the performance robustness requirements, we elaborate on the term J^{pr} of the inequalities (11) and (17). For (11) we get, straightforwardly,

$$J_{H_\infty}^{pr} \triangleq \|W(\frac{1}{1+PC} - \frac{1}{1+\widehat{P}C})\|_\infty = \|W\frac{(P-\widehat{P})C}{(1+\widehat{P}C)(1+PC)}\|_\infty. \tag{21}$$

---

[6] The term 'local' refers to the fact that, at each iteration, the controller design (resp. the identification design) is performed on the basis of some present (i.e. local) plant model (resp. presently operating (i.e. local)) controller.

The computations for (17) are more complicated. An LQG control design minimizing (4) leads to a two-degree-of-freedom controller as shown in Figure 2.4, where n_t is the input to the reference model, $r_t = R(z)n_t$: see e.g. [BGW90].

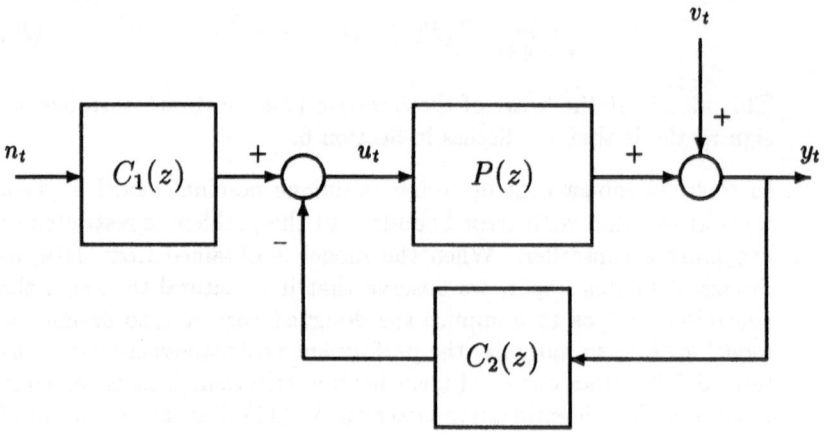

Figure 2.4: Actual LQG controlled system

Assume now that we take $v'_t \equiv 0$ in the corresponding design loop: see Figure 2.5. It is shown in [ZBG91] that the second right hand term of (17) can then be expressed as follows.

$$
\begin{aligned}
J_{LQG}^{pr} &\triangleq \lim_{N \to \infty} \frac{1}{N} \sum_{t=1}^{N} [(y_t - y_t^c)^2 + \lambda(u_t - u_t^c)^2] \\
&= \frac{1}{2\pi} \int_{-\pi}^{\pi} \left\{ \frac{|(P - \widehat{P})C_1|^2(1 + \lambda|C_2|^2)}{|(1 + PC_2)(1 + \widehat{P}C_2)|^2} \Phi_n \right. \\
&\qquad \left. + \frac{(1 + \lambda|C_2|^2)}{|1 + PC_2|^2} \Phi_v \right\} d\omega
\end{aligned}
\tag{22}
$$

In these expressions, y_t and u_t are defined on the actual LQG controlled plant of Figure 2.4, while y_t^c and u_t^c are defined from the design LQG loop of Figure 2.5, and Φ_n and Φ_v are the spectra of the signals n_t and v_t, respectively. Note that the two loops are driven by the same external signal, n_t, but that the loop of Figure 2.5 is noise free. Assuming that the spectrum of the input to the reference model, n_t, dominates that of the noise, v_t, within the passband of the closed loop system, then the model fit obtained by the minimization of this Least Squares criterion will be essentially determined by the first term.

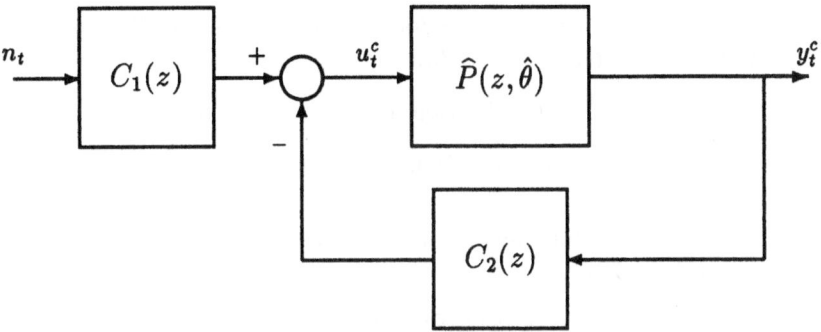

Figure 2.5: Nominal (or design) LQG controlled system

In both cases (H_∞ and LQG design) we observe that the model fit that is imposed by the minimization of J^{pr} is one in which the error between $P(e^{j\omega})$ and $\widehat{P}(e^{j\omega})$ must be made small in a frequency weighted sense, where the frequency weighting contains the product of the actual and designed sensitivities. Thus, the model errors must be small where these sensitivities are large, and in particular around the crossover frequency of the actual and the designed closed loop systems. Comparing with the robust stability criterion (10), we note that we have an additional weighting by the sensitivity function of the *actual* closed loop system. In Section 5, we examine how these performance robustness criteria can be minimized by identification design.

3 Questions raised by the interplay

The expressions of the previous section have shown that the control design and the identification design are closely intertwined. For example, the robust stability inequality (10) shows that the model error can be large in frequency areas where the mixed sensitivity function of the designed closed loop system is small, but must be small where this sensitivity function is large. Similarly, the robust performance considerations have shown that what really matters for performance is not the error between the open loop transfer functions $P(z)$ and $\widehat{P}(z)$, but the error between the closed loop transfer functions (or, equivalently, the closed loop sensitivities) of the achieved and designed loops of Figures 2.3 and 2.2.

These expressions, and the interplay they reveal, raise several questions, leading to several research topics.

1. Can we estimate the model error $P(e^{j\omega}) - \widehat{P}(e^{j\omega})$, or a bound on this error, for every ω, when that model has been obtained by identification on the basis of input/output data information? This is the question of *model error quantification.*

2. Can we design the identification in such a way that

   - the controller designed from that model will stabilize the actual plant,

   - and the performance robustness criterion J^{pr} is minimized in (21) or (22)?

   This is the question of *identification for control.*

3. Can we jointly optimize the identification and the control design, in order to maximize the achieved control performance? This is the question of *synergistic identification and control design.*

As stated in the introduction, the first question is where most of the research effort is presently concentrated. Although most of this work is said to be 'motivated' by robust control, the control design or the control performance criterion is rarely mentioned in the papers that deal with 'identification for robust control'. The effort is spent on trying to develop identification methods whose sole merit is to deliver error bounds, without much attention paid to the question of whether the identified model is good for the control design problem at hand. Our analysis above has clearly indicated that the identification method must take account of the control objective. *Given that with a restricted complexity model set a certain amount of plant/model mismatch is inevitable, the control objective must determine the distribution of this plant/model error.* The point of the remark above is to stress that the identification methods will be useful for robust control design only if they do take account of the control objective.

The quantification of the error on identified models is of course a very important objective in itself: it has historically been a trademark of engineers that they should be able to produce an evaluation of the quality of the product that they deliver. In addition, the estimation of the error on an identified model is indeed an important ingredient for robust control design. However, it is not central to the design of the identification for robust control or to the identification/control interplay. Therefore we do not attempt here to cover the huge amount of literature that is published on this problem. Let it suffice to say that the methods are essentially distinguished by the following ingredients.

- The type of prior knowledge that is assumed about the unmodeled dynamics and the noise: this prior knowledge can be in the form of stochastic descriptions or hard bounds, it can be unstructured or parametric.

- The form in which the data are assumed to be injected in the algorithms, such as time series data, or Fourier transform estimates.

- The methods that are used to propagate the prior uncertainty using data information. Examples are least squares, recursive least squares with bounding ellipsoids, or worst case constraints leading to large size linear programming problems.

- The norms, criteria and algorithms that are used to formulate the approximation problems: H_∞, H_2, l_1, etc.

It is fair to say that, even though the objective is to arrive at bounds on transfer function errors, the imagination of our research community to get there is essentially unbounded.

Recent surveys of available methods can be found in [WL92], [Tem93], [MV91], [GK92], [Gev91]. Finally, we like to mention that [GGN92] is probably the only contribution in which the prior assumptions on unmodeled dynamics and noise are of a qualitative rather than a quantitative nature: the prior uncertainty descriptions are parametric functions whose parameters are subsequently estimated from the data.

4   What can identification do for you?

We shall from now on adopt a performance enhancement objective. We have seen in Section 2 that a reasonable scheme for the minimization of the *achieved performance* by the combined design of identification and control is to perform a succession of model-based controller designs and controller-based identification designs: for a given (local) plant model, a designed control criterion, J^{des}, is minimized over the class of admissible controllers, and with a given (local) controller operating on the plant, an identification criterion, J^{pr}, is minimized over the set of plant models. This odd-looking identification criterion consists of a measure of the difference between the actual and the design loop, this measure being derived from and compatible with the global control performance criterion.

We now examine whether these bizarre control-performance-based identification criteria can indeed be minimized by classical identification methods. We shall do the development for the LQG criterion (an H_2 control criterion) because, as we will show, the corresponding criterion J^{pr}_{LQG} can be naturally connected to a prediction error identification criterion (an H_2 identification criterion). Similarly, the H_∞ criterion $J^{pr}_{H_\infty}$ naturally leads to an H_∞ identification criterion.

Least squares prediction error identification

We first recall the basic ingredients of prediction error identification. Remember that the true plant is assumed to be representable by (1). We consider that the model set takes the form

$$y_t = P(z, \theta)u_t + H(z)e_t. \tag{23}$$

Here $P(z, \theta)$ is a proper rational transfer function parametrized by some real vector θ, e_t is a zero mean white noise sequence, while $H(z)$ is, for simplicity, assumed to be some fixed noise model chosen by the user. From the model set (23) it is easy to write the one-step ahead prediction for y_t:

$$\hat{y}_{t|t-1}(\theta) = H^{-1}(z)P(z, \theta)u_t + [1 - H^{-1}(z)]y_t. \tag{24}$$

The one-step ahead prediction error is

$$
\begin{aligned}
\epsilon_t(\theta) &\triangleq y_t - \hat{y}_{t|t-1}(\theta) \\
&= H^{-1}(z)[(P(z) - P(z, \theta))u_t + v_t].
\end{aligned}
\tag{25}
$$

In Least Squares prediction error identification, the estimation of the parameter vector θ on the basis of N input-output data is obtained by minimizing the sum of the squares of the prediction errors $\{\epsilon_t(\theta), t = 1, \ldots, N\}$. However, for reasons that will soon become transparent, it is often desirable to minimize a frequency weighted sum or, equivalently, to filter the errors by some stable filter with transfer function $D(z)$. We denote by $\epsilon_t^f(\theta)$ the filtered errors :

$$\epsilon_t^f(\theta) \triangleq D(z)\epsilon_t(\theta). \tag{26}$$

Least-squares prediction error identification amounts to estimating θ that minimizes

$$V_N(\theta) \triangleq \frac{1}{N} \sum_{t=1}^{N} [\epsilon_t^f(\theta)]^2. \tag{27}$$

The parameter estimate is then defined as

$$\hat{\theta}_N = \arg \min_{\theta \in D_\theta} V_N(\theta), \tag{28}$$

where D_θ is a predefined set of admissible values. The parameter vector $\hat{\theta}_N$ then defines an estimated input-output model $P(z, \hat{\theta}_N)$.

Under reasonable conditions on the data and the model structure (see [Lju87]), $\hat{\theta}_N$ converges as $N \to \infty$ to

$$\theta^* = \arg \min_{\theta \in D_\theta} \bar{V}(\theta), \tag{29}$$

where

$$\bar{V}(\theta) = \lim_{N \to \infty} EV_N(\theta). \tag{30}$$

If the data are a realization of a stationary stochastic process, then $\bar{V}(\theta) = E[\epsilon_t^f(\theta)]^2$, the variance of the filtered prediction errors. Expressing these filtered prediction errors as a function of the 'true system' and the model transfer functions, and using Parseval's identity,

$$E[\epsilon_t^f(\theta)]^2 = \frac{1}{2\pi} \int_{-\pi}^{\pi} \Phi_{\epsilon^f}(\omega) d\omega,$$

allows one to obtain an expression for the frequency distribution of the asymptotic model error. To make this exercise useful, we shall successively derive the expressions of ϵ_t^f in the case of open loop and closed loop identification.

The filtered prediction error, ϵ_t^f, can be written, using (25) and (26),

$$\epsilon_t^f(\theta) = D(z)H^{-1}(z)[(P(z) - P(z,\theta))u_t + v_t]. \tag{31}$$

Identification in open loop

Assume first that the data have been collected while the process operates in open loop. In such case, the signals u_t and v_s are uncorrelated for all t and s. It then follows from (31) that,

$$\bar{V}(\theta) \triangleq E[\epsilon_t^f(\theta)]^2 = \frac{1}{2\pi} \int_{-\pi}^{\pi} \{|P(e^{j\omega}) - P(e^{j\omega},\theta)|^2 \Phi_u(\omega) + \Phi_v(\omega)\}$$
$$\times \frac{|D(e^{j\omega})|^2}{|H(e^{j\omega})|^2} d\omega \tag{32}$$

Since $\hat{\theta}_N$ converges to θ^*, and $\theta^* \triangleq \arg\min_{\theta \in D_\theta} \bar{V}(\theta)$, this integral expression gives an implicit characterization of the model $P(e^{j\omega}, \theta^*)$ to which $P(e^{j\omega}, \hat{\theta}_N)$ will converge if the number of data tends to infinity. In other words, it gives an implicit characterization of the asymptotic bias error.

The expression (32) shows that, when identification is performed on data collected in open loop operation, $P(e^{j\omega}, \hat{\theta}_N)$ converges to that model within the model set that minimizes a frequency weighted integral of the square error between the true transfer and the model transfer function, with a frequency weighting $\frac{\Phi_u(\omega)|D(e^{j\omega})|^2}{|H(e^{j\omega})|^2}$. With our assumption of a fixed noise model (i.e. $H(z)$ is θ-independent), the convergence point of $\hat{\theta}_N$ is independent of the actual noise distribution. It depends on the noise model $H(e^{j\omega})$, but only through the combined weighting $\frac{\Phi_u|D|^2}{|H|^2}$.

The formula (32) is useful because it shows that, in the situation where some restricted complexity model structure has been chosen for $P(z,\theta)$, one can still manipulate the frequency distribution of the plant/model error to a certain extent by playing with the design variables Φ_u, D and H. Since the whole interplay between identification and robust control design is based on obtaining a frequency distribution of the plant/model error that satisfies performance constraints, we will come back to this design issue and examine, on the basis of formula (32), whether open loop identification with the required choices of Φ_u, D and H can help us obtain robust performance. But first, we derive a similar expression for the frequency distribution of the plant/model error in the case of closed loop identification.

Identification in closed loop

We now consider that the data have been collected on the true system when some controller was operating, and we compute the expression of the filtered one step ahead prediction error. To make our derivation more general, and because we shall return to the LQG controller in Sections 5 and 6, we consider that the system operates under a two-degree-of-freedom controller as shown in Figure 2.4. First we compute ϵ_t^f. Substituting the expression for u_t derived from Figure 2.4 into (31) yields:

$$\epsilon_t^f(\theta) = \frac{D(z)}{H(z)[1 + P(z)C_2(z)]}[(P(z)-P(z,\theta))C_1(z)n_t+(1+\hat{P}(z,\theta)C_2(z))v_t].$$

(33)

To get a better insight into some properties of closed loop identification, we shall further assume that the noise v_t on the true system can be modeled as $v_t = H_0(z)e_t$, where e_t is zero mean white noise. The expression for $\epsilon_t^f(\theta)$ can then be rewritten as (dropping the dependence on z for simplicity of notation):[7]

$$\epsilon_t^f(\theta) = \frac{D}{H(1+PC_2)}[(P-\hat{P}(\theta))C_1n_t+(H_0(1+\hat{P}(\theta)C_2)-H(1+PC_2))e_t]+De_t.$$

(34)

When identification is performed on closed loop data collected on the process operating under a two-degree-of-freedom controller, the estimate $\hat{\theta}_N$ converges to the minimum of the cost function $\bar{V}(\theta) \triangleq E[\epsilon_t^f(\theta)]^2$. From (33) we get, using Parseval's theorem:

$$E[\epsilon_t^f(\theta)]^2 = \frac{1}{2\pi}\int_\pi^\pi \left\{ \left| \frac{(P(e^{j\omega}) - \hat{P}(e^{j\omega},\theta))}{1 + P(e^{j\omega})C_2(e^{j\omega})} \right|^2 |C_1(e^{j\omega})|^2 \Phi_n(\omega) \right.$$

---

[7]This alternative calculation results from an insightful discussion with R. Hakvoort and P. Van den Hof.

$$+ \left| \frac{1 + \widehat{P}(e^{j\omega}, \theta)C_2(e^{j\omega})}{1 + P(e^{j\omega})C_2(e^{j\omega})} \right|^2 \Phi_v(\omega) \right\} \times \frac{|D(e^{j\omega})|^2}{|H(e^{j\omega})|^2} d\omega \ (35)$$

An interesting alternative expression of the minimizing value of $\bar{V}(\theta)$ can be obtained using (34). Since the last term in (34) is independent of θ, the minimization of $\bar{V}(\theta)$ with respect to θ is equivalent with the minimization of[8]

$$V^*(\theta) = \frac{1}{2\pi} \int_\pi^\pi \left\{ \left| \frac{(P - \widehat{P}(\theta))}{1 + PC_2)} \right|^2 |C_1|^2 \Phi_n(\omega) \right.$$

$$\left. + \left| \frac{H_0(1 + \widehat{P}(\theta)C_2) - H(1 + PC_2)}{1 + PC_2} \right|^2 \sigma_e^2 \right\} \times \frac{|D|^2}{|H|^2} d\omega \ (36)$$

The expressions (35) and (36) describe in an implicit way the asymptotic distribution of the error between the true system $P(e^{j\omega})$ and the estimated model $\widehat{P}(e^{j\omega}, \hat{\theta})$ when the identification is performed on data collected in closed loop using a reduced complexity model set. The following remarks are worth making.

- The model fit is definitely influenced by the controller: the component C_1 shapes the spectrum $|C_1|^2 \Phi_n$ that enters the loop, while the component C_2 exerts its influence through the sensitivity $\frac{1}{1+PC_2}$ of the actual closed loop. The weighting on both terms of the integrand will be large where this sensitivity is large, namely around the crossover frequency of the closed loop system.

- External excitation is definitely needed for closed loop identification. The model $\widehat{P}$ will approximate P (in the frequency weighted sense determined by the formula (35)) only if the reference signal spectrum $|C_1|^2 \Phi_n$ that enters the loop dominates the noise spectrum Φ_v within the closed loop bandwidth.

- Without external reference, the model will attempt to approximate the inverse of the controller, $C_2^{-1}(e^{j\omega})$.

- The data filter $D(e^{j\omega})$ can again be used to shape the fit globally.

- The expression (36) shows that, even if the model set $\mathcal{M} = \{\widehat{P}(z, \theta)\}$ is able to represent the true system $P(z)$, closed loop identification using a direct prediction error method will lead to a biased estimate of $P(z)$ if the noise model is incorrect, that is if $H(z) \neq H_0(z)$. This

---

[8] We drop the dependence of the transfer functions on $e^{j\omega}$ here to simplify notation.

is a serious drawback which has led Hansen [Han89] and Schrama [Sch92a] to propose an alternative indirect scheme that transforms the closed loop identification problem into an open loop scheme.

The Hansen-Schrama scheme

We present the Hansen scheme, as modified by Schrama, to perform closed loop identification using open loop methods. The scheme is based on the idea that if a compensator $C(z)$ stabilizes the plant $P(z)$, then $P(z)$ can be represented in the Youla parametrization of all plants stabilized by the compensator $C(z)$: this is the dual of the classical Youla parametrization. Thus, consider the loop of Figure 2.3, and assume that the noise v_t can be modeled as $v_t = H(z)e_t$, with $H(z)$ a rational transfer function and e_t white noise of zero mean. Let P_0 be an auxiliary model (possibly an estimate of P) such that the closed loop (P_0, C) is stable, and let P_0 and C have right coprime factorizations $P_0 = N_0(D_0)^{-1}$ and $C = N_c(D_c)^{-1}$, respectively, where N_0, D_0, N_c, D_c are all stable transfer functions. It can then be shown [Sch92a] that the feedback system of Figure 2.3, with $v_t = H(z)e_t$, is stable if and only if $[H \quad P]$ has a right coprime factorization of the form

$$[H \quad P] = [D_c S \quad N_0 + D_c R] \begin{bmatrix} I & 0 \\ N_c S & D_0 - N_c R \end{bmatrix}^{-1}, \qquad (37)$$

where $R(z)$ and $S(z)$ are stable transfer functions. We note, in particular, that $P(z) = (N_0 + D_c R)(D_0 - N_c R)^{-1}$. For future use, we define:

$$N^a \triangleq N_0 + D_c R \qquad D^a \triangleq D_0 - N_c R. \qquad (38)$$

Using this coprime factor representation of P and H, the feedback system of Figure 2.3 can be redrawn as in Figure 4.1.

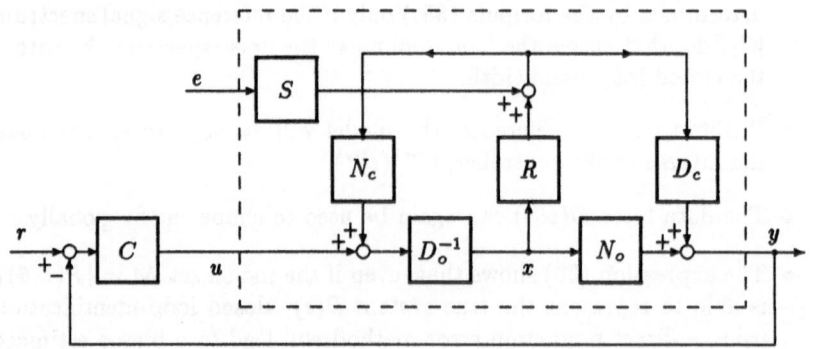

Figure 4.1: Coprime factor representation of P and H.

Now it is easy to show that the signal x_t of Figure 4.1 can be reconstructed from u_t and y_t through

$$x_t = (D_0 + CN_0)^{-1}(u_t + Cy_t). \tag{39}$$

In addition, it follows immediately from the figure that $u_t + Cy_t = r_t$. Therefore, x_t is uncorrelated with e_t, and hence with the noise v_t.

There are a number of ways to extract an estimate of the plant from this coprime factor representation. They are all based on the observation, immediately derived from (38) and the block-diagram, that

$$\begin{pmatrix} u_t \\ y_t \end{pmatrix} = \begin{pmatrix} D^a \\ N^a \end{pmatrix} x_t + \begin{pmatrix} -N_c S \\ D_c S \end{pmatrix} e_t. \tag{40}$$

Thus, the signal vector $(u \;\; y)^T$ is expressed as the output of an unknown system driven by the known signal x_t, plus a noise term that is uncorrelated with x_t. These equations serve as a basis for the identification of the transfer functions N^a and D^a, thus yielding an estimate $\widehat{P} = \widehat{N}^a(\widehat{D}^a)^{-1}$ of the unknown plant P. Notice that, unlike the estimate obtained by direct identification (see (36)), the present estimate is not biased by the incorrectness of the noise model. Finally, we mention that this coprime factor representation has also been used by Zhu et al. [ZS92] for the identification of closed loop systems using spectral estimates.

We have briefly presented the bare essentials of prediction error identification theory, and we have given a frequency domain characterization of the criteria that are minimized by prediction error methods when identification is performed in open loop and in closed loop. These characterizations can also be seen as an implicit description of the frequency distribution of the asymptotic model error. We have also shown how the bias effects introduced by the noise in closed loop identification can be circumvented.

5   Identification for control

We now examine how the identification theory of the previous section can be used to tune the identification criterion towards the satisfaction of a control performance criterion, as suggested in Section 2. We thus return to the LQG design problem posed in Section 2. Ideally, one would like to design the identification in such a way that the difference between the cost achieved using the model-dependent controller $\widehat{C}$ on the real plant P achieves a performance, J^{ach}, that is as close as possible to J^{opt}.

Given that the global control performance criterion is given by J_{LQG} (see (4)), this strategy should, in principle, be pursued by designing an identification criterion that minimizes

$$V_N^{opt}(\theta) = \frac{1}{N} \sum_{t=1}^{N} [(y_t - y_t^{opt})^2 + \lambda(u_t - u_t^{opt})^2]. \tag{41}$$

For the case of minimum variance control ($\lambda = 0$), this was precisely the experiment design criterion adopted in [GL86], and it led to the conclusion that the identification should be performed in closed loop under minimum variance control. Although an exact implementation of this result is of course impossible because the minimum variance controller is a function of the unknown plant, it suggests an iterative design strategy in which the identification is pursued under feedback control with a minimum variance controller computed from the present model estimate. Adaptive minimum variance control is a fast implementation of this idea.

In the minimum variance control case, an optimal experiment design can be derived because the controllers C^{opt} and $\widehat{C}$ are an explicit function of the system and model parameters, respectively. The same is true for model reference control: see [Lju87]. In our present LQG design problem, the controller is a nonlinear function of the plant, and this seems to rule out a design based on a comparison of the optimal loop and the achieved loop. Thus, we return to a comparison between the achieved and the design loop, equipped with (or reinforced by) the inequalities (19). It was suggested in Section 2 that one could minimize the first term of the upper bound, J^{des}, by control design (this is a standard LQG control design problem) and the second term by identification design. The question was raised as to whether the non-standard criterion (18) can be minimized by identification techniques.

We first recall that the criterion (18) has been reformulated in the frequency domain as (22), and we now compare (22) with the prediction error criteria (32) and (35) for open loop and closed loop identification, respectively. It is immediately obvious that the criterion (22) cannot be minimized as a result of open loop identification. However, it can be minimized by a classical least squares prediction error method provided,

- the identification is performed in closed loop;

- with a data filter $D(z)$ obtained as the solution of

$$|D(z,\theta)|^2 = \frac{|H(z)|^2(1 + \lambda|C_2(z)|^2)}{|1 + \widehat{P}(z,\theta)|^2}. \tag{42}$$

We thus have the following remarkable result, that was derived in [ZBG91].

'Remarkable?' result (denoted R?R)

Assume that some two-degree-of-freedom controller $[C_1(z), C_2(z)]$ operates on the true system as in Figure 2.4, and on a simulated design loop containing a θ-dependent model set $\widehat{P}(z, \theta)$ as in Figure 2.5. Let the two loops be driven by the same reference signal source, n_t. Then the criterion

$$J_{LQG}^{pr,N} = \frac{1}{N} \sum_{t=1}^{N} [(y_t - y_t^c)^2 + \lambda(u_t - u_t^c)^2], \qquad (43)$$

which expresses the 'distance' between the two loops in a measure that is determined by the global LQG criterion, can be minimized over this model set by a classical least squares prediction error method, provided the identification is performed in closed loop with the data filter $D(z, \theta)$ defined by (42).

This result is remarkable because it is a priori not obvious that the criterion (43), which is in fact a 'control performance error criterion', can be made identical to a classical prediction error criterion, given the proper experimental set-up and the proper data filter. This equivalence allows one to estimate the model $\widehat{P}(z, \hat{\theta})$ that minimizes (43) using standard identification algorithms. We now make a few comments about this result.

Comments

1. One of the important consequences of result R?R is that, if one identifies a model for control design, then the identification should be performed in closed loop. The intuition behind this result is that a model will be good for control design if its closed loop properties are close to the closed loop properties of the actual system under the same feedback control, that is if the closed loop transfer functions of the loops of Figure 2.3 and 2 are close. This is the message of the inequalities (11) and (17).

2. The problem is that the controller that should ideally operate during the collection of data for identification is precisely the optimal controller that is to be designed from the identified model. This is a classical case of the design method biting its own tail, and is the main reason for introducing iterative design methods. In these iterative designs, each control design step is followed by an identification design step, and vice versa. The rationale of these iterative designs, in the light of our present identification analysis, is that, even if the presently acting controller, say $[C_{1,i}, C_{2,i}]$, is not the optimal one in the class of admissible reduced complexity controllers, it will produce a frequency distribution of the input signal, $\frac{|C_{1,i}|^2}{|1+PC_{2,i}|^2}\Phi_n$, that will force the next identified model, say $\widehat{P}_{i+1}(z, \hat{\theta})$, to have closed loop properties close to those of the true system.

3. A number of other approaches to the 'identification for control' problem have also led to the conclusion that the identification should be performed in closed loop and with appropriate data filters, and this is now widely recognized as a key ingredient for the success of control design based on reduced complexity identified models.

   - In [Han89] (see also [HFK89]) the idea of [GL86] has been extended from an H_2 measure of the performance degradation between the optimal and the actual system to an H_∞ measure of the difference between the closed loop transfer functions of these two feedback systems.

   - In [BGW90] the H_∞ stability and performance robustness constraints detailed in Section 2 were used to motivate the use of closed loop prediction error identification in the context of modeling for LQG control design. Indeed, a comparison between the robust performance criterion (21) and the closed loop identification criterion (35) shows that, by proper choice of the data filter $D(z)$ and by closed loop identification, the robust performance expression $|\frac{(P-\widehat{P})C}{(1+PC)(1+\widehat{P}C)}|$ can be made small in an H_2 sense. Even though the H_2 minimization of this quantity cannot guarantee a bound on its H_∞ norm, the idea of making that quantity small within the bandwidth of the closed loop system to enhance performance robustness was advocated.

   - The Delft group, Hakvoort, Schrama, and Van den Hof, have brought important contributions to the identification for control problem (see e.g. [Hak90], [Sch91], [Sch92a], [Sch92b], [SvdH92]). In his remarkable thesis [Sch92a], Schrama examined many facets of the 'identification for control' and 'iterative identification and control design' problems. In particular, he demonstrated convincingly with a dramatic simulation example that an open loop model that would pass all standard model validation tests can result in a disastrous controller. Conversely, the best model for control design can be so poor as an open loop model of the plant that it would fail most classical model validation tests.

   - The equivalence between a performance robustness criterion and a prediction error criterion established in [ZBG91] for an LQG control criterion has been extended to a pole placement control design by Åström [Åst93]. For this particular control design, he shows that the 'control performance error' (i.e. the error between the outputs y and y^c of the loops of Figures 2.3 and 2.2) can be made identical to the 'identification error' (i.e. the filtered prediction error) by performing closed loop identification with a specific data filter.

- Mäkilä and Partington have advocated closed loop identification, both as a way of identifying open loop unstable but stabilizable systems [MP92a] and as a procedure for enhancing control performance robustness when the model is used for control design [MP92b]. In [MP92a] H_∞ identification of the closed loop transfer function is performed, while parameter bounding techniques using l_∞-stable coprime factor descriptions are used in [MP92b].

- Liu and Skelton [LK90] pointed to the need for closed loop identification and proposed an iterative design scheme using Skelton's q-Markov Cover models.

We have demonstrated in this section that the performance robustness criterion J^{pr}_{LQG}, defined for some given controller operating both on the plant and on the model, can be minimized over a set of parametrized models by least squares prediction error identification in closed loop with an appropriate data filter. A similar conclusion can be drawn for the minimization of $J^{pr}_{H_\infty}$.

We now examine methods for the synergistic identification and control design that are based on the iterative minimization of J^{des} by controller manipulation and of J^{pr} by model manipulation.

6   Iterative identification and control design

Suppose that, for some plant P to be controlled, a preliminary analysis has led to the choice of some model set, $\mathcal{M} = \{\widehat{P}(z,\theta), \theta \in D_\theta\}$, and some control performance criterion, $J(P,C)$, with the property that the minimization of $J(\widehat{P},C)$ with respect to C, for any $\widehat{P} \in \mathcal{M}$, uniquely determines a designed controller, $\widehat{C} = \widehat{C}(\widehat{P})$. The choice of an adequate model set and of an adequate control objective is of course very much part of the control design. In particular, one must choose a control performance objective that is compatible with the achievable closed loop bandwidth, etc. However, we adopt these assumptions (or play this game) to illustrate the central features of the synergistic design problem. Ideally, the problem of joint identification and control could then be reformulated as a parameter estimation problem as follows:

$$\min_{\theta \in D_\theta} J(P, \widehat{C}(\widehat{P}(\theta))). \tag{44}$$

To make things concrete, assume for example that $J(P,C)$ is the LQG criterion (4) and that $\mathcal{M}$ is a parametrized set of third order output error models with a parameter vector $\theta \in D_\theta = \mathbb{R}^6$. The optimal control problem has thus been turned into an identification problem.

The direct minimization of such a global control performance criterion over a set of restricted complexity models is typically intractable, but one way to attack the problem is to perform a succession of local identification steps and local control design steps in an iterative way. Several motivations can be given to rationalize such iterative procedures.

- A feasibility motivation: the intractable joint optimization problem alluded to above is replaced by a sequence of tractable closed loop identification problems with fixed controller, and controller design problems with a given model.

- A theoretical motivation: the triangle inequalities (11) and (19) are one way of giving theoretical credibility to the the idea that by performing small controller changes that minimize J^{des} for a given $\widehat{P}$, followed by small model changes that minimize J^{pr} for a given $\widehat{C}$, then this iterative procedure may tend to jointly minimize the upper bound on the achieved cost. For the moment, no hard results are yet available.

- A practical motivation: in the process control industry, it is common practice to design a controller, let it operate for a while, collect data on the controlled process, and then use these data to perform a new design in order to ameliorate performance. The novel contribution of the iterative design schemes is to provide a systematic and theoretically justified framework to perform these successive designs. Thus, they should really be seen for what they are, namely performance enhancement schemes.

There are many variants to the iterative design schemes - and we shall discuss some of them - but the basic idea is as follows.

1. **Step** 0: Identify an open loop model, $\widehat{P}_0$, from input-output data, and design a controller, $\widehat{C}_0$, that stabilizes both the true plant P and the estimated model $\widehat{P}_0$. Apply this controller to the plant and collect new input-output data.

2. **Step** i: Using the closed loop data collected on the plant while the controller $\widehat{C}_{i-1}$ operates, identify a new model $\widehat{P}_i$ by minimizing a local identification criterion. Using this new identified model $\widehat{P}_i$, design a new controller $\widehat{C}_i$ that stabilizes both P and $\widehat{P}_i$, by minimizing a local control design criterion. Apply this controller to the plant and collect new input-output data.

3. **Step** ∞: Do not iterate until convergence (who would?), for two good reasons: convergence has not been proved for any of these schemes, and which practical control engineer would want to redesign his controllers every day anyway?

In [ZBG91] such a procedure was proposed, first for the case where the global criterion is a classical H_∞ criterion, then for the case of an even more classical LQG criterion. In the case of an H_∞ criterion, the iteration of identification steps and control design steps can be formulated, and it was proved that the achieved performance criterion decreases at every step. However, no feasible algorithm is presently available for the H_∞ identification step of this joint design. The same idea of iterative minimization of an H_∞ criterion was developed independently in [BYM92], where the exact same conclusion was reached, a rather fortunate coincidence.

We therefore turn to the LQG criterion, and present the algorithm known in the process control industry as the Zangscheme.[9]

The Zangscheme[10]

The Zangscheme of [ZBG91] uses the LQG criterion (4) as the global criterion to be minimized. To simplify the notation, we shall often use the vector notation $C(z)$ to denote the two-degree-of-freedom controller $C_1(z), C_2(z)$. Thus, $C(z) \triangleq [C_1(z) \quad C_2(z)]$.

Consider now that we are at i-th iteration of the design (see above), and that the two-degree-of-freedom controller $\widehat{C}_{i-1}$, designed at the previous iteration, is operating on the real plant P. Thus, $C(z)$ is replaced by $\widehat{C}_{i-1}(z)$ in the loop of Figure 2.4, and N data y_t and u_t are being collected on that actual closed loop system.

The identification step is performed by minimizing the local prediction error identification criterion,

$$J^{id,N} = \frac{1}{N} \sum_{1}^{N} [D_i(z,\theta)\epsilon_t(\theta)]^2, \tag{45}$$

over the model set $\mathcal{M}$, where $\epsilon_t(\theta) \triangleq y_t - \hat{y}_t(\theta)$, and where the data filter is computed from

$$|D_i(z,\theta)|^2 = \frac{|H(z)|^2(1 + \lambda|\widehat{C}_{2,i-1}(z)|^2)}{|1 + P(z,\theta)|^2}. \tag{46}$$

The criterion is 'local' only through its dependence on the present controller, $\widehat{C}_{2,i-1}$, acting in the loop. As shown in Section 5, this local identification criterion is identical to the robust performance criterion J^{pr} of LQG defined in (18), where the signals u_t^c and y_t^c are those that would be generated by

---

[9] ... and better known in Australia and Belgium as the Zangstuff.

[10] We present here a slightly improved version of the Zangscheme, taking account of modifications introduced by Partanen and Bitmead [PB93] and by the author: progress just cannot be stopped.

applying the same signal n_t to the loop of Figure 2.5 controlled by the same controller $\widehat{C}_{i-1}$ and with no noise input. The new model that results from the identification step is denoted $\widehat{P}_i(z)$:

$$\widehat{P}_i(z) = \arg \min_{P(z,\theta) \in \mathcal{M}} J^{id,N} \tag{47}$$

Comments

We note that the data filter depends on the model that is being identified. There are several ways to cope with this problem.

- The first solution is to replace the unknown $\widehat{P}(z,\theta)$ in the data filter by the most recent estimate, $\widehat{P}_{i-1}(z)$; this is the solution proposed in [ZBG91]. However, the equivalence between J^{pr} and J^{id} breaks down with this solution, as pointed out by Hakvoort and Van den Hof [HV93].

- A better solution is to let the filter $D_i(z)$ be θ-dependent. This effectively corresponds to solving a prediction error problem with a modified model structure. It will, however, typically lead to a more complicated minimization problem for $J^{id,N}$.

We now turn to the i-th control design iteration. The certainty equivalence control design criterion would be to minimize the following performance criterion J^{des}:

$$J^{des} = \lim_{N\to\infty} \frac{1}{N} \sum_{t=1}^{N} \{(y_t^c - r_t)^2 + \lambda(u_t^c)^2\}, \tag{48}$$

where u_t^c is the designed control signal and y_t^c is the output of the identified model, $\widehat{P}_i$, driven by u_t^c.

Instead of following the certainty equivalence route of minimizing (48), the Zangscheme performs a controller design that takes account of the present plant/model uncertainty in the following way.

- A closed loop simulation is performed with the controller $\widehat{C}_{i-1}(z)$ acting on the present plant model $\widehat{P}_i(z)$. The actual closed loop system of Figure 2.4, and the simulation loop of Figure 2.5, with the same controller $\widehat{C}_{i-1}(z)$, are driven by the same signal n_t[11], thus generating the signals u_t and y_t, respectively u_t^c and y_t^c.

- With these experimental and simulated data sets, low order (typically third order AR models) are fitted to the signals $(y - r)$, u, $(y^c - r)$ and u^c, thus yielding spectral estimates $\widehat{\Phi}_{y-r}^{\frac{1}{2}}(z)$, $\widehat{\Phi}_u^{\frac{1}{2}}(z)$, $\widehat{\Phi}_{y^c-r}^{\frac{1}{2}}(z)$ and $\widehat{\Phi}_{u^c}^{\frac{1}{2}}(z)$.

---

[11] In addition, the actual system is also driven by the noise source v_t.

- The following frequency weighted *local control criterion* is minimized to compute $C_i(z)$:

$$J^c = \lim_{N \to \infty} \frac{1}{N} \sum_{t=1}^{N} \{[F_1(z)(y_t^c - r_t)]^2 + \lambda[F_2(z)u_t^c]^2\}, \qquad (49)$$

where F_1 and F_2 are weighting functions (linear filters) that are chosen as the following ratios of the estimated spectra:

$$F_1 = \left(\frac{\widehat{\Phi}_{y-r}}{\widehat{\Phi}_{y^c-r}}\right)^{1/2}, \quad F_2 = \left(\frac{\widehat{\Phi}_u}{\widehat{\Phi}_{u^c}}\right)^{1/2}. \qquad (50)$$

We comment that all the signals necessary for the computations of the filters $F_1(z)$ and $F_2(z)$ are readily available at every iteration step. The effect of the frequency weightings is to make the filtered tracking error signal and control signal, respectively, in (49) have the same spectra as the corresponding signals in the global (ideal) performance criterion J_{LQG} of (4). Thus, the frequency weightings are a distortion of the certainty equivalence criterion that takes account of plant/model mismatch in order to reflect the global criterion. The plant/model mismatch information is injected in the design on the basis of signal information only.

Besides forcing the local control objective to mimic the global one, as explained above, the effects of the frequency weightings in (49) have entirely logical and intuitive interpretations. If at some frequency Φ_{y-r} is larger than Φ_{y^c-r}, it means that at that frequency the model fit is poor with the consequence that the achieved tracking performance (with the presently active controller) is worse than expected from the designed system. Hence, more emphasis should be put on the tracking penalty at that frequency at the next control design stage, which is reflected by the weighting being larger than 1. If at some frequency Φ_{y-r} is smaller than Φ_{y^c-r}, it also means that at that frequency the model fit is poor, but in such a way that the presently active controller actually achieves a better tracking performance on the true plant than on the model. The emphasis on the tracking penalty at that frequency should therefore be decreased at the next control design stage to provide scope for improvement at other frequencies. Similar astute and entirely intuitive observations can be made by the reader as regards the frequency weighting on the control.

The i-th iteration of the Zangscheme (at least in one of its variants) thus involves the following steps:

- Apply the controller $\widehat{C}_{i-1}(z)$ to the true plant $P(z)$ with an external reference r_t, as in Figure 2.4, to generate a data set $\{y_t, u_t\}$ of length N.

- Compute the data filter $D_i(z, \theta)$ using (46).

- With the data set $\{y_t, u_t\}$ identify $\widehat{P}_i(z, \hat{\theta})$ using $D_i(z, \theta)$.

- Perform a closed loop simulation, driven by the same external reference r_t, with the controller $\widehat{C}_{i-1}(z)$ acting upon the plant model $\widehat{P}_i(z, \hat{\theta})$, as in Figure 2.5, to generate a data set $\{y_t^c, u_t^c\}$ of length N.

- With the data sets from the experiment and the simulation, identify AR models of $(y - r)$, u, $(y^c - r)$, u^c to yield $\widehat{\Phi}_{y-r}^{\frac{1}{2}}(z)$, $\widehat{\Phi}_{y^c-r}^{\frac{1}{2}}(z)$, $\widehat{\Phi}_u^{\frac{1}{2}}(z)$ and $\widehat{\Phi}_{u^c}^{\frac{1}{2}}(z)$.

- Calculate the frequency weightings F_1 and F_2 using (50).

- Design a new frequency weighted feedback controller $\widehat{C}_i(z)$ based on $\widehat{P}_i(z, \hat{\theta})$ and the identified signal spectra.

A number of other variants have been proposed [PB93], [HSV92], and a large number of simulations have been performed (see [ZBG91]). The disturbance rejection properties of the Zangscheme have been examined in [ZBG92]. The simulations typically exhibit an improvement of the *achieved cost* during the first three or four iterations, with no significant improvement thereafter. They also show that the model that is obtained after these few iterations can be very different from the best open loop model: this last finding is corroborated by the simulations performed with all other iterative design methods, again confirming that the best model for control design is definitely not the best open loop model.

Alternative iterative design schemes

We now briefly describe the key features of some of the other iterative identification and control design schemes. They are all based on trying to make the design loop of Figure 2.2 close to the achieved loop of Figure 2.3 in some sense.

- In Liu and Skelton [LS90], the q-Markov Cover theory of Skelton et al., which is a method for model reduction, is used in the identification step to identify a model of the closed loop system with the previously designed controller operating in the loop. Since the controller is known, a model of the open loop plant can be derived. A minimum energy controller with output variance constraint is used in the control design step.

- In [BYM92] Bayard et al. formulate an H_∞ robust performance criterion for the joint optimization of control design and identification design. A relaxation algorithm is proposed for solving the joint optimization problem, based on alternating between curve fitting and control design steps. This strategy yields a monotonically improving achieved performance, but is presently not implementable. A numerical example using an approximate implementation, for which no descent property can be proved, shows the usefulness of the approach.

- The Delft group has done a thorough analysis of the iterative design scheme. Although several alternative methods and variants have been examined, the basic approach in [Sch92a], [Sch92b] and [SV92] is to use, as global performance measure, an H_∞ norm of the feedback system linking the exogenous signals to the $(u \quad y)^T$ vector. This encompasses most H_∞ control design criteria. The triangle inequality (11) is used to justify the iterative design schema. Both the control and the identification design are performed using coprime stable factor representations of the plant model and the controller, as explained in Section 4. This use of coprime factor representations guarantees that the designed controller is optimally robust against perturbations of the coprime factors, as shown by Vidyasagar [Vid85]. The closed loop identification step is based on the open loop scheme detailed in Section 4, using the auxiliary signal x_t defined in Figure 4.1. For lack of a satisfactory H_∞-algorithm, the minimization of the performance robustness criterion $J_{H_\infty}^{pr}$ in the identification step is replaced by an H_2 minimization. The control design step is a certainty equivalence minimization of the nominal H_∞ criterion. One interesting feature of the Schrama scheme is that it can 'predict' the achieved performance; this feature is used to update the performance requirements as the closed loop model becomes better.

- The idea of improving the performance requirements as the closed loop model becomes closer to the actual closed loop system is central to the philosophy of the scheme developed by Lee et al. [LAKM92], who call this idea the windsurfer approach to adaptive control. The techniques used by Lee et al. are based on the Hansen representation of the closed loop system, and are therefore close to those of Schrama et al., but the global objective is different. It is formulated as the minimization of the H_∞ norm of the difference between the achieved closed loop transfer function and that of a reference model. The emphasis is put on how to update the reference model (i.e. the performance specifications) as the model and the controller improve. The identification step, which is still not fully resolved, attempts to estimate the stable factor R in the representation (38), rather than the factors N^a and D^a in the Schrama scheme.

7   Conclusions

Two o'clock in the morning. What more can I say? The joint design of identification and control is a fresh field, ripe with ideas and probably a few misconceptions as well. Some convergent streams are emerging from the array of different approaches that have been applied to the identification and control design problem. These streams and guidelines bear the names 'iterative designs', 'closed loop identification', 'control-determined identification criterion', 'appropriate data filter'. However, the evidence is still circumstantial, the methods ad hoc, and the hard proofs scarce. So far, it appears that the benefits to be drawn from this research area are more attuned towards the development of performance enhancement design schemes for industrial process control than they are towards the fine-tuning of controllers that need to stabilize high performance aircraft.

However, beneath the surface lie fascinating and deep theoretical questions that have information theoretic significance going back to the dual control ideas of Fel'dbaum, and enough hard mathematical and control-theoretic questions to keep a few generations of PhD students busy.

Acknowledgements

The ideas developed in this paper are in large part the result of enthusiastic work and heated discussions with Bob Bitmead and Zhuquan Zang, who will have recognized many of their own devilish thoughts. I am also deeply indebted to my Delft colleagues Paul Van den Hof, Ruud Schrama and Richard Hakvoort for the many insightful discussions we have had on the problems discussed in this paper.

References

[Åst93] K.J. Åström, "Matching criteria for control and identification", *2nd European Control Conference*, Groningen, Holland, 1993.

[ÅB65] K.J. Åström and T. Bohlin, "Numerical identification of linear dynamic systems from normal operating records", *IFAC Symposium on Self-Adaptive Systems*, Teddington, UK, 1965.

[ÅW71] K.J. Åström and B. Wittenmark, "Problems of identification and control", *Journal of Mathematical Analysis and Applications*, Vol. 34, pp. 90-113, 1971.

[BGW90] R.R. Bitmead, M. Gevers, V. Wertz, "Adaptive Optimal Control - The Thinking Man's GPC", Prentice Hall International, Series in Systems and Control Engineering, 1990.

[**BYM92**] D.S. Bayard, Y. Yam and E. Mettler, "A criterion for joint optimization and robust control", *IEEE Transactions on Automatic Control*, Vol. 37, No 7, pp. 986-991, July 1992.

[**Doy92**] J. Doyle, not so private communication, Santa Barbara Workshop, June 1992.

[**Fel60**] A.A. Fel'dbaum, "The theory of dual control", Parts 1-4, *Automation and Remote Control*, 1960-1961.

[**Gev91**] M. Gevers, "Connecting identification and robust control: a new challenge", Technical Report 91.48, CESAME, Louvain University, 1991; also *Proc. 9th IFAC Symp. on Identification and System Parameter Estimation*, Budapest, Hungary, pp. 1-10, July 1991.

[**GGN92**] G.C. Goodwin, M. Gevers, B. Ninness, "Quantifying the error in estimated transfer functions with application to model order selection", *IEEE Transactions on Automatic Control*, Vol. 37, No 7, pp. 913-929, July 1992.

[**GK92**] G. Gu, P. Khargonekar, "Linear and Nonlinear Algorithms for Identification in H_∞ with Error Bounds", *IEEE Transactions on Automatic Control*, Vol. 37, No 37, pp. 953-963, July 1992.

[**GL86**] M. Gevers and L. Ljung, "Optimal experiment designs with respect to the intended model application", *Automatica*, Vol.22, pp. 543-554, 1986.

[**Hak90**] R.G. Hakvoort, "Optimal experiment design for prediction error identification in view of feedback design", *Selected Topics in Identification, Modeling and Control*, Delft University Press, Vol. 2, pp. 71-78, 1990.

[**Han89**] F.R.Hansen, "A fractional representation approach to closed-loop system identification and experiment design", PhD Thesis, Stanford University, CA, USA, March 1989.

[**HFK89**] F.R. Hansen, G.F. Franklin and R. Kosut, "Closed-loop identification via fractional representation: experiment design", *Proc. American Control Conference*, Pittsburgh, PA, pp. 1422-1427, 1989.

[**HSV92**] R.G. Hakvoort, R.J.P. Schrama, P.M.J. Van den Hof, "Approximate identification with closed loop performance criterion and application to LQG feedback design", Technical Report, Mech. Engin. Systems and Control Group, Delft University of Technology, 1992.

[**HV93**] R.G. Hakvoort and P. Van den Hof, private communication.

[Kal60] R.E. Kalman, "Contributions to the theory of optimal control", *Bol. Soc. Mathematica Mexicana*, pp. 102-119, 1960.

[LAKM92] W.S. Lee, B.D.O. Anderson, R.L. Kosut and I.M.Y. Mareels, "On adaptive robust control and control-relevant system identification", Technical Report, Australian National University, 1992.

[Lju87] L. Ljung, "System Identification: Theory for the User", Prentice Hall US, 1987.

[LS90] K. Liu and R.E. Skelton, "Closed-loop identification and iterative controller design", *Proc. 29th IEEE Conf. on Decision and Control*, Honolulu, Hawaii, pp. 482-487, December 1990.

[Mäk91] P.M. Mäkilä, "Identification of stabilizable systems: closed loop approximation", *Int. J. Control*, Vol. 54, No 3, pp. 577-592, 1991.

[MP92a] P.M. Mäkilä and J.R. Partington, "Robust identification of strongly stabilizable systems", *IEEE Transactions on Automatic Control*, Vol. 37, No 11, pp. 1709-1716, November 1992.

[MP92b] P.M. Mäkilä and J.R. Partington, "On bounded error identification of feedback systems", Report 92-5, Åbo Akademi University, Dept of Chemical Engineering, Finland, submitted for publication.

[MV91] M. Milanese, A. Vicino, "Optimal estimation theory for dynamic systems with set membership uncertainty : an overview", *Automatica*, Vol. 27, No 6, pp. 997-1009, 1991.

[PB93] A.G. Partanen and R.R. Bitmead, "Two stage iterative identification/controller design and direct experimental controller refinement", submitted to CDC 1993.

[Sch91] R. Schrama, "Control-oriented approximate closed-loop identification via fractional representations", *Proc. American Control Conference*, Boston, MA, pp. 719-720, 1991.

[Sch92a] R. Schrama, "Approximate identification and control design", PhD Thesis, Delft University of Technology, 1992.

[Sch92b] R. Schrama, "Accurate identification for control: the necessity of an iterative scheme", *IEEE Transactions on Automatic Control*, Vol. 37, No 7, pp. 991-994, July 1992.

[SS89] T.S. Söderström and P. Stoica, "System Identification"", Prentice Hall International, 1989.

[SV92] R.J.P. Schrama and P.M.J. Van den Hof, "An iterative scheme for identification and control design based on coprime factorizations",*Proc. American Control Conference*, pp. 2842-2846, 1992.

[Tem93] R. Tempo, "Robust and Optimal Algorithms for Worst-Case Parametric System Identification", Research report, CENS-CNR, Politecnico di Torino, 1993.

[Vid85] M. Vidyasagar, "Control System Synthesis: A Factorization Approach", MIT Press, Cambridge, MA, USA, 1985.

[WL86] B. Wahlberg and L. Ljung, "Design variables for bias distribution in transfer function estimation", *IEEE Transactions on Automatic Control*, Vol. AC-31, pp. 134-144, February 1986.

[WL92] B. Wahlberg and L. Ljung, "Hard frequency-domain model error bounds from least-squares like identification techniques", *IEEE Transactions on Automatic Control*, Vol.37, No 7, pp. 900-912, July 1992.

[Zam81] G. Zames, "Feedback and optimal sensitivity: model reference transformations, multiplicative seminorms, and approximate inverses", *IEEE Transactions on Automatic Control*, Vol. AC-26, pp. 301-320, April 1981.

[ZBG91] Z. Zang, R.R. Bitmead and M. Gevers, "Iterative Model Refinement and Control Robustness Enhancement", submitted for publication to *IEEE Transactions on Automatic Control*; a preliminary version was presented at the 30th IEEE Conference on Decision and Control, Brighton, UK, pp. 279-284, December 1991.

[ZBG92] Z. Zang, R.R. Bitmead and M. Gevers, "Disturbance Rejection: On-Line Refinement of Controllers by Closed Loop Modeling", *Proc. American Control Conference*, Vol.4, pp. 1829-1833, Chicago, Illinois, June 1992.

[ZS92] Y.C. Zhu and A.A. Stoorvogel, "Closed Loop Identification of Coprime Factors", *Proc. 31st Conference on Decision and Control*, pp. 453-545, December 1992.

6. Nonlinear State Space $\mathcal{H}_\infty$ Control Theory

A.J. van der Schaft *

1 Introduction

Although the $\mathcal{H}_\infty$ control problem was originally formulated [55] as a *linear* design problem in the *frequency domain* (in fact, $\mathcal{H}_\infty$ stands for the Hardy space of complex functions bounded and analytic in the open right-half complex plane), it can be naturally translated to the *time-domain* and extended to *nonlinear* state-space systems. Indeed, the standard $\mathcal{H}_\infty$ control problem can be equivalently formulated as the optimal attenuation of the L_2-induced norm from exogenous inputs (inputs with unknown power spectrum) to the to-be-controlled outputs, under the constraint of internal stability. Also, although early research in $\mathcal{H}_\infty$ control was conducted solely using frequency domain methods, a satisfactory state space solution to the linear $\mathcal{H}_\infty$ (sub-)optimal control problem was reached by the end of the eighties (see especially [12], [31], [17], [46], [16], [30], [48], [49], [47]). Moreover, this state space solution relies on tools familiar from LQ and LQG theory, in particular Riccati equations and Hamiltonian matrices. In the classical paper by Willems on LQ control [52] the relations of these tools with the underlying notion of *dissipativity* were being stressed; while in [53] dissipativity was defined for general nonlinear systems, encompassing notions of passivity of physical systems and input-output stability of nonlinear (feedback) systems. The resulting *dissipation inequalities* were fruitfully explored in e.g. [35], [20], [21], also linking them to the Hamilton-Jacobi equation from classical nonlinear optimal control (see also [36]).

Summarizing, the $\mathcal{H}_\infty$ control problem from linear control theory can be naturally extended to nonlinear systems, and the linear state space suboptimal solution relies on tools which have proper nonlinear extensions. Also, by the end of the eighties it was realized that the linear state space solution admits a natural *differential-game* interpretation ("control" playing versus "disturbance"); while the theory of differential games has been

*Department of Applied Mathematics, University of Twente, P.O. Box 217, Enschede, The Netherlands

also developed for nonlinear systems, centering around a Hamilton-Jacobi equation called Isaacs equation, see e.g. [14], [7].

With this in mind it is not surprising that during the last couple of years some promising work has been performed towards a solution of the *nonlinear* $\mathcal{H}_\infty$ control problem, see in particular the work of Ball, Helton & Walker [2], [3], [4], [5], Basar & Bernhard [6], Isidori & Astolfi [25], [26], [23], [24] and myself [40], [41], [43], [44], [45]. The present paper does *not* intend to give a survey of this young area of research. Instead, I will basically follow the approach of my previous papers [40], [43], [44], and then try to incorporate some results of other authors.

The main novelty of this paper is in Section 5, where the nonlinear *central controller* is derived based on the differential-game -theoretic approach of Basar and Bernhard [6]. Unfortunately this nonlinear central controller is in general (apart from e.g. the *linear* case) *infinite-dimensional*. In fact, the situation resembles very much the solution to the *nonlinear filtering* problem, and further research is needed to clarify many issues. Before we come to this, after a brief statement of the nonlinear $\mathcal{H}_\infty$ (sub-)optimal control problem in Section 2, we give a concise treatment of L_2-gain analysis of nonlinear systems in Section 3, largely based on [43], [40]. In Section 4 we treat the *state feedback* $\mathcal{H}_\infty$ suboptimal control problem, mainly following [43], [40], with a few new observations scattered throughout the text. Section 5 about the dynamic measurement feedback $\mathcal{H}_\infty$ problem is mainly based on [5], [44], and offers, apart from the derivation of the nonlinear central controller, an analysis of *necessary* conditions for the solvability of the problem and a generalized version of a beautiful *separation principle* obtained in [5]. Furthermore, following some unpublished work of Doyle [11], an explicit solution to the $\mathcal{H}_\infty$ optimal control problem for nonlinear *lossless* systems is discussed.

A key element in our approach is the strict relation between Hamilton-Jacobi equations and *invariant manifolds* of (hyperbolic) Hamiltonian vector fields. This relation provides insight about the global solvability of Hamilton-Jacobi equations, and is a very powerful tool for proving *local* existence of solutions. In fact, this will allow us to solve, in a very simple manner, the *local* nonlinear $\mathcal{H}_\infty$ sub-optimal control problem, based on the solution of the linear $\mathcal{H}_\infty$ problem for the *linearized* system.

Notation: The notation used is fairly standard. We denote by $z^T z$ or $\| z \|^2$ the squared norm of a vector $z \in \mathbf{R}^k$. The notation $L_2[0,T]$ will be also used for vector-valued functions, i.e., we say that $z : [0,T] \to \mathbf{R}^k$ is in $L_2[0,T]$ if $\int_0^T \| z(t) \|^2 \, dt < \infty$. For a differentiable function $V : \mathbf{R}^n \to \mathbf{R}$ we denote by $V_x(x)$ the row-vector of partial derivatives, and by $V_x^T(x)$ the transposed (column-)vector. By $\mathbf{C}^-$ we denote the open left-half complex plane, and by $\sigma(A)$ the set of eigenvalues of a square matrix A. $\mathbf{R}^+$ is the set of non-negative reals.

2   The standard $\mathcal{H}_\infty$ optimal control problem for nonlinear systems

In analogy with the linear case, see e.g. [13], [12], we formulate the standard $\mathcal{H}_\infty$ problem for nonlinear systems as follows. Consider a nonlinear state space system

$$
\begin{aligned}
\dot{x} &= f(x, u, d) \\
y &= g(x, u, d) \\
z &= h(x, u, d)
\end{aligned}
\tag{1}
$$

with two sets of inputs u and d, and two sets of outputs y and z, and state vector x. Here u denotes the vector of *control inputs*, d are the *exogenous inputs* (disturbances to be rejected and/or reference signals to be tracked), y are the *measured outputs* and z are the *to-be-controlled outputs* (tracking errors, cost variables). Throughout we assume the existence of a fixed equilibrium x_0, u_0, d_0, i.e. $f(x_0, u_0, d_0) = 0$, and after a suitable coordinate shift we may assume that $x_0 = 0$, $u_0 = 0$, $d_0 = 0$. Also without loss of generality we assume that $g(x_0, u_0, d_0) = 0$ and $h(x_0, u_0, d_0) = 0$.

Now let γ be a fixed positive constant. Then the nonlinear $\mathcal{H}_\infty$ *suboptimal control problem* (for disturbance attenuation level γ) is to find a nonlinear compensator

$$
\begin{aligned}
\dot{\xi} &= k(\xi, y) \\
u &= m(\xi, y)
\end{aligned}
\tag{2}
$$

with state vector ξ, and satisfying $k(0,0) = 0$, $m(0,0) = 0$, such that the closed-loop system (1), (2) has L_2-*gain less than or equal to* γ, in the sense that

$$
\int_0^T \| z(t) \|^2 \, dt \le \gamma^2 \int_0^T \| d(t) \|^2 \, dt
\tag{3}
$$

for all functions $d(\cdot)$ and all $T \ge 0$, where $z(\cdot)$ denotes the response of the closed-loop system (1), (2) for $d(\cdot)$ and *initial condition* $x(0) = 0$, $\xi(0) = 0$. Actually the above definition constitutes a slight departure from the common definition in linear $\mathcal{H}_\infty$ control, where (3) is considered only for $T = \infty$ while it is required that the closed-loop system is *asymptotically stable*. It can be verified that these requirements *imply* that (3) holds for arbitrary $T \ge 0$. On the other hand, also in the nonlinear case some kind of stability of the closed-loop system is very desirable; however we will see that generically some sort of asymptotic stability will be *implied* by the finite

L_2-gain condition (3), and we will find it easier to formulate the nonlinear $\mathcal{H}_\infty$ control problem without any specific a priori stability requirements.

Furthermore, for nonlinear systems it is not quite natural to consider only the *zero-state* response as in (3); therefore we will actually strenghten (3) by requiring that for all initial conditions $x(0)$, $z(0)$ of the closed-loop system (1), (2) there exists a nonnegative constant K (depending on $x(0)$, $z(0)$ and zero for $x(0) = 0, z(0) = 0$) such that (see [10], [19])

$$\int_0^T \| z(t) \|^2 \, dt \leq \gamma^2 \int_0^T \| d(t) \|^2 \, dt + K \tag{4}$$

for all $d(\cdot)$ and $T \geq 0$, with $z(\cdot)$ denoting the response for initial condition $x(0), \xi(0)$. This strengthening of the notion of finite L_2-gain will also enable us to make a smooth transition to the theory of *dissipative systems*.

Finally, the $\mathcal{H}_\infty$ *optimal control problem* is to find the smallest $\gamma^* \geq 0$ such that the $\mathcal{H}_\infty$ suboptimal control problem is solvable for *all* $\gamma > \gamma^*$. However even in the *linear* state space $\mathcal{H}_\infty$ theory the optimal $\mathcal{H}_\infty$ problem is quite delicate, and thus for nonlinear systems we will mostly restrict ourselves in the present paper to the (strictly) suboptimal case. Furthermore we will confine ourselves to nonlinear systems which are *affine* in the input variables u and d, and where also the output equations for y and z have a more special structure. Largely this is done because of clarity of exposition; however there are some inherent problems if the dependence on u and d is completely general, see e.g. [45]. Moreover, a main assumption in the present paper is that all control inputs u appear non-trivially in the to-be-controlled outputs z, and that all measurements y are non-trivially affected by the disturbances d; relaxing this assumption leads to *singular* nonlinear $\mathcal{H}_\infty$ control problems (see for the linear case, e.g. [48], [47]).

3 L_2-gain analysis of nonlinear systems

As has become clear from Section 2 the notion of finite L_2-gain is essential in the nonlinear $\mathcal{H}_\infty$ control problem. Actually this notion has been intensively studied in the sixties and seventies, mainly for stability investigations, and has been incorporated into a general notion of dissipativity [53], [35], [20], [21]. We will briefly recapitulate the main ingredients of this theory in our context, and then reinterpret and extend them using the geometric approach based on invariant manifolds of Hamiltonian vectorfields as put forward in [43], [40].

Consider a nonlinear system with a single set of input variables u and a single set of output variables y:

$$\begin{aligned} \dot{x} &= f(x) + g(x)u \quad, \quad u \in \mathbf{R}^m \quad, \quad f(0) = 0 \\ y &= h(x) \qquad\qquad, \quad y \in \mathbf{R}^p \quad, \quad h(0) = 0 \end{aligned} \tag{5}$$

Here $x = (x_1, \cdots, x_n)$ are local coordinates for a C^∞ state space manifold M and $g(x)$ is an $n \times m$ matrix with elements depending on x. We assume throughout that all data, i.e. f, g, h, are C^k, with $k \geq 1$ (at least continuously differentiable).

Definition 3.1 Let γ be a fixed non-negative constant. The nonlinear system (5) has L_2-gain $\leq \gamma$ if for all $x \in M$ there exists a constant $K(x)$, $0 \leq K(x) < \infty$, with $K(0) = 0$, such that

$$\int_0^T \| y(t) \|^2 \, dt \leq \gamma^2 \int_0^T \| u(t) \|^2 \, dt + K(x), \qquad (6)$$

for all $u \in L_2[0, T]$ and all $T \geq 0$, with $y(t)$ denoting the response of (5) for initial condition $x(0) = x$. The nonlinear system (5) has L_2-gain $< \gamma$ if there exists a $\tilde{\gamma} \geq 0$ with $\tilde{\gamma} < \gamma$ such that (5) has L_2-gain $\leq \tilde{\gamma}$, and finally the system has L_2-gain $= \gamma$ if it has L_2-gain $\leq \gamma$ and there does *not* exist $0 \leq \tilde{\gamma} < \gamma$ such that it has L_2-gain $\leq \tilde{\gamma}$.

Remark 3.2 For a very interesting interpretation of L_2-gain in terms of *periodic* input functions we refer to [26].

On the other hand, define for every x the function $V_a : M \to \mathbf{R}^+ \cup \{\infty\}$, called *available storage* in [53], as

$$V_a(x) = \sup_u \frac{1}{2} \int_0^T (\| y(t) \| - \gamma^2 \| u(t) \|^2) dt, \qquad x(0) = x, \qquad (7)$$

where the supremum is taken over all $u \in L_2[0, T]$ and all $T \geq 0$, and where $y(t)$ denotes the response for initial condition $x(0) = x$. It is immediate that condition (6) is equivalent to the condition that V_a is *finite* for every $x \in M$, i.e. $V_a : M \to \mathbf{R}^+$, while $V_a(0) = 0$. Following the theory of dissipative systems [53] V_a satisfies the *integral dissipation inequality*

$$V(x(t_1)) - V(x(t_0)) \leq \tfrac{1}{2} \int_{t_0}^{t_1} (\gamma^2 \| u(t) \|^2 - \| y(t) \|^2) dt,$$
$$V(0) = 0, \qquad (8)$$

for all $t_1 \geq t_0$ and all $u \in L_2[t_0, t_1]$ (with $x(t_1)$ denoting the solution at time t_1 for initial condition $x(t_0)$ at time t_0). In fact, see [53], V_a is the *minimal* function $V : M \to \mathbf{R}^+$ satisfying the integral dissipation inequality, and for every x the constant $2V_a(x)$ equals the *minimal* value of $K(x)$ for which (6) holds. (Thus $2V_a(x)$ measures, for a system with L_2-gain $\leq \gamma$, to what extent the L_2-norm of the output signal can be larger than γ^2 times the L_2-norm of the input signal, depending on the initial state x.) A function $V : M \to \mathbf{R}^+$ satisfying (8) is called a *storage function* in [53], [35], and if there exists such a storage function then the system (5) is called *dissipative* with respect to the supply rate $\frac{1}{2}\gamma^2 \parallel u \parallel^2 - \frac{1}{2} \parallel y \parallel^2$.

Now assume that there exists a storage function $V : M \to \mathbf{R}^+$ which is *continuously differentiable*, then it is immediately seen that (8) can be rewritten as the *differential dissipation inequality*

$$V_x(x)f(x) + V_x(x)g(x)u \leq \frac{1}{2}\gamma^2 \parallel u \parallel^2 - \frac{1}{2} \parallel y \parallel^2, \qquad V(0) = 0, \quad (9)$$

for all $u \in \mathbf{R}^m$, with $y = h(x)$. Furthermore by a simple completion of the square argument [40] it is seen that (9) is equivalent to the *Hamilton-Jacobi inequality*

$$V_x(x)f(x) + \frac{1}{2}\frac{1}{\gamma^2}V_x(x)g(x)g^T(x)V_x^T(x) + \frac{1}{2}h^T(x)h(x) \leq 0,$$

$$V(0) = 0,$$
(10)

for all $x \in M$. Summarizing [43]:

Theorem 3.3 *System (5) has L_2-gain $\leq \gamma$ if and only if there exists a solution $V : M \to \mathbf{R}^+$ to the integral dissipation inequality (8), while there exists a non-negative C^1 solution to (8) if and only if there exists a non-negative C^1 solution to the differential dissipation inequality (9), if and only if there exists a non-negative C^1 solution to the Hamilton-Jacobi inequality (10).*

Our approach will be primarily based on the Hamilton-Jacobi inequality (10), and thus our focus is on storage functions which are at least C^1. Actually we will show, under appropriate conditions, that at least on a neighborhood of 0 the *minimal* storage function V_a has the same degree of differentiability as the system (5), i.e. C^k, with $k \geq 1$. An interesting alternative to our restriction to C^1 storage functions is provided by considering *viscosity solutions* to the Hamilton-Jacobi inequality (10), see e.g. [28], [29]; however we will not go into this in the present paper.

There is a close connection between (internal) asymptotic stability of (5) and the property of finite L_2-gain. In fact, storage functions serve as *Lyapunov* functions [20], [21], [51]:

Proposition 3.4 *Suppose there exists a nonnegative C^1 solution V to (10). Assume the system (5) is zero-state observable (i.e. $y(t) = 0, \forall t \geq 0$, implies $x(0) = 0$). Then $V(x) > 0$ for all $x \neq 0$, and 0 is a locally asymptotically stable equilibrium of (5). Moreover, if V is proper (i.e., for each $c > 0$ the set $\{x \in M \mid 0 \leq V(x) \leq c\}$ is compact), then 0 is a globally asymptotically stable equilibrium. Conversely, if 0 is a globally asymptotically stable equilibrium of (5) then every C^1 solution of (10) is nonnegative.*

Now we pass on to the *Hamiltonian* point of view. First of all we note that the differential dissipation inequality (9) can be written as

$$K_\gamma(x, V_x^T(x), u) \leq 0, \qquad V(0) = 0, \qquad \text{for all } u, \tag{11}$$

where the *pre-Hamiltonian* $K_\gamma : T^*M \times \mathbf{R}^m \to \mathbf{R}$ is defined as

$$K_\gamma(x, p, u) = p^T f(x) + p^T g(x)u - \frac{1}{2}\gamma^2 \parallel u \parallel^2 + \frac{1}{2}h^T(x)h(x) \tag{12}$$

with $x = (x_1, \cdots, x_n), p = (p_1, \cdots, p_n)$ natural coordinates for the $2n$-dimensional cotangent bundle T^*M (p denoting the co-state variables). Substituting the maximizing input $u^* = \frac{1}{\gamma^2}g^T(x)p$ into (11) yields the Hamilton-Jacobi inequality (10), which we succinctly write as

$$H_\gamma(x, V_x^T(x)) \leq 0, \qquad V(0) = 0, \tag{13}$$

for the Hamiltonian $H_\gamma : T^*M \to \mathbf{R}$ given as $K_\gamma(x, p, u^*)$, i.e.,

$$H_\gamma(x, p) = p^T f(x) + \frac{1}{2}\frac{1}{\gamma^2}p^T g(x)g^T(x)p + \frac{1}{2}h^T(x)h(x) \tag{14}$$

We will now momentarily forget that we are interested in *nonnegative* solutions V of (13), and furthermore we will restrict ourselves to solutions of the Hamilton-Jacobi *equality* or equation

$$H_\gamma(x, V_x^T(x)) = 0, \qquad V(0) = 0, \qquad V_x(0) = 0. \tag{15}$$

(Notice that the extra condition $V_x(0) = 0$ is automatically satisfied if V is non-negative).

Also we consider, as in classical mechanics, the *Hamiltonian vectorfield* X_{H_γ} on T^*M corresponding to the Hamiltonian function H_γ, given by the canonical equations [1]

$$
\begin{aligned}
\dot{x}_i &= \frac{\partial H_\gamma}{\partial p_i}(x, p) \\
&\qquad\qquad\qquad\qquad i = 1, \cdots, n \\
\dot{p}_i &= -\frac{\partial H_\gamma}{\partial x_i}(x, p)
\end{aligned}
\tag{16}
$$

with equilibrium $(x, p) = (0, 0) \in T^*M$. It now can be verified ([43], [40]) that $V : M \to \mathbf{R}$ is a solution of (15) if and only if the following n-dimensional submanifold of T^*M

$$N = \{(x, p) \in T^*M \mid p^T = V_x(x)\} \tag{17}$$

is an *invariant* manifold of X_{H_γ} through $(0, 0)$, meaning that $X_{H_\gamma}(x, p)$ is tangent to N at every point $(x, p) \in N$. (Furthermore, N is a special type of submanifold of T^*M, since it is n-dimensional and parametrized by the x-coordinates, and since it is a *Lagrangian* submanifold of T^*M, cf. [1]).

Summarizing, solutions of the Hamilton-Jacobi equality (15) correspond to special n-dimensional invariant manifolds of the Hamiltonian vectorfield X_{H_γ}. Now, assume that the linearization of X_{H_γ} at $(0,0)$ has no purely imaginary eigenvalues (such a vectorfield is called *hyperbolic*). From the theory of nonlinear differential equations [38] we then know there exist two invariant manifolds for X_{H_γ}, namely the *stable invariant manifold N^-* consisting of all points in T^*M converging to $(0,0)$ along the flow of X_{H_γ}, and the *unstable invariant* manifold N^+ consisting of all points converging to $(0,0)$ along the flow of $-X_{H_\gamma}$. Moreover, the degree of smoothness of N^- and N^+ is the same as of the vectorfield X_{H_γ}.

In fact, using the Hamiltonian structure we can prove

Proposition 3.5 *[40] Consider the nonlinear system (5), with f, g, h C^k, $k \geq 2$. Assume the C^{k-1} vectorfield X_{H_γ} is hyperbolic in $(0,0) \in T^*M$. Then the stable and unstable invariant manifolds N^- and N^+ are n-dimensional C^{k-1} immersed submanifolds of T^*M, which are both Lagrangian.*

Thus, N^- and N^+ will generate solutions of the Hamilton-Jacobi equation (15) if and only if N^- and N^+ can be *parametrized by the x-coordinates*. Let $\pi : T^*M \to M$ denote the canonical projection $(x, p) \mapsto x$.

Corollary 3.6 *[43], [44] Take the same assumptions as in Proposition 3.5. Additionally, assume that $\pi : N^- \subset T^*M \to M$ is a C^{k-1} diffeomorphism, then there exists a C^k function $V^- : M \to \mathbf{R}$ such that*

$$N^- = \{(x, p) \mid p^T = V_x^-(x)\} \tag{18}$$

and V^- is the unique solution of the Hamilton-Jacobi equation (15) with the property that the vectorfield

$$f + \frac{1}{\gamma^2} g g^T (V_x^-)^T \quad \text{is globally asymptotically stable on } M \tag{19}$$

(for the equilibrium $0 \in M$). Similarly, if $\pi : N^+ \subset T^*M \to M$ is a C^{k-1} diffeomorphism then there exists a unique C^k solution $V^+ : M \to \mathbf{R}$ of the Hamilton-Jacobi equation (15) with the property that

$$-(f + \frac{1}{\gamma^2} g g^T (V_x^+)^T) \quad \text{is globally asymptotically stable.} \tag{20}$$

Moreover, every C^1 solution to the Hamilton-Jacobi inequality (10) satisfies $V^- \leq V \leq V^+$.

Remark 3.7 The dynamics on the invariant manifold N^- (resp. N^+), projected onto M by π, is given as $\dot{x} = f(x) + \frac{1}{\gamma^2} g(x) g^T(x) (V_x^-)^T(x)$ (resp. $\dot{x} = f(x) + \frac{1}{\gamma^2} g(x) g^T(x) (V_x^+)^T(x)$).

In fact, the solutions V^- and V^+ have an immediate interpretation in terms of dissipativity whenever they are nonnegative:

Proposition 3.8 *[43] Take the same assumptions as in Corollary 3.6. Assume additionally that $V^- \geq 0$ (For example, see Proposition 3.4, this is the case if f is globally asymptotically stable.) Then the available storage V_a as defined in (7) is equal to V^-, and in particular is finite for every x and is C^k. Hence the system has L_2-gain $\leq \gamma$, and in fact (by hyperbolicity of X_{H_γ}) the system has L_2-gain $< \gamma$. Furthermore, assume additionally that the system (5) is reachable from 0, then $V^+(x)$ is given as*

$$\inf \frac{1}{2} \int_0^T (\gamma^2 \parallel u(t) \parallel^2 - \parallel y(t) \parallel^2) dt, \qquad x(0) = 0, \tag{21}$$

where the infimum is taken over all $u \in L_2[0,T]$ and $T \geq 0$ such that $x(T) = x$. (The expression (21) is called the required supply in [53]).

Of course, the condition that N^- and/or N^+ is parametrizable by the x-coordinates is in general not easily checkable. Instead, on the basis of the *linearization* of (5) at 0, i.e.

$$\begin{aligned}
\dot{\bar{x}} &= F\bar{x} + G\bar{u} \ , & \bar{u} \in \mathbf{R}^m, \quad \bar{x} \in \mathbf{R}^n, \\
\bar{y} &= H\bar{x} & , \quad \bar{y} \in \mathbf{R}^p,
\end{aligned} \tag{22}$$

with $F = \frac{\partial f}{\partial x}(x_0), G = g(x_0), H = \frac{\partial h}{\partial x}(x_0)$ we will now develop easily verifiable conditions for *local* parametrization of N^- or N^+ by the x-coordinates, and thus for *local* finite L_2-gain of the nonlinear system. First observation is that the *linearization* of X_{H_γ} at $(0,0)$ is given by the Hamiltonian matrix

$$Ham_\gamma = \begin{pmatrix} F & \frac{1}{\gamma^2} GG^T \\ -H^T H & -F^T \end{pmatrix}, \tag{23}$$

and thus hyperbolicity of X_{H_γ} means that Ham_γ does not have purely imaginary eigenvalues. Furthermore from the theory of nonlinear differential equations [38] we know that the tangent spaces to N^-, respectively N^+, at $(0,0)$ are given by the (generalized) stable, respectively unstable, eigenspaces of Ham_γ.

Let us now concentrate on N^- and the stable eigenspace of Ham_γ (see for N^+ [13], [40]). From the theory of Riccati equations (e.g. [13]) it is well-known that if (F, G) is stabilizable then the stable eigenspace of Ham_γ is of the form $span \begin{pmatrix} I_n \\ P \end{pmatrix}$, where $P = P^T$ is the unique solution of the Riccati equation

$$F^T P + PF + \frac{1}{\gamma^2} PGG^T P + H^T H = 0 \tag{24}$$

satisfying

$$\sigma(F + \frac{1}{\gamma^2}GG^T P) \subset \mathbf{C}^- \tag{25}$$

Thus stabilizability of (F,G) implies that the stable eigenspace of Ham_γ can be parametrized by the $\bar{x}$-coordinates, and therefore, by the above mentioned property of N^- being tangent to the stable eigenspace, there exists a *neighborhood* W of 0 in M where N^- can be parametrized (in a C^{k-1} manner) by the x-coordinates. It thus follows that locally about $(0,0) \in T^*M$ the manifold N^- is given as

$$N^- = \{(x,p) \mid p^T = V_x^-(x), \ x \in W\} \tag{26}$$

for some C^k function $V^- : W \to \mathbf{R}$ satisfying the Hamilton-Jacobi equation (15) *on* W, and such that the vectorfield $f + \frac{1}{\gamma^2}gg^T(V_x^-)^T$ is asymptotically stable *on* W. Thus if $V^- : W \to \mathbf{R}$ is nonnegative (which for instance, cf. Proposition 3.4, is implied by asymptotic stability of f *on* W) then the system *restricted* to the neighborhood W of 0 will have L_2-gain $\leq \gamma$. Since we can replace γ by another constant $\tilde{\gamma} < \gamma$ with $\tilde{\gamma}$ sufficiently close to γ we have obtained :

Theorem 3.9 *Consider the nonlinear system (5) with its linearization (22). Assume (F,G) is stabilizable and Ham_γ in (23) does not have purely imaginary eigenvalues, while f is locally asymptotically stable. Then the system has locally L_2-gain $< \gamma$ in the sense that there exists a constant $\tilde{\gamma} < \gamma$ and a neighborhood $W \subset M$ of 0 and $K(x), 0 \leq K(x) < \infty$, with $K(0) = 0$, such that (cf. (6))*

$$\int_0^T \| y(t) \|^2 \, dt \leq \tilde{\gamma}^2 \int_0^T \| u(t) \|^2 \, dt + K(x) \tag{27}$$

for all $u \in L_2[0,T]$, all $T \geq 0$, and all $x = x(0) \in W$ such that the state space trajectories do not leave the neighborhood W. Furthermore V_a as defined in (7) is finite for every $x \in W$, and is C^k on W.

On the other hand, assume F is asymptotically stable, then it is well-known (e.g. [46]) that the existence of a symmetric solution P of (24), (25) is equivalent to the linearized system (22) having L_2-gain $< \gamma$. In fact, we can prove the following equivalence.

Theorem 3.10 *[43] Consider the nonlinear system (5) with its linearization (22). Assume F is asymptotically stable. Then the following statements are equivalent:*
(a) The linearization (22) has L_2-gain $< \gamma$.
(b) $\exists P = P^T \geq 0$ solution of (24), (25).
(c) The nonlinear system (5) has locally L_2-gain $< \gamma$.

Remark 3.11 We may also replace in (a) and (c) L_2-gain $< \gamma$ by *zero-state L_2-gain* $< \gamma$, i.e. the weaker property that (6) or (27) only holds for $x = 0$.

In relation with the implication (a) $\rightarrow$ (c) we observe that if there exists a $C^k(k \geq 2)$ solution V to the Hamilton-Jacobi equation (15), then the Hessian matrix $P := \frac{\partial^2 V}{\partial x^2}(0)$ satisfies the equation

$$x^T P F x + \frac{1}{2}\frac{1}{\gamma^2}x^T PGG^T P x + \frac{1}{2}x^T H^T H x = 0, \tag{28}$$

or equivalently, P is a symmetric solution of the Riccati equation (24). Also we note that the Riccati equation (24) is equivalent to the Hamilton-Jacobi equation (15) if the nonlinear system (5) *equals* the linear system (22).

We note that the *optimal* case, i.e. L_2-gain being *equal* to γ^*, corresponds to the case that Ham_{γ^*} has purely imaginary eigenvalues. In this case the geometric analysis as described above becomes much more delicate, since the stable and unstable invariant manifolds of X_{H_γ} (partly) deteriorate, for $\gamma \downarrow \gamma^*$, into the *center* manifold of $X_{H_\gamma^*}$; and in fact we need to split the center manifold of $X_{H_\gamma^*}$ into two equally dimensioned invariant manifold in order to obtain solutions of the Hamilton-Jacobi equation (15).

4   State feedback $\mathcal{H}_\infty$ control

As announced in Section 2 we will study the $\mathcal{H}_\infty$ problem for a somewhat special class of nonlinear systems, namely affine systems of the form

$$\dot{x} = a(x) + b(x)u + g(x)d_1 \quad , \quad a(0) = 0,$$

$$y = c(x) + d_2 \qquad\qquad , \quad c(0) = 0, \tag{29}$$

$$z = \begin{pmatrix} h(x) \\ u \end{pmatrix} \qquad\qquad , \quad h(0) = 0,$$

where $x = (x_1, \cdots, x_n)$ are local coordinates for a C^∞ state space manifold M, $u \in \mathbf{R}^m$ are the control inputs, $d = \begin{pmatrix} d_1 \\ d_2 \end{pmatrix} \in \mathbf{R}^r$ the exogenous inputs, $y \in \mathbf{R}^p$ the measured outputs and $z \in \mathbf{R}^s$ the to-be-controlled outputs. All data will be assumed to be C^k, with $k \geq 2$. It is not very difficult, like in the linear case (see e.g. [17], [48], [47]), to generalize everything to affine systems having a more general dependence on u and d (as long as we remain within the "non-singular case"); only the resulting formulae will become more complicated (see e.g. [5], [26], [24]). For some extensions to nonlinear systems which are not affine in u and d we refer to [5], [45].

First we will consider the *state feedback $\mathcal{H}_\infty$ suboptimal control problem*, where we assume that all components of the state vector are available for noiseless measurements (i.e. $y = x, d_2$ void and $d = d_1$), and where we seek for a nonlinear static state feedback

$$u = \alpha(x), \qquad \alpha(0) = 0, \tag{30}$$

such that the closed-loop system (29), (30) has L_2-gain $\leq \gamma$ from d to z, i.e., cf. (6), for every $x \in M$ there exists a constant $K(x), 0 \leq K(x) < \infty$, with $K(0) = 0$, such that

$$\int_0^T \| z(t) \|^2 \, dt \leq \gamma^2 \int_0^T \| d(t) \|^2 \, dt + K(x) \tag{31}$$

for all $d \in L_2[0, T]$ and all $T \geq 0$, with $z(t)$ denoting the response of (29), (30) for initial condition $x(0) = x$. If this can be done for some $\tilde{\gamma} < \gamma$ then the state feedback $\mathcal{H}_\infty$ *strictly* suboptimal control problem (for γ) is said to be solvable. In analogy with the previous section, cf. formula (12), we define the pre-Hamiltonian $K_\gamma : T^*M \times \mathbf{R}^r \times \mathbf{R}^m$ as

$$K_\gamma(x, p, d, u) = p^T(a(x) + b(x)u + g(x)d)$$
$$-\tfrac{1}{2}\gamma^2 \| d \|^2 + \tfrac{1}{2} \| z \|^2 \tag{32}$$

with $\| z \|^2 = h^T(x)h(x) + \| u \|^2$. ¿From the equations $\frac{\partial K_\gamma}{\partial d} = 0$ and $\frac{\partial K_\gamma}{\partial u} = 0$ we determine respectively

$$d^* = \frac{1}{\gamma^2} g^T(x)p, \qquad u^* = -b^T(x)p, \tag{33}$$

which have the *saddle point property* [14], [7], [25]

$$K_\gamma(x, p, d, u^*) \leq K_\gamma(x, p, d^*, u^*) \leq K_\gamma(x, p, d^*, u) \tag{34}$$

for every (d, u) and (x, p). This leads to the *Hamiltonian* $H_\gamma : T^*M \to \mathbf{R}$ given as $K_\gamma(x, p, d^*, u^*)$, i.e.

$$H_\gamma(x, p) = p^T a(x) + \tfrac{1}{2}p^T \left[\tfrac{1}{\gamma^2} g(x)g^T(x) - b(x)b^T(x) \right] p$$
$$+\tfrac{1}{2}h^T(x)h(x) \tag{35}$$

It now immediately follows from the theory of differential games [14], [7] (see for clear exposé's in this context [6], [23], [25]), that the solution of the state feedback $\mathcal{H}_\infty$ suboptimal control centers around finding solutions $P : M \to \mathbf{R}^+$ to the Hamilton-Jacobi inequality $H_\gamma(x, P_x^T(x)) \leq 0$, i.e.

$$P_x(x)a(x) + \tfrac{1}{2}P_x(x) \left[\tfrac{1}{\gamma^2} g(x)g^T(x) - b(x)b^T(x) \right] P_x^T(x)$$
$$+\tfrac{1}{2}h^T(x)h(x) \leq 0, \quad P(0) = 0 \tag{36}$$

(In differential game theory (36) with equality is called the *Hamilton-Jacobi-Isaacs equation*.) In fact we have

Theorem 4.1 *[40], [43] Consider the nonlinear system (29) with $y = x$. Let $\gamma > 0$. Suppose there exists a $C^r(k \geq r \geq 1)$ solution $P \geq 0$ to the Hamilton-Jacobi inequality (36), then the C^{r-1} state feedback*

$$u = -b^T(x)P_x^T(x) \tag{37}$$

solves the state feedback $\mathcal{H}_\infty$ suboptimal control problem. Conversely, suppose there exists a state feedback (30) which solves the state feedback $\mathcal{H}_\infty$ suboptimal control problem in the sense that there exists a C^1 solution $V \geq 0$ to the Hamilton-Jacobi inequality (10) for the closed-loop system (29), (30) (for the L_2-gain from d to z), then V is also a solution of (36).

The proof of Theorem 4.1 is very simple; let us only treat the first part of it. Indeed, rewrite (36) as

$$P_x(x)[a(x) - b(x)b^T(x)P_x^T(x)] + \frac{1}{2}\frac{1}{\gamma^2}P_x(x)g(x)g^T(x)P_x^T(x)$$

$$+ \frac{1}{2}P_x(x)b(x)b^T(x)P_x^T(x) + \frac{1}{2}h^T(x)h(x) \leq 0, \tag{38}$$

then the result immediately follows from Theorem 3.3. We see that P is a candidate *Lyapunov function* for the closed-loop system, and in fact we have the following corollary of Proposition 3.4.

Proposition 4.2 *[43] Suppose there exists a C^1 solution $P \geq 0$ to (36). Assume the system $\dot{x} = a(x)$ with outputs $z = \begin{bmatrix} h(x) \\ -b^T(x)P_x^T(x) \end{bmatrix}$ is zero-state observable (cf. Proposition 3.4), then the closed-loop system (29), (37) is locally asymptotically stable, and globally asymptotically stable if P happens to be proper.*

The geometric approach to the Hamilton-Jacobi inequality (36) is again based on the Hamiltonian H_γ given in (35) and its corresponding Hamiltonian vectorfield X_{H_γ} on T^*M. We have the following analog of Corollary 3.6 and Proposition 3.8. Let $\pi : T^*M \to M$ denote again the canonical projection $(x, p) \mapsto x$.

Proposition 4.3 *[43] Let H_γ be the C^k Hamiltonian given in (35). Assume the C^{k-1} vectorfield X_{H_γ} is hyperbolic in $(0, 0) \in T^*M$. Suppose that $\pi : N^- \subset T^*M \to M$ is a C^{k-1} diffeomorphism, where N^- is the stable invariant manifold of X_{H_γ}. Then there exists a C^k function $P^- : M \to \mathbf{R}$ such that*

$$N^- = \{(x, p) \in T^*M \mid p^T = P_x^-(x)\} \tag{39}$$

and P^- is the unique solution of the Hamilton-Jacobi equation

$$H_\gamma(x, P_x^T(x)) = 0, \qquad P(0) = 0, P_x(0) = 0, \tag{40}$$

with the property that the vectorfield

$$a - bb^T P_x^{-T} + \frac{1}{\gamma^2} gg^T P_x^{-T} \tag{41}$$

is globally asymptotically stable (in $0 \in M$). Assume additionally that $P^- \geq 0$ (for example, see Proposition 3.4, this is the case if $a - bb^T P_x^{-T}$ is globally asymptotically stable), then the state feedback $\mathcal{H}_\infty$ suboptimal control problem is solvable.

Remark 4.4 Under extra technical conditions it is shown in [43] that P^- is the minimal nonnegative solution of the Hamilton-Jacobi inequality (36).

We now again turn attention to the linearization of (29) at 0, i.e.

$$\begin{aligned}
\dot{\bar{x}} &= A\bar{x} + B\bar{u} + G\bar{d}_1 \\[4pt]
\bar{y} &= C\bar{x} + \bar{d}_2 \\[4pt]
\bar{z} &= \begin{pmatrix} H\bar{x} \\ \bar{u} \end{pmatrix}
\end{aligned} \tag{42}$$

with $A = \frac{\partial a}{\partial x}(0), B = b(0), G = g(0), C = \frac{\partial c}{\partial x}(0), H = \frac{\partial h}{\partial x}(0)$, in order to derive conditions for at least local existence of a nonnegative solution of the Hamilton-Jacobi inequality (36). Since we are dealing with state feedback we first take $\bar{y} = \bar{x}, \bar{d}_2$ void, and $\bar{d} = \bar{d}_1$. We obtain the following analog of Theorem 3.10.

Theorem 4.5 [43], [40] Consider the nonlinear system (29) with $y = x$ and $d = d_1$, and its linearization (42) with $\bar{y} = \bar{x}$ and $\bar{d} = \bar{d}_1$. Assume (A, H) is detectable. Then the following statements are equivalent.
(a) There exists a linear state feedback

$$\bar{u} = F\bar{x} \tag{43}$$

such that the closed-loop system (42), (43) (with inputs $\bar{d}$ and outputs $\bar{z}$) is asymptotically stable and has L_2-gain $< \gamma$ (i.e. the linear state feedback $\mathcal{H}_\infty$ strictly suboptimal control problem is solvable).
(b) There exists a symmetric solution $X = X^T \geq 0$ to the Riccati equation

$$A^T X + XA + X(\frac{1}{\gamma^2}GG^T - BB^T)X + H^T H = 0 \tag{44}$$

satisfying additionally

$$\sigma(A - BB^T X + \frac{1}{\gamma^2}GG^T X) \subset \mathbf{C}^- \tag{45}$$

(c) There exists a neighborhood $W \subset M$ of 0, and a nonlinear state feedback $u = \alpha(x)$ as in (30) defined on W, such that $A + BF$, with $F = \frac{\partial \alpha}{\partial x}(0)$, is asymptotically stable and the closed-loop system (29), (30) (with inputs d and outputs z) has locally L_2-gain $< \gamma$ on W, cf. Theorem 3.9 (i.e., the nonlinear state feedback $\mathcal{H}_\infty$ strictly suboptimal control problem is solvable on W).

Of course, the equivalence between (a) and (b) is well-known from linear state space $\mathcal{H}_\infty$ theory. Theorem 4.5 primarily tells us that the solvability of the $\mathcal{H}_\infty$ strictly suboptimal problem of the *linearized* system implies at least the *local* solvability of the nonlinear $\mathcal{H}_\infty$ strictly suboptimal problem (both by state feedback). Indeed, the existence of a solution $X = X^T \geq 0$ to (44), (45) implies the local existence of a solution $P \geq 0$ to the Hamilton-Jacobi equality (36) (with ≤ 0 replaced by $= 0$), such that $a - bb^T P_x^T + \frac{1}{\gamma_2}gg^T P_x^T$ is locally asymptotically stable. Furthermore there is a strict relation between the matrix X and the function P, namely

$$X = \frac{\partial^2 P}{\partial x^2}(0). \tag{46}$$

It follows that the *linear* part of the nonlinear feedback $u = -b^T(x)P_x^T(x)$ prescribed by Theorem 4.1 is equal to $-B^T Xx$, which is precisely the linear feedback which solves the state feedback $\mathcal{H}_\infty$ strictly suboptimal control problem for the *linearized* system! On the other hand, using Theorem 3.10 ((a) $\rightarrow$ (c)), it follows ([43]) that also this *linear* feedback $-B^T Xx$ will solve the *nonlinear* state feedback $\mathcal{H}_\infty$ strictly suboptimal control problem on some neighborhood $\tilde{W}$ of 0. We *conjecture* that always $\tilde{W} \subseteq W$, so that the nonlinear feedback will always have a *larger domain of validity* than its linear part, but a proof of this conjecture has not yet been found.

The main limitation to the use of Theorem 4.5 is that it does not give us any clue about the *size* of the neighborhood W of 0 where the nonlinear $\mathcal{H}_\infty$ problem is solvable. Indeed, Theorem 4.5 only guarantees that locally near 0 the map $\pi : N^- \subset T^*M \rightarrow M$ is a C^{k-1} diffeomorphism, and a really nonlinear analysis has to tell us *if* and *where* problems with the parametrization of the stable invariant manifold N^- by the x-coordinates arise.

Example 4.6 Consider the scalar system

$$\begin{aligned} \dot{x} &= u + (\arctan x)d \\ y &= x \\ z &= \begin{bmatrix} x \\ u \end{bmatrix} \end{aligned} \tag{47}$$

The Hamilton-Jacobi inequality (36) for this system can be written as

$$[P_x(x)]^2 \left[1 - \frac{1}{\gamma^2} arctan^2 x\right] \geq x^2, \qquad P(0) = 0 \tag{48}$$

Clearly, (48) has a solution $P : \mathbf{R} \to \mathbf{R}^+$ if and only if $| \, arctan \, x \, | \leq \gamma$ for all $x \in \mathbf{R}$, implying that the optimal level of disturbance attenuation is $\gamma^* = \frac{\pi}{2}$. In fact, for any $\gamma > \gamma^*$ the $\mathcal{H}_\infty$ suboptimal control problem is solved by the state feedback

$$u = \alpha_\gamma(x) = -x(1 - \frac{1}{\gamma^2} arctan^2 x)^{-\frac{1}{2}} \tag{49}$$

On the other hand, the linearized system is given as

$$\bar{x} = \bar{u}, \qquad \bar{y} = \bar{x}, \qquad \bar{z} = \begin{bmatrix} \bar{x} \\ \bar{u} \end{bmatrix}, \tag{50}$$

and clearly has the optimal level of disturbance attenuation equal to 0. Hence by Theorem 4.5 for every $\gamma > 0$ the Hamilton-Jacobi inequality (48) should have at least a *local* solution. Indeed, for every $\gamma > 0$ (48) has a solution $P \geq 0$ on the neighborhood

$$W_\gamma = \{x \in \mathbf{R} \mid \; | \, arctan \, x \, | < \gamma\} \tag{51}$$

yielding the state feedback $u = \alpha_\gamma(x)$ *defined on* W_γ. (Notice that $\alpha_\gamma(x)$ becomes unbounded for $| \, x \, | \uparrow \tan \gamma$, and that $W_\gamma \to \{0\}$ for $\gamma \downarrow 0$.)

Now, let us apply instead of the nonlinear feedback $u = \alpha_\gamma(x)$ its linear part given by $u = -x$, yielding the closed-loop system

$$\begin{aligned} \dot{x} &= -x + (arctan \, x)d \\ z &= \begin{bmatrix} x \\ -x \end{bmatrix} \end{aligned} \tag{52}$$

Then the Hamilton-Jacobi inequality (10) is given as

$$- V_x(x)x + \frac{1}{2}\frac{1}{\gamma^2}V_x^2(x) \, arctan^2 x + \frac{1}{2}x^2 + \frac{1}{2}x^2 \leq 0, V(0) = 0, \tag{53}$$

having a solution $V \geq 0$ for all x satisfying $| \, arctan \, x \, | < \frac{1}{2}\sqrt{2}\gamma$, i.e. a domain which is indeed *smaller* than the domain of validity W_γ obtained for the nonlinear controller (49). $\qquad\square$

It is clear that Theorem 4.5 is quite important in solving the state feedback $\mathcal{H}_\infty$ optimal or near-optimal control problem. Indeed, using linear $\mathcal{H}_\infty$ theory one may compute γ^*_{lin} for the *linearized* system, and this gives a useful lower bound for γ^* of the nonlinear system. Also in some applications one may be satisfied with solving the nonlinear $\mathcal{H}_\infty$ problem on a sufficiently large neighborhood of the equilibrium, and thus one may consider γ with

$\gamma_{lin}^* \le \gamma \le \gamma^*$. The actual computation of solutions $P \ge 0$ to the Hamilton-Jacobi inequality (36) for given values of γ is not an easy task; analytic solutions will often not exist. It is useful to observe that for $\gamma \to \infty$ the Hamilton-Jacobi *equality* (40) tends to the Hamilton-Jacobi-Bellman equation of the optimal control problem for the system $\dot{x} = a(x) + b(x)u$ with cost functional $\int_0^\infty (\frac{1}{2}h^T(x(t))h(x(t)) + \frac{1}{2}u^T(t)u(t))dt$, i.e.

$$P_x(x)a(x) - \frac{1}{2}P_x(x)b(x)b^T(x)P_x^T(x) + \frac{1}{2}h^T(x)h(x) = 0 \qquad (54)$$

for which local solutions always exist under the assumption of stabilizability of the linearized system [9], [32], [42].

Various methods have been proposed in the literature for obtaining *approximate* solutions of (54), some of which are also directly applicable to the Hamilton-Jacobi equation (40). Perhaps the most useful method is the power series method as given in Lukes [32] (see also [9], [15]), extended to the Hamilton-Jacobi equation (40) in [43].

The idea is as follows. Let $X = X^T \ge 0$ be a solution of (44), (45). Then expand the solution $P(x)$ of (40), (41) as

$$P(x) = \frac{1}{2}x^T X x + P^{(3)}(x) + P^{(4)}(x) + \cdots \qquad (55)$$

with $P^{(m)}(x)$ denoting the m-th order terms of $P(x)$ (with $m \le k$ if P is C^k). The crucial fact is now that $P^{(m)}(x)$ can be computed *inductively* from $X, P^{(3)}(x), \cdots, P^{(m-1)}(x), m = 3, 4, \cdots$. Truncation of (55) will yield an approximate or even *exact* solution of the Hamilton-Jacobi inequality (36).

Note furthermore that for the computation of the suboptimal state feedback (37) it is sufficient to compute $P_x(x)$ instead of $P(x)$; indeed one may look at the equation (40) (or (36)) for every x as a quadratic equation in the unknown $P_x(x)$ (which however has to be the gradient of some function $P(x)$!)

With regards to *robustness* properties of the suboptimal $\mathcal{H}_\infty$ state feedback control (37) we mention the following immediate extension of a simple robustness property of the nonlinear optimal (state feedback) control obtained from the Hamilton-Jacobi-Bellman equation (54), see [15]. We consider *static perturbations* in the inputs

$$u = \varphi(u_{nom}) \qquad (56)$$

for some nonlinear mapping $\varphi : \mathbf{R}^m \to \mathbf{R}^m, \varphi(0) = 0$, where u_{nom} is the nominal input value provided by the state feedback controller.

Proposition 4.7 *Consider the suboptimal $\mathcal{H}_\infty$ state feedback $u_{nom} = -b^T(x)P_x^T(x)$, where $P \ge 0$ is a C^1 solution to the Hamilton-Jacobi inequality (36). Take the same assumptions as in Proposition 4.2, implying* *that the closed-loop system $\dot{x} = a(x) - b(x)b^T(x)P_x^T(x)$ is globally asymptotically stable. Then the closed-loop system will remain asymptotically stable for all perturbations (56) satisfying for some $\varepsilon > 0$*

$$z^T\varphi(z) \geq \frac{1}{2}(1+\varepsilon)z^T z, \qquad \text{for all } z \in \mathbf{R}^m \tag{57}$$

Remark 4.8 This can be interpreted [15] as "infinite gain margin-50% gain reduction tolerance", like in linear quadratic (state feedback) control.

Proof By (36), (57) we obtain (omitting for simplicity the argument x)

$$P_x\left[a + b\varphi(-b^T P_x^T)\right] \;\leq\; -\tfrac{1}{2}h^T h - \tfrac{1}{2}\tfrac{1}{\gamma^2}P_x gg^T P_x^T +$$

$$\tfrac{1}{2}P_x bb^T P_x^T + P_x b\varphi(-b^T P_x^T)$$

$$\leq\; -\tfrac{1}{2}h^T h - \tfrac{1}{2}\tfrac{1}{\gamma^2}P_x gg^T P_x^T - \tfrac{\varepsilon}{2}P_x bb^T P_x^T$$

and the result follows from Lyapunov theory as in Proposition 4.2. □

For later use we will now briefly sketch how the solvability of the state feedback $\mathcal{H}_\infty$ suboptimal control problem can be checked for systems having a *direct feedthrough* from disturbance d_2 to to-be-controlled outputs z, and for *time-varying* nonlinear systems. Consider the nonlinear system (29) with additional feedthrough term in the z-equation, i.e. (29) with

$$z = \begin{bmatrix} h(x) + f(x)d_2 \\ u \end{bmatrix}, \qquad h(0) = 0, \tag{58}$$

where $d_1 \in \mathbf{R}^{r_1}, d_2 \in \mathbf{R}^{r_2}$. Then a sufficient condition for the solvability of the state feedback $\mathcal{H}_\infty$ suboptimal control problem for $\gamma > 0$ consists of two parts, namely

$$D(x) := \gamma^2 I_{r_2} - f^T(x)f(x) > 0, \tag{59}$$

together with the existence of a C^1 solution $P \geq 0$ to the modified Hamilton-Jacobi inequality

$$P_x(x)a(x) + \tfrac{1}{2}P_x(x)\left[\tfrac{1}{\gamma^2}g(x)g^T(x) - b(x)b^T(x)\right]P_x^T(x)+$$

$$\tfrac{1}{2}h^T(x)h(x) + \tfrac{1}{2}h^T(x)f(x)D^{-1}(x)f^T(x)h(x) \leq 0, \quad P(0) = 0 \tag{60}$$

(Notice that for $f(x) = 0$ this is precisely (36).) Furthermore the suboptimal control is again given by (37). The interpretation of (59) is quite clear, while (60) is obtained by considering $K_\gamma(x, p, d, u)$ as in (32) with $\| z \|^2 = \| h(x) + f(x)d_2 \|^2 + \| u \|^2$. As before, cf. (33), one computes $u^* = -b^T(x)p$ and $d_1^* = \tfrac{1}{\gamma^2}g^T(x)p$, while $\frac{\partial K_\gamma}{\partial d_2} = 0$ yields $d_2^* =$

$[\gamma^2 I_{r_2} - f^T(x)f(x)]^{-1} f^T(x)h(x)$. Substitution of these functions in K_γ then yields the Hamiltonian $H_\gamma(x,p)$ corresponding to (60). (Note that this procedure can be easily extended to other generalizations of (29), e.g. z directly depending on d_1).

Next, consider a nonlinear system (29), which is *time-varying*, i.e.

$$\dot{x} = a(x,t) + b(x,t)u + g(x,t)d_1 \quad , \quad a(0,t) = 0$$

$$y = c(x,t) + d_2 \qquad\qquad , \quad c(0,t) = 0 \qquad\qquad (61)$$

$$z = \begin{bmatrix} h(x,t) \\ u \end{bmatrix} \qquad\qquad , \quad h(0,t) = 0$$

Then the state feedback $\mathcal{H}_\infty$ suboptimal control problem (for $\gamma > 0$) is solvable if there exists a C^1 solution $P : M \times \mathbf{R} \to \mathbf{R}^+$ to the *non-stationary* Hamilton-Jacobi inequality generalizing (36):

$$P_t(x,t) + P_x(x,t)a(x,t) + \tfrac{1}{2}h^T(x,t)h(x,t) +$$

$$\tfrac{1}{2}P_x(x,t)\left[\tfrac{1}{\gamma^2}g(x,t)g^T(x,t) - b(x,t)b^T(x,t)\right]P_x^T(x,t) \le 0, \qquad (62)$$

$$P(0,0) = 0,$$

while the suboptimal $\mathcal{H}_\infty$ control is given as the *time-varying* nonlinear feedback

$$u(t) = -b^T(x(t),t)P_x^T(x,t), \quad t \in \mathbf{R} \qquad\qquad (63)$$

For the (simple) proof we refer to [45].

5 $\mathcal{H}_\infty$ control by dynamic output feedback

Let us now study the full, i.e., dynamic output feedback, $\mathcal{H}_\infty$ *suboptimal control* problem as formulated in Section 2. We will consider affine nonlinear systems (29), and in order to stay within the framework of affine nonlinear systems without feedthrough term we will restrict attention to C^k compensators of the form (compare with (2))

$$\dot{\xi} = k(\xi) + \ell(\xi)y \quad , \quad k(0) = 0$$
$$\qquad\qquad\qquad\qquad\qquad\qquad\qquad\qquad (64)$$
$$u = m(\xi) \qquad\qquad , \quad m(0) = 0$$

with $\xi = (\xi_1, \cdots, \xi_\nu)$ local coordinates for the compensator state space manifold M_c. Let now $\gamma > 0$. First we want to derive *necessary* conditions for the existence of a compensator (64) which solves the $\mathcal{H}_\infty$ suboptimal control problem (for disturbance attenuation level γ).

Thus suppose there exists a compensator (64) such that the closed-loop system (29), (64) has L_2-gain $\leq \gamma$ (from d to z), in the sense that there exists a C^1 solution $V(x,\xi) \geq 0$ to the Hamilton-Jacobi inequality (10) for the closed-loop system, which takes the form

$$V_x(x,\xi)\left[a(x) + b(x)m(\xi)\right] + V_\xi \left[k(\xi) + \ell(\xi)c(x)\right] +$$

$$\frac{1}{2}\frac{1}{\gamma^2}V_x(x,\xi)g(x)g^T(x)V_x^T(x,\xi) + \frac{1}{2}\frac{1}{\gamma^2}V_\xi(x,\xi)\ell(\xi)\ell^T(\xi)V_\xi^T(x,\xi)$$

$$+ \frac{1}{2}h^T(x)h(x) + \frac{1}{2}m^T(\xi)m(\xi) \leq 0, \quad V(0,0) = 0. \tag{65}$$

There are two ways to obtain from (65) an equation only involving the x-variables. Let us consider the equation

$$V_\xi(x,\xi) = 0 \tag{66}$$

and suppose this equation has a C^1 solution $\xi = F(x)$, with $F(0) = 0$. (By the implicit function theorem this will locally be the case if the partial Hessian matrix $V_{\xi\xi}$ is non-singular.) Define

$$P(x) := V(x, F(x)) \geq 0, \tag{67}$$

then substitution of $\xi = F(x)$ into (65) and completing the squares yields (note that $P_x(x) = V_x(x, F(x))$, since $V_\xi(x, F(x)) = 0$)

$$P_x(x)a(x) + \frac{1}{2}\frac{1}{\gamma^2}P_x(x)g(x)g^T(x)P_x^T(x) + \frac{1}{2}h^T(x)h(x)$$

$$+ \frac{1}{2}\parallel b^T(x)P_x^T(x) + m(F(x)) \parallel^2 - \frac{1}{2}P_x(x)b(x)b^T(x)P_x^T(x) \leq 0 \tag{68}$$

while $P(0) = V(0,0) = 0$. Thus we see that $P(x)$ as defined above is a nonnegative C^1 solution to the Hamilton-Jacobi inequality (36) governing the state feedback $\mathcal{H}_\infty$ suboptimal control problem!

Alternatively, we may define

$$R(x) := V(x,0) \geq 0 \tag{69}$$

and substitution of $\xi = 0$ into (65) and completing the squares yields

$$R_x(x)a(x) + \frac{1}{2}\frac{1}{\gamma^2}R_x(x)g(x)g^T(x)R_x^T(x) + \frac{1}{2}h^T(x)h(x)$$

$$+ \frac{1}{2}\frac{1}{\gamma^2}\parallel \gamma^2 c(x) + \ell^T(0)V_\xi^T(x,0) \parallel^2 - \frac{1}{2}\gamma^2 c^T(x)c(x) \leq 0 \tag{70}$$

with $R(0) = 0$, and consequently $R \geq 0$ is a solution of

$$R_x(x)a(x) + \frac{1}{2}\frac{1}{\gamma^2}R_x(x)g(x)g^T(x)R_x^T(x) + \frac{1}{2}h^T(x)h(x)$$

$$- \frac{1}{2}\gamma^2 c^T(x)c(x) \leq 0, \quad R(0) = 0. \tag{71}$$

In fact this latter equation (71) is well-known from linear state space $\mathcal{H}_\infty$ theory: for a linear system (42) $R(x)$ is of the form $\frac{1}{2}x^T Rx$ for some matrix $R \geq 0$, and it is easily checked that $Y := \gamma^2 \, R^{-1}$ (assuming R to be nonsingular!) is a solution of the *dual* Riccati inequality

$$AY + YA^T + GG^T + Y(\frac{1}{\gamma^2}H^T H - C^T C)Y \leq 0 \qquad (72)$$

as obtained in [12]. We shall therefore call (71) the *dual Hamilton-Jacobi inequality*.

Actually the dual Hamilton-Jacobi inequality has an immediate interpretation (a similar observation has been made for the linear case in [30]). Consider the closed-loop system (29), (64) having L_2-gain $\leq \gamma$ for initial condition $(x(0), \xi(0)) = (0,0)$ and let the output y of (29) be identically zero. Then also u as delivered by the compensator (64) is identically zero, and thus the system (29) with $u = 0$ *satisfying the constraint* $y = 0$ necessarily has L_2-gain $\leq \gamma$! More explicitly, necessarily the system

$$\begin{aligned} \dot{x} &= a(x) + g(x)d_1 \\[2mm] y &= c(x) + d_2 \\[2mm] z &= h(x) \end{aligned} \qquad (73)$$

with inputs $d = \begin{bmatrix} d_1 \\ d_2 \end{bmatrix}$ and outputs has L_2-gain $\leq \gamma$ *for all d_2 such that* $c(x) + d_2 = 0$, i.e. $d_2 = -c(x)$. Thus the pre-Hamiltonian (see (12)) of (73) is

$$p^T a(x) + p^T g(x)d_1 - \frac{1}{2}\gamma^2 \parallel d_1 \parallel^2 - \frac{1}{2}\gamma^2 \parallel d_2 \parallel^2 + \frac{1}{2}h^T(x)h(x) =$$

$$p^T a(x) + p^T g(x)d_1 - \frac{1}{2}\gamma^2 \parallel d_1 \parallel^2 - \frac{1}{2}\gamma^2 c^T(x)c(x) + \frac{1}{2}h^T(x)h(x) \quad (74)$$

which by substitution of $d_1^* = \frac{1}{\gamma^2}g^T(x)p$ leads to the Hamiltonian $H_\gamma(x,p) = p^T a(x) + \frac{1}{2}\frac{1}{\gamma^2}p^T g(x)g^T(x)p + \frac{1}{2}h^T(x)h(x) - \frac{1}{2}\gamma^2 c^T(x)c(x)$ corresponding to the Hamilton-Jacobi inequality (71).

Finally, we observe that since $V(x,\xi) \geq 0$ and $V_\xi(x, F(x)) = 0$ necessarily $P(x) = V(x, F(x)) \leq V(x, 0) = R(x)$, at least for x near zero.

Summarizing we have the following theorem, independently obtained in [5] and [44] (with the cooperation of C. Scherer):

Theorem 5.1 *Suppose the $\mathcal{H}_\infty$ suboptimal control problem for $\gamma \geq 0$ is solvable by a compensator (64) in the sense that there exists a C^1 solution $V(x,\xi) \geq 0$ to the Hamilton-Jacobi inequality (65). Assume that the equation $V_\xi(x,\xi) = 0$ has a C^1 solution $\xi = F(x), F(0) = 0$, with $F : M \to M_c$.*

Then the nonnegative functions $P(x)$ and $R(x)$ defined in (67), (69) satisfy the Hamilton-Jacobi inequality (36), respectively the dual Hamilton-Jacobi inequality (71), as well as the coupling condition

$$P(x) \le R(x), \quad \text{for all } x \text{ near } 0 \tag{75}$$

Remark 5.2 In the linear case the existence of nonnegative solutions $P(x)$, $R(x)$ to (36), (71) and (75) basically reduces to the famous necessary *and* sufficient conditions for solvability of the $\mathcal{H}_\infty$ suboptimal problem obtained in [12], i.e. the existence of $X \ge 0, Y \ge 0$ to (44), respectively (72), with $X \le \gamma^2 Y^{-1}$.

As in the linear case, one would hope that the existence of nonnegative solutions $P(x), R(x)$ to (36), (71), (75) is also (almost) *sufficient* for the solvability of the $\mathcal{H}_\infty$ suboptimal control problem, and leads to the construction of a compensator (64) solving the $\mathcal{H}_\infty$ suboptimal control problem. Unfortunately, some problems arise which do not seem to have easy solutions. A first indication of the problems to be expected is that, as explained above, the *dual* Hamilton-Jacobi inequality gives information about the system (29) *only* for the output value *zero*, which in principle is quite limited information for a nonlinear system.

We notice that the inequalities (68) and (70) suggest a natural choice of the function $m(\xi)$ in the compensator, while the choice of $\ell(\xi)$ is much more problematic. Indeed, suppose there exist $P \ge 0, R \ge 0$ satisfying (36) and (71) *with equality*. Then the proof of Theorem 5.1 suggests to construct a function $V(x,\xi)$ such that $V(x,0) = R(x)$ and $V(x,F(x)) = P(x)$ for some function $\xi = F(x)$. Then it follows from (68), respectively (70), that $m(\xi)$ and $\ell(\xi)$ have to be constructed in such a way that

$$m(F(x)) = -b^T(x)P_x^T(x) \tag{76}$$

$$\ell^T(0)V_\xi^T(x,0) = -\gamma^2 c(x), \text{ for all } x \tag{77}$$

In fact, choosing $m(\xi)$ as in (76) corresponds to "maximizing the dissipation rate of the closed-loop system", cf. [5]. Notice that (77) only yields information about the choice of $\ell(0)$, i.e., about the *linearization* of the to-be-constructed compensator (64), while, on the other hand, putting a severe condition on the existence of the storage function $V(x,\xi)$.

We conclude that the choice of $\ell(\xi)$ is not very clear. On the other hand, for the construction of $k(\xi)$ we have the following result, generalizing a very appealing theorem obtained in [5]:

Theorem 5.3 *Take the same assumptions as in Theorem 5.1. Moreover assume that the partial Hessian $V_{\xi\xi}(x, F(x))$ is non-singular for every x. Suppose $m(\xi)$ is given as in (76), and suppose that the Hamilton-Jacobi* *inequality (65) is fulfilled with equality at all points* $(x, F(x)) \in M \times M_c$. *Then* $k(\xi)$ *satisfies*

$$k(F(x)) = F_x(x) \left[a(x) - b(x)b^T(x)P_x^T(x) + \frac{1}{\gamma^2}g(x)g^T(x)P_x^T(x) \right]$$

$$-\ell(F(x))c(x) \qquad (78)$$

Proof Rewrite (see Section 3) (65) as

$$V_x(x,\xi)\left[a(x) + b(x)m(\xi) + g(x)d_1^*\right] +$$

$$V_\xi(x,\xi)\left[k(\xi) + \ell(\xi)c(x) + \ell(\xi)d_2^*\right] - \qquad (79)$$

$$\tfrac{1}{2}\gamma^2 \parallel d_1^* \parallel^2 - \tfrac{1}{2}\gamma^2 \parallel d_2^* \parallel^2 + \tfrac{1}{2}h^T(x)h(x) + \tfrac{1}{2}m^T(\xi)m(\xi) \le 0$$

with $d_1^* = \frac{1}{\gamma^2}g^T(x)V_x^T(x,\xi), d_2^* = \frac{1}{\gamma^2}\ell^T(\xi)V_\xi^T(x,\xi)$. Now by assumption (79) holds with equality at all points $(x, F(x)) \in M \times M_c$, and thus differentiation of (79) with respect to ξ at all points $(x, F(x))$ yields zero. On the other hand, differentiation of (79) to ξ via d_1^* and d_2^* yields zero because of the maximizing property (w.r.t. to the pre-Hamiltonian K_γ, cf. (12)) of d_1^* and d_2^*, and the same holds for differentiation to ξ via $m(\xi)$ since we choose $m(\xi)$ satisfying (76), and thus minimizing (79), cf. (68). Thus what remains is (note also that $d_2^* = 0$ at $(x, F(x))$)

$$V_{x\xi}(x, F(x))\left[a(x) + b(x)m(F(x)) + g(x)d_1^*\right] +$$

$$V_{\xi\xi}(x, F(x))\left[k(F(x)) + \ell(F(x)c(x)\right] = 0, \qquad (80)$$

with $m(F(x))$ given by (76). Furthermore since $V_\xi(x, F(x)) = 0$ for all x, differentiation with respect to x yields

$$V_{x\xi}(x, F(x)) + V_{\xi\xi}(x, F(x))F_x(x) = 0, \qquad (81)$$

and (78) follows from substitution of (81) in (80), and by nonsingularity of $V_{\xi\xi}(x, F(x))$. □

Note that if $F : M \to M_c$ is a *diffeomorphism*, then we may choose coordinates ξ for the compensator such that $\xi = F(x)$ is the *identity* mapping, and (78) yields

$$k(\xi) = \left[a(\xi) - b(\xi)b^T(\xi)P_\xi^T(\xi) + \frac{1}{\gamma^2}g(\xi)g^T(\xi)P_\xi^T(\xi) \right] - \ell(\xi)c(\xi) \quad (82)$$

This is precisely the result proved in [5] (actually for non-affine nonlinear systems!), and is called a *separation principle* in [5] since the resulting compensator (64) can be rewritten as

$$\dot{\xi} = a^*(\xi) + \ell(\xi)(y - c(\xi)) \qquad (83)$$

$$u = m(\xi) = -b^T(\xi)P_\xi^T(\xi) \tag{84}$$

with $a^*(\xi) = a(\xi) - b(\xi)b^T(\xi)P_\xi^T(\xi) + \frac{1}{\gamma^2}g(\xi)g^T(\xi)P_\xi^T(\xi)$. Hence the control u is given as the suboptimal *state* feedback $-b^T(x)P_x^T(x)$ with x replaced by its *estimate* ξ as given by the identity-observer (83).

Notice that the general case (78) admits a similar interpretation; however the nonlinear observer dictated by (78) is in general a *reduced-order observer*, see [39]. (It follows from (78) that the compensator state space M_c can be always reduced to $F(M)$, i.e. the *image* of F in M_c, and thus has dimension $\leq n$!)

Summarizing, if we have nonnegative solutions P and R to the Hamilton-Jacobi inequalities (36) and (71) then a natural choice for a compensator (64) is to take $M_c = M$ (with $F : M \to M_c$ the identity mapping), $m(\xi) = -b^T(\xi)P_\xi^T(\xi)$ and $k(\xi)$ given as in (82). Thus the construction of $\ell(\xi)$ is the main stumbling block, and already much effort has been devoted [5], [25], [26], [24] in investigating if (or when) it is possible to construct $\ell(\xi)$ in such a way that the resulting compensator (64) solves the $\mathcal{H}_\infty$ suboptimal control problem, or almost equivalently, if there exists a solution $V(x, \xi) \geq 0$ to the Hamilton-Jacobi inequality (65). So far, these investigations did not yield a decisive answer. Let us just make two remarks before we will go on to another line of reasoning which sheds a new light on the problem.

First of all, it is quite natural to construct the compensator (64) in such a way that its *linearization* at $\xi = 0$ equals the *central controller* for the linearized system (42). In fact, it is easy to see that the choice of $k(\xi)$ and $m(\xi)$ as in (76), (82) is in accordance with this idea, while it leads to a value of $\ell(0)$ given as

$$\ell(0) = \gamma^2 [S_{xx}(0)]^{-1} C^T \tag{85}$$

with $S(x) := R(x) - P(x)$ (provided the Hessian $S_{xx}(0)$ is positive definite). (Indeed, since $X = P_{xx}(0)$ and $Y = \gamma^2 R_{xx}^{-1}(0)$ is a solution of the Riccati inequalities (44) and (72), $\ell(0)$ as given in (85) equals $(\gamma^2 Y^{-1} - X)C^T$ which is the gain of the central controller [12]).

Secondly, the following idea has proved to be useful in *linear* $\mathcal{H}_\infty$ theory [12]. Consider a solution $P \geq 0$ to the Hamilton-Jacobi inequality (36). Then substitution of (36) into $\frac{d}{dt}P = P_x(x)a(x) + P_x(x)b(x)u + P_x(x)g(x)d_1$ and completion of the squares yields ([44]) $2\frac{d}{dt}P + \| z \|^2 - \gamma^2 \| d_1 \|^2 \leq \| u + b^T(x)P_x^T(x) \|^2 - \gamma^2 \| d_1 - \frac{1}{\gamma^2}g^T(x)P_x^T(x) \|^2$, and integrating from $t = 0$ to $t = T$, using $x(0) = 0$ and $P(x(T)) \geq 0$,

$$\int_0^T (\| z(t) \|^2 - \gamma^2 \| d(t) \|^2)dt \leq$$

$$\int_0^T (\| \tilde{z}(t) \|^2 - \gamma^2 \| \tilde{d}(t) \|^2)dt \tag{86}$$

where $\tilde{z} := u + b^T(x)P_x^T(x)$, $\tilde{d}_1 := d_1 - \frac{1}{\gamma^2}g^T(x)P_x^T(x)$, and $\tilde{d} = \begin{bmatrix} \tilde{d}_1 \\ d_2 \end{bmatrix}$.

Now $\tilde{z}$ and $\tilde{d}$ can be viewed as to-be-controlled outputs and disturbances for the *transformed system* (with $\tilde{a}(x) = a(x) + \frac{1}{\gamma^2}g(x)g^T(x)P_x^T(x)$)

$$
\begin{aligned}
\dot{x} &= \tilde{a}(x) + b(x)u + g(x)\tilde{d}_1 \\
y &= c(x) + d_2 \\
\tilde{z} &= b^T(x)P_x^T(x) + u
\end{aligned}
\tag{87}
$$

while it follows from (86) that if we can construct a compensator which solves the $\mathcal{H}_\infty$ suboptimal control problem for (87) then the *same* compensator will also solve the problem for the original system (29)! The point is that since we have already used the knowledge of the existence of a $P \geq 0$ to (36), we may *expect* that the $\mathcal{H}_\infty$ problem for (87) will be more tractable than for (29). Indeed, if we apply a compensator (64) to (87), and assume that the closed-loop system has L_2-gain $\leq \gamma$ then instead of (65) we have to consider the following Hamilton-Jacobi inequality for the closed-loop system

$$
\tilde{V}_x(x,\xi)\left[a(x) + \frac{1}{\gamma^2}g(x)g^T(x)P_x^T(x) + b(x)m(\xi)\right] +
$$

$$
\tilde{V}_\xi(x,\xi)\left[k(\xi) + \ell(\xi)c(x)\right] + \frac{1}{2}\frac{1}{\gamma^2}\tilde{V}_x(x,\xi)g(x)g^T(x)\tilde{V}_x^T(x,\xi)
\tag{88}
$$

$$
+\frac{1}{2}\frac{1}{\gamma^2}\tilde{V}_\xi(x,\xi)\ell(\xi)\ell^T(\xi)\tilde{V}_\xi^T(x,\xi) + \frac{1}{2}\parallel b^T(x)P_x^T(x) + m(\xi) \parallel^2 \leq 0
$$

Now if we choose $m(\xi)$ as in (76), i.e. $m(F(x)) = -b^T(x)P_x^T(x)$ for some $F : M \to M_c$, then it is immediately seen that we can take $\tilde{V}(x,\xi)$ to be such that $\tilde{V}(x, F(x)) \equiv 0$, while by completing the squares as in (70) $S(x) := \tilde{V}(x,0) \geq 0$ satisfies

$$
S_x(x)\left[a(x) + \frac{1}{\gamma^2}g(x)g^T(x)P_x^T(x)\right] + \frac{1}{2}\frac{1}{\gamma^2}S_x(x)g(x)g^T(x)S_x^T(x)
$$

$$
-\frac{1}{2}\gamma^2 c^T(x)c(x) + \frac{1}{2}P_x^T(x)b(x)b^T(x)P_x^T(x) \leq 0, \quad S(0) = 0
\tag{89}
$$

Actually this last equation has a clear interpretation: by subtracting (36) from (71) we see that (89) is satisfied by

$$
S(x) := R(x) - P(x)
\tag{90}
$$

which underlines the need for the coupling condition $P(x) \leq R(x)$, for all x. The resulting Hamilton-Jacobi inequality (89) can be regarded as a crucial tool in the investigations on the $\mathcal{H}_\infty$ control problem made in [25], [26], [24].

Now we will look at the $\mathcal{H}_\infty$ suboptimal control problem from a somewhat different perspective, using the theory of differential games. Indeed, we will first briefly recapitulate the crucial observations made in [[6], Chapter 5]. Let us consider the *finite horizon* $\mathcal{H}_\infty$ suboptimal control problem, i.e. we consider the L_2-gain on some *finite*, fixed time-interval $[T_1, T_2]$. (Thus in (31) we consider integrals $\int_{T_1}^{T} () dt$ with $\tau \le T_2$.)

Following the theory of differential games ([14], [7]) we look at the min-max solution (with respect to u and d) of the performance criterion (cf. [6])

$$N_1(x(T_1)) + \frac{1}{2} \int_{T_1}^{T_2} (u^T u + h^T h - \gamma^2 d_1^T d_1 - \gamma^2 d_2^T d_2) dt + N_2(x(T_2)) \quad (91)$$

where the control $u(t), t \in [T_1, T_2]$, is only allowed to depend on $y(\tau)$, with $T_1 \le \tau \le t$. (For completeness we have included an initial and terminal cost N_1 and N_2 with $N_1(0) = N_2(0) = 0$; these can be set equal to zero, for example.) As shown by Başar & Bernhard [6] , this problem can be split into two parts.

First there is the finite horizon *state feedback* $\mathcal{H}_\infty$ suboptimal control problem concerned with the min-max solution of the performance criterion

$$\frac{1}{2} \int_{T_1}^{T_2} (u^T u + h^T h - \gamma^2 d_1^T d_1 - \gamma^2 d_2^T d_2) dt + N_2(x(T_2)) \quad (92)$$

where, for every $t \in [T_1, T_2]$, $u(t)$ is allowed to depend on the *full state* $x(t)$, leading, as in the time-varying case (see (62)), to the non-stationary Hamilton-Jacobi equation

$$P_t(x, t) + P_x(x, t) a(x) + \frac{1}{2} h^T(x) h(x) +$$

$$\frac{1}{2} P_x(x) \left[\frac{1}{\gamma^2} g(x) g^T(x) - b(x) b^T(x) \right] P_x^T(x, t) = 0, \quad (93)$$

$$P(x, T_2) = N_2(x)$$

with resulting suboptimal state feedback $u(t) = -b^T(x(t)) P_x^T(x, t)$, cf. (63).

Secondly, let $\tau \in [T_1, T_2]$, and let $\bar{u}(t), t \in [T_1, \tau]$ and $\bar{y}(t), t \in [T_1, \tau]$ be a given pair of control input and measured output trajectories of the system (29). Then we formulate the *auxiliary problem* of maximizing the performance criterion

$$N_1(x(T_1)) + \frac{1}{2} \int_{T_1}^{\tau} (\bar{u}^T \bar{u} + h^T h - \gamma^2 d_1^T d_1 - \gamma^2 d_2^T d_2) dt + P(x(\tau), \tau) \quad (94)$$

with respect to $x(T_1)$ and $d(t), t \in [T_1, \tau]$, under the *constraint* that the measured output of the system (29) *equals* $\bar{y}(t), t \in [T_1, \tau]$. (Here $P(x, t)$ is the solution of (93).)

Assume that this auxiliary problem admits a unique maximizing solution $\hat{x}(T_1), \hat{d}(t), t \in [T_1, \tau]$. Then let $\hat{x}(t), t \in [T_1, \tau]$, be the state trajectory generated by $\hat{x}(T_1), \hat{d}(t)$ and $\bar{u}(t), t \in [T_1, \tau]$. Define

$$\tilde{u}(\tau) = -b^T(\hat{x}(\tau))P_x^T(\hat{x}(\tau), \tau) \qquad (95)$$

Assume the auxiliary problem can be solved for every $\tau \in [T_1, T_2]$, then, since $\hat{x}(\cdot)$ depends on $\bar{u}(\cdot)$, we have defined by (95) a causal mapping [6]

$$\bar{u}(t), t \in [T_1, T_2] \mapsto \tilde{u}(t), t \in [T_1, T_2] \qquad (96)$$

and the *fixed point* of this mapping will be denoted by $\hat{u}(t), t \in [T_1, T_2]$. By construction $\hat{u}(\tau)$ only depends on $\bar{y}(t), t \in [T_1, \tau]$, and it is shown in [6] that $\hat{u}(t)$ is the solution of the optimization problem with performance criterion (91). This is called a *worst case certainty equivalence principle* in [6], since by (95) $\hat{u}(t)$ is given as

$$\hat{u}(t) = -b^T(\hat{x}(t)))P_x^T(\hat{x}(t), t), t \in [T_1, T_2], \qquad (97)$$

where $\hat{x}(\tau)$ is the state at time τ resulting from $\hat{u}(t), t \in [T_1, \tau]$, and the *worst case* initial state and disturbance $\hat{x}(T_1), \hat{d}(t), t \in [T_1, \tau]$, *compatible* with the available information at time τ, i.e. $\bar{y}(t), t \in [T_1, \tau]$.

We attack the auxiliary problem for every τ in the following classical way. First we maximize (under the constraint that the measured output equals $\bar{y}(t), t \in [T_1, \tau]$) the criterion (94) under the *additional* constraint that $x(\tau) = x$; denote this value by $-S(x, \tau)$. Then we maximize $-S(x, \tau)$ with respect to x. Note that $-S(x, \tau)$ can be written as $P(x, \tau) - R(x, \tau)$, where $-R(x, \tau)$ denotes the maximum of the performance index

$$N_1(x(T_1)) + \frac{1}{2}\int_{T_1}^{\tau}(\bar{u}^T\bar{u} + h^Th - \gamma^2 d_1^T d_1 - \gamma^2 d_2^T d_2)dt \qquad (98)$$

with respect to $x(T_1)$ and $d(t), t \in [T_1, \tau]$, under the constraint of given measured output $\bar{y}(t), t \in [T_1, \tau]$, and final state $x(\tau) = x$. Since $y = c(x) + d_2$ this reduces to the *unconstrained* optimization problem

$$N_1(x(T_1)) + \frac{1}{2}\int_{T_1}^{\tau}\left[\bar{u}^T(t)\bar{u}(t) + h^T(x(t))h(x(t)) - \gamma^2 d_1^T(t)d_1(t)\right.$$

$$\left. - \gamma^2 \parallel \bar{y}(t) - c(x(t)) \parallel^2\right)dt \qquad (99)$$

with respect to $x(T_1)$ and $d_1(t), t \in [T_1, \tau]$. It follows (see e.g. [18], [34]) that $R(x, t)$ satisfies

$$R_t(x, t) + R_x(x, t)\left[a(x) + b(x)\bar{u}(t)\right] + \frac{1}{2}\frac{1}{\gamma^2}R_x(x, t)g(x)g^T(x)R_x^T(x, t)$$

$$+\frac{1}{2}h^T(x)h(x) - \frac{1}{2}\gamma^2 c^T(x)c(x) + \gamma^2 c^T(x)\bar{y}(t)$$

$$-\frac{1}{2}\gamma^2 \parallel \bar{y}(t) \parallel^2 + \frac{1}{2} \parallel \bar{u}(t) \parallel^2 = 0, \quad R(x,T_1) = -N_1(x). \tag{100}$$

Summarizing, let $\hat{x}(t)$ be the minimum of $S(x,t) = R(x,t) - P(x,t)$, with P and R given by (93) and (100), then the suboptimal feedback is given as

$$\hat{u}(t) = -b^T(\hat{x}(t))P_x^T(\hat{x}(t),t) \tag{101}$$

where *we have substituted* $\bar{u}(t) = \hat{u}(t)$ *in* (100)!

Now we proceed as follows. *Assume* that the minimum $\hat{x}(t)$ of $S(x,t)$ is determined by the equation $S_x(\hat{x}(t),t) = 0$. Then differentiation to t, and *assuming* that $S_{xx}(\hat{x}(t),t)$ is invertible, yields (see [18])

$$\dot{\hat{x}}(t) = -[S_{xx}(\hat{x}(t),t)]^{-1} S_{tx}(\hat{x}(t),t) \tag{102}$$

Rewrite (93) as (omitting arguments and boundary conditions)

$$P_t + P_x a + P_x b u^* + P_x g d_1^* + \frac{1}{2}u^{*^T}u^* - \frac{1}{2}\gamma^2 d_1^{*^T}d_1^* + \frac{1}{2}h^T h = 0 \tag{103}$$

with (see Section 4) $u^* = -b^T P_x^T, d_1^* = \frac{1}{\gamma^2}g^T P_x^T$, and rewrite (100) with $\bar{u} = \hat{u}$ as

$$R_t + R_x a + R_x b\hat{u} + \frac{1}{2}\hat{u}^T\hat{u} - \frac{1}{2}\gamma^2 \hat{d}_1^T\hat{d}_1 + R_x g\hat{d}_1 + \frac{1}{2}h^T h$$

$$-\frac{1}{2}\gamma^2 c^T c + \gamma^2 c^T \bar{y} - \frac{1}{2}\gamma^2 \parallel \bar{y} \parallel^2 = 0 \tag{104}$$

with $\hat{u}$ given by (101) and $\hat{d}_1 = \frac{1}{\gamma^2}g^T R_x^T$. Subtracting (103) from (104) yields

$$S_t + S_x a + R_x b\hat{u} + R_x g\hat{d}_1 - P_x bu^* - P_x g d_1^* + \frac{1}{2}\hat{u}^T\hat{u} - \frac{1}{2}u^{*^T}u^*$$

$$-\frac{1}{2}\gamma^2 \hat{d}_1^T\hat{d}_1 + \frac{1}{2}\gamma^2 d_1^{*^T}d_1^* - \frac{1}{2}\gamma^2 c^T c + \gamma^2 c^T\bar{y} - \frac{1}{2}\gamma^2 \parallel \bar{y} \parallel^2 = 0 \tag{105}$$

Finally, differentiation of (105) with respect to x in $x = \hat{x}$ yields (note that $R_x(\hat{x},t) = P_x(\hat{x},t)$, and if $x = \hat{x}$, then $u^* = \hat{u}, d_1^* = \hat{d}_1$!)

$$S_{tx} + S_{xx}a + S_{xx}bu^* + S_{xx}g d_1^* + \gamma^2 \frac{\partial c^T}{\partial x}\bar{y} - \gamma^2 \frac{\partial c^T}{\partial x}c = 0 \tag{106}$$

and thus solving S_{tx} from (106) and substituting into (102) yields the compensator

$$\dot{\hat{x}} = \left[a(\hat{x}) - b(\hat{x}) b^T(\hat{x}) P_x^T(\hat{x}, t) + \tfrac{1}{\gamma^2} g(\hat{x}) g^T(\hat{x}) P_x^T(\hat{x}, t) \right]$$

$$+ \gamma^2 \left[S_{xx}(\hat{x}, t) \right]^{-1} \tfrac{\partial c^T}{\partial x}(\hat{x}) \left[y(t) - c(\hat{x}) \right] \tag{107}$$

$$u = -b^T(\hat{x}) P_x^T(\hat{x}, t)$$

solving the finite-horizon $\mathcal{H}_\infty$ suboptimal control problem.

The transition to the infinite-horizon $\mathcal{H}_\infty$ problem is rather straightforward (see for similar reasoning in the linear case [6]). Indeed, let $T_2 \to \infty$ in (93), while imposing the boundary condition $x(t) \to 0$ for $t \to \infty$. Then instead of $P(x, t)$ solving (93) we are looking for the solution $P(x)$ of (40) (i.e., (36) with equality) satisfying (41). Similarly, let $T_1 \to -\infty$ in (100), (101) (with $P(x, t)$ replaced by $P(x)$!), while imposing the boundary condition $x(t) \to 0$ for $t \to -\infty$. This means that we are looking for solutions $R(x, t)$ of (100), (101), with $T_1 = -\infty$, such that the origin is an asymptotically stable equilibrium for the vectorfield $-(a(x) + \tfrac{1}{\gamma^2} g(x) g^T(x) R_x^T(x, t))$. Note however that because of the dependence on $\bar{y}(t)$ the partial differential equation (100) is inherently time-dependent, and so are its solutions $R(x, t)$!

We will call (107), with $P(x, t)$ and $S(x, t) = R(x, t) - P(x, t)$ replaced by these infinite-horizon versions, the *nonlinear central controller* for the $\mathcal{H}_\infty$ suboptimal control problem. Note that it is in general an *infinite-dimensional* controller, since $R(x, t)$ and thus the controller gain $[S_{xx}(\hat{x}, t)]^{-1}$ is time-varying and generated by (the infinite-horizon version of) the time-dependent partial differential equation (100), (101). In the *linear* case it can be easily checked that $[S_{xx}(\hat{x}, t)]^{-1}$ does not depend on t and that (107) is precisely the linear central controller as obtained in [12].

Thus the situation is very much like in nonlinear Kalman filtering or nonlinear deterministic filtering, cf. [18], [34]. However the fact that the nonlinear central controller is generally infinite-dimensional does *not* necessarily rule out the possible existence of a *finite-dimensional* compensator which solves the $\mathcal{H}_\infty$ suboptimal control problem, especially in the strictly suboptimal case.

We have noted that (93) is the non-stationary version of (40), while the stationary version of (100) for $\bar{u}(t) = 0, \bar{y}(t) = 0$ is precisely (71) with equality. Hence a logical *finite-dimensional approximation* to the nonlinear central controller (107) is obtained by replacing $S_{xx}(\hat{x}, t)$ in (107) by $S_{xx}(x)$, where $S(x) = R(x) - P(x)$, with $P(x)$ the stabilizing solution of (36) and $R(x)$ the anti-stabilizing solution of (71) with equality. This leads to the compensator

$$\dot{\xi} = \quad a(\xi) - b(\xi)b^T(\xi)P_\xi^T(\xi) + \tfrac{1}{\gamma^2}g(\xi)g^T(\xi)P_\xi^T(\xi)$$

$$+\gamma^2 \left[R_{\xi\xi}(\xi) - P_{\xi\xi}(\xi) \right]^{-1} \tfrac{\partial c^T}{\partial \xi}(\xi) \left[y(t) - c(\xi) \right] \tag{108}$$

$$u = \quad -b^T(\xi)P_\xi^T(\xi)$$

where $P \geq 0$ and $R \geq 0$ are solutions to (36), respectively (71), with equality, such that

$$a - bb^T P_x^T + \tfrac{1}{\gamma^2}gg^T P_x^T \quad \text{is asymptotically stable}$$

$$-(a + \tfrac{1}{\gamma^2}gg^T R_x^T) \quad \text{is asymptotically stable} \tag{109}$$

$$R_{xx}(x) > P_{xx}(x), \forall x$$

Proposition 5.4 *Consider the system (29), and suppose there exist solutions $P \geq 0, R \geq 0$ to (36) and (71) with equality, satisfying (109). Then the closed-loop system (29), (108) has locally L_2-gain $< \gamma$ (i.e., the compensator (108) solves locally the $\mathcal{H}_\infty$ strictly suboptimal control problem).*

Proof The linearization of (108) is the central controller [12] for the linearized system (42). Now apply Theorem 3.10.   □

Remark 5.5 Note that (108) is of the previously obtained form (83), (84), with $\ell(\xi) = \gamma^2 \left[R_{\xi\xi}(\xi) - P_{\xi\xi}(\xi) \right]^{-1} \tfrac{\partial c^T}{\partial \xi}(\xi)$. The same compensator (108) has been proposed before in [5], using a different motivation, and in [[5], Theorem 5.3] a result similar to Proposition 5.4 has been obtained. For some interesting related results we refer to [24], [25].

We remark that the construction of the finite-dimensional compensator (108) is not really satisfactory. Indeed we could have taken *any* compensator whose linearization equals the linearization of (108) in order that Proposition 5.4 continues to hold. In fact, already in [43] such a compensator (different from (108)) has been proposed. The important issue is therefore to construct a compensator with a *maximal* domain of validity. Since the compensator (108) *resembles* the nonlinear central controller there is reason to believe that it has favorable properties.

Finally, we close this section with an example of a class of nonlinear systems which *does* admit a finite-dimensional compensator which *globally* solves the $\mathcal{H}_\infty$ control problem (even for the *optimal* case). Note that the constructed compensator is *not* the central controller.

Example 5.6 A nonlinear system $\dot{x} = a(x) + b(x)u$, $y = c(x)$, $x \in M$, $u \in \mathbf{R}^m$, $y \in \mathbf{R}^m$ is called *lossless* [53] if there exists a nonnegative function $H : M \to \mathbf{R}$ (the internal energy) such that

$$
\begin{aligned}
H_x(x)a(x) &= 0 \\
H_x(x)b(x) &= c^T(x)
\end{aligned}
\tag{110}
$$

(This is equivalent to the requirement $\frac{d}{dt}H = y^T u$.) Following [11] we will now add to the control input $u \in \mathbf{R}^m$ an exogenous input $d_1 \in \mathbf{R}^m$, while the measured output y is additively contaminated with the exogenous input $d_2 \in \mathbf{R}^m$. Furthermore, the to-be-controlled outputs are given as $\begin{bmatrix} y \\ u \end{bmatrix}$, i.e., we consider the system

$$
\begin{aligned}
\dot{x} &= a(x) + b(x)u + b(x)d_1 &, \quad x \in M, u \in \mathbf{R}^m, d_1 \in \mathbf{R}^m \\
y &= c(x) + d_2 &, \quad y \in \mathbf{R}^m, d_2 \in \mathbf{R}^m \\
z &= \begin{bmatrix} c(x) + d_2 \\ u \end{bmatrix}
\end{aligned}
\tag{111}
$$

with a, b, c satisfying (110). Suppose $a(0) = 0$, and, without loss of generality, set $c(0) = 0$ and $H(0) = 0$. For the state feedback $\mathcal{H}_\infty$ control problem we notice that (111) is of the generalized form (58), and thus (see (59)) $\gamma^* \geq 1$, and the solution of the state feedback $\mathcal{H}_\infty$ suboptimal control problem is concerned with the Hamilton-Jacobi inequality (60), which takes the form (for $\gamma > 1$)

$$
P_x(x)a(x) - \tfrac{1}{2}\tfrac{\gamma^2 - 1}{\gamma^2} P_x(x)b(x)b^T(x)P_x^T(x)
$$

$$
+ \tfrac{1}{2}\tfrac{\gamma^2}{\gamma^2 - 1}c^T(x)c(x) \leq 0, \quad P(0) = 0.
\tag{112}
$$

Assume now that $\dot{x} = a(x)$, $y = c(x)$ is zero-state observable. It follows that $a(x) - \alpha b(x)b^T(x)P_x^T(x)$, for $\alpha > 0$, is a locally asymptotically stable vectorfield. (This vectorfield results from applying the feedback $u = -\alpha y$ to the undisturbed lossless system; use then $\frac{d}{dt}H = y^T u = -\alpha \| y \|^2$ and zero-state observability as in Proposition 4.2.) By using (110) it is seen that the *minimal* solution $P \geq 0$ to (112) for $\gamma > 1$ is given as

$$
P(x) = \frac{\gamma^2}{\gamma^2 - 1}H(x),
\tag{113}
$$

leading to the suboptimal stabilizing state feedback $u = -\frac{\gamma^2}{\gamma^2 - 1}b^T(x)H_x^T(x)$.

The *dual* Hamilton-Jacobi inequality (71) is again computed by setting $u = 0$ and $y = 0$, and thus $d_2 = -c(x)$. This leads to the pre-Hamiltonian (see (74)) $K_\gamma = p^T a(x) + p^T g(x) d_1 - \frac{1}{2}\gamma^2 d_1^T d_1 - \frac{1}{2}\gamma^2 c^T(x)c(x)$, and thus to the Hamilton-Jacobi inequality

$$R_x(x)a(x) + \frac{1}{2}\frac{1}{\gamma^2} R_x(x)b(x)b^T(x)R_x^T(x)$$
$$-\frac{1}{2}\gamma^2 c^T(x)c(x) \le 0, \quad R(0) = 0. \tag{114}$$

Using again (110) it is seen that the *maximal* solution of (114) is

$$R(x) = \gamma^2 H(x) \tag{115}$$

(Note that this is the anti-stabilizing solution of (114), since the vectorfield $a(x) + \frac{1}{\gamma^2}b(x)b^T(x) \cdot \gamma^2 H_x^T(x)$ is anti-stable!.) Now the *coupling condition* (75) tells us that $P(x) \le R(x)$, and thus by (113) and (115) a *necessary* condition for the solvability of the $\mathcal{H}_\infty$ suboptimal control problem is that $\gamma^2 \ge 2$! On the other hand, the *static output feedback*

$$u = -y \tag{116}$$

yields a closed-loop system which has L_2-gain $\le \sqrt{2}$. This can be checked either by using the methods of Section 3 for arbitrary $\gamma > \sqrt{2}$, or more simply (as has been done in [11]) by noting that the closed-loop system (111), (116) satisfies

$$\| d_1 \|^2 + \| d_2 \|^2 = \| d_1 + d_2 - y \|^2 + (y - d_2)^T(d_1 - y) +$$
$$+ \| y \|^2 = \| y \|^2 + \| d_1 + d_2 - y \|^2 + (y - d_2)^T(d_1 + u) \tag{117}$$

and thus by integrating from 0 to T, and observing that, because of loss-lessness, $\int_0^T (y - d_2)^T(d_1 + u)dt = H(x(T)) - H(x(0))$

$$\int_0^T \| d \|^2 dt \ge \int_0^T \| y \|^2 dt \left(= \int_0^T \| u \|^2 dt\right) - H(x(0)), \tag{118}$$

so that $\int_0^T \| z \|^2 dt \le 2 \int_0^T \| d \|^2 dt + 2H(x(0))$.

We conclude that γ^* *equals* $\sqrt{2}$, and that (116) is an $\mathcal{H}_\infty$ optimal (output) feedback. (This same result has been obtained before in [11] by different methods.) Furthermore, cf. Proposition 3.4, if the internal energy H is *proper*, then the closed-loop system is globally asymptotically stable.

We remark that $u = -y$ also solves the *nonlinear optimal control* problem for $d_1 = d_2 = 0$ and cost criterion $\frac{1}{2} \| z \|^2 = \frac{1}{2} \| y \|^2 + \frac{1}{2} \| u \|^2$, leading to the Hamilton-Jacobi-Bellman equation

$$V_x(x)a(x) - \frac{1}{2}V_x(x)b(x)b^T(x)V_x^T(x) + \frac{1}{2}c^T(x)c(x) = 0, \quad V(0) = 0 \tag{119}$$

(obtained from (112) by letting $\gamma \to \infty$). In fact, because of (110) this equation has the solution $V(x) = H(x)$! (In the context of robot manipulators this fact has been noted before in [50].)

6   Conclusions

It has become apparent that nonlinear $\mathcal{H}_\infty$ control entails some interesting problems. From a general point of view the *state feedback* $\mathcal{H}_\infty$ suboptimal control problem is reasonably well-understood. Important problems remain with regard to a priori information on the *size* of the neighborhood where the local state feedback $\mathcal{H}_\infty$ problem is solvable, and with regard to the nature of solutions P to the Hamilton-Jacobi inequality (36) (differentiability issues, viscosity solutions, properness of P as candidate Lyapunov function).

On the other hand, the full (dynamic output feedback) $\mathcal{H}_\infty$ control problem is still largely open. Appealing necessary conditions have been found generalizing the necessary *and sufficient* conditions of the linear $\mathcal{H}_\infty$ control problem. Furthermore there is insight into the structure of possible compensators, e.g. there exists a separation principle (Theorem 5.3). In the present paper the nonlinear *central controller* has been derived, which however has the severe drawback of being *infinite-dimensional*. A priori this does not rule out the possibility of existence of *finite*-dimensional compensators solving the (sub-)optimal problem; in fact the $\mathcal{H}_\infty$ optimal control problem for lossless systems has been shown to admit a finite-dimensional (actually, *static*) solution. Much effort has already been put into obtaining finite-dimensional controllers in general [5], [24], [26]; so far without decisive answers. Also, further work needs to be done on the nonlinear $\mathcal{H}_\infty$ *optimal* problem, on the *singular* nonlinear $\mathcal{H}_\infty$ (sub-)optimal control problem (see for a partial attempt [33]), and on the relation with classical nonlinear optimal control.

A very crucial question is the usefulness of nonlinear $\mathcal{H}_\infty$ control as a nonlinear design method. In the linear case many control problems, e.g. the mixed sensitivity problem, can be recast into a standard $\mathcal{H}_\infty$ problem for a *generalized* plant; this needs close attention in the nonlinear case (see for some remarks in this direction [5]). Of course, to a first approximation, one could base the choice of weighting functions on the linearized system. Especially the usefulness of the nonlinear $\mathcal{H}_\infty$ problem with regard to *robust* nonlinear control needs to be investigated.

Another main issue for the applicability of nonlinear $\mathcal{H}_\infty$ control is its computational complexity; hopefully todays computing power can manage solving the Hamilton-Jacobi inequalities.

Finally, the relation with modern nonlinear geometric control theory, see e.g. the textbooks [22], [37], deserves further study. It would be nice to bring these ideas together; after all, the two benchmark papers by Zames [55] and Isidori et al. [27] were published just 10 pages apart from each other!

Acknowledgements It is a pleasure to acknowledge the help of all colleagues who lent me a willing ear for questions on linear $\mathcal{H}_\infty$ control during the last couple of years. Especially I would like to thank S. Weiland, C. Scherer and G. Meinsma for very helpful discussions. Also I like to thank J. Grizzle, P. Khargonekar, A. Stoorvogel, T. Başar, A. Isidori, J. Ball and J.W. Helton for useful comments. I thank C. Scherer for a fruitful cooperation leading to the results of Theorem 5.1, J. Doyle for pointing out the nice example of $\mathcal{H}_\infty$ control of lossless systems, and J. de Does and J.M. Schumacher for stimulating conversations on the nonlinear central controller as derived in Section 5.

References

[1] R.A. Abraham and J.E. Marsden, *Foundations of Mechanics*, 2nd ed. Reading, MA: Benjamin/Cummings, 1978.

[2] J.A. Ball and J.W. Helton, "Factorization of nonlinear systems: Towards a theory for nonlinear $\mathcal{H}^\infty$ control," in *Proc. 27th CDC*, Austin, TX, 1988, pp. 2376-2381.

[3] J.A. Ball and J.W. Helton, "$\mathcal{H}^\infty$ control for nonlinear plants: Connections with differential games," in *Proc. 28th CDC*, Tampa, FL, 1989, pp. 956-962.

[4] J. Ball, J.W. Helton, "$\mathcal{H}^\infty$ control for stable nonlinear plants," *Math. Contr. Sign. Syst.*, vol. 5, pp. 233-262, 1992.

[5] J. Ball, J.W. Helton, M. Walker, "$\mathcal{H}_\infty$ control for nonlinear systems via output feedback," Preprint University of California at San Diego, August 1991.

[6] T. Başar and P. Bernhard, $\mathcal{H}_\infty$-*optimal control and related minimax design problems*, Birkhauser, 1990.

[7] T. Başar and G.J. Olsder, *Dynamic Noncooperative Game Theory*, New York: Academic, 1982.

[8] R.W. Brockett, *Finite Dimensional Linear Systems*, New York: Wiley, 1970.

[9] P. Brunovsky, "On the optimal stabilization of nonlinear systems," *Czech. Math. J.*, vol. 18, pp. 278-293, 1968.

[10] C.A. Desoer and M. Vidyasagar, *Feedback Systems: Input-Output Properties*. New York: Academic, 1975.

[11] J.C. Doyle, Unpublished notes

[12] J.C. Doyle, K. Glover, P.P. Khargonekar and B.A. Francis, "State s-pace solutions to standard $\mathcal{H}_2$ and $\mathcal{H}_\infty$ control problems," *IEEE Trans. Automat. Contr.*, vol. 34, pp. 831-846, 1989.

[13] B.A. Francis, *A Course in $\mathcal{H}_\infty$ Control Theory* (Lect. Notes Contr. Inf. Sci., Vol. 88), Berlin: Springer-Verlag, 1987.

[14] A. Friedman, *Differential games*, Wiley-Interscience, 1971.

[15] S.T. Glad, "Robustness of nonlinear state feedback — A survey," *Automatica*, vol. 23, pp. 425-435, 1987.

[16] K. Glover and J.C. Doyle, "A state space approach to $\mathcal{H}_\infty$ optimal control," in *Three Decades of Mathematical System Theory*, H. Nijmeijer and J.M. Schumacher, Eds. (Lect. Notes Contr. Inf. Sci., Vol. 135). Berlin: Springer-Verlag, 1989, pp. 179-218.

[17] K. Glover and J.C. Doyle, "State-space formulaes for all stabilizing controllers that satisfy an $\mathcal{H}_\infty$ norm bound and relations to risk sensitivity", *Syst. Contr. Lett.*, Vol. 11, pp. 167-172, 1988.

[18] O.B. Hijab, *Minimum Energy Estimation*, Doctoral dissertation, University of California, Berkeley, 1980.

[19] D.J. Hill, "Dissipativeness, stability theory and some remaining problems," in *Analysis and Control of Nonlinear Systems* (eds. C.I. Byrnes, C.F. Martin, R.E. Saeks), North-Holland, Amsterdam, pp. 443-452, 1988.

[20] D. Hill and P. Moylan, "The stability of nonlinear dissipative systems," *IEEE Trans. Automat. Contr.*, vol. AC-21, pp. 708-711, 1976.

[21] D. Hill and P. Moylan, "Connections between finite gain and asymptotic stability," *IEEE Trans. Automat. Contr.*, vol. AC-25, pp. 931-936, 1980.

[22] A. Isidori, *Nonlinear Control Systems*, 2nd ed. Berlin: Springer-Verlag, 1989.

[23] A. Isidori, *Feedback control of nonlinear systems*, Proc. 1st ECC, Grenoble, July 2-5, 1991, Hermes, Paris, pp. 1001-1012.

[24] A. Isidori, "$\mathcal{H}_\infty$ control via measurement feedback for affine nonlinear systems", Dept. of Systems Science and Mathematics, Washington University, St. Louis, November '92.

[25] A. Isidori, A. Astolfi, *Nonlinear $\mathcal{H}_\infty$-control via measurement feedback*, *J. Math. Systems, Estimation, Contr.*, 2, 1992, pp. 31-44.

[26] A. Isidori, A. Astolfi, "Disturbance attenuation and $\mathcal{H}_\infty$ control via measurement feedback in nonlinear systems, *IEEE Trans. Automat. Contr.*, AC-37, pp. 1283-1293, 1992.

[27] A. Isidori, A.J. Krener, C. Gori-Giorgi, S. Monaco, "Nonlinear decoupling via feedback: a differential geometric approach", *IEEE Trans. Automat. Contr.*, vol. AC-26, pp. 331-345, 1981.

[28] M.R. James, "A partial differential inequality for dissipative systems," preprint 1992.

[29] M.R. James, "Computing the $\mathcal{H}_\infty$ norm for nonlinear systems", preprint 1992.

[30] P.P. Khargonekar, "State-space $\mathcal{H}_\infty$ control theory and the *LQG*-problem, in *Mathematical System Theory – The influence of R.E. Kalman* (ed. A.C. Antoulas), Springer: Berlin, 1991.

[31] P.P. Khargonekar, I.R. Petersen and M.A. Rotea, "$\mathcal{H}^\infty$ optimal control with state feedback," *IEEE Trans. Automat. Contr.*, vol. AC-33, pp. 786-788, 1988.

[32] D.L. Lukes, "Optimal regulation of nonlinear dynamical systems," *SIAM J. Contr.*, vol. 7, pp. 75-100, 1969.

[33] R. Marino, W. Respondek, A.J. van der Schaft, P. Tomei, "Almost $\mathcal{H}_\infty$ disturbance decoupling," University of Twente, TW-Memorandum 1066, 1992.

[34] R.E. Mortensen, "Maximum likelihood recursive nonlinear filtering," *J. Optimization Theory and Applic.*, 2, pp. 386-394, 1968.

[35] P.J. Moylan, "Implications of passivity in a class of nonlinear systems," *IEEE Trans. Automat. Contr.*, vol. AC-19, pp. 373-381, 1974.

[36] P.J. Moylan and B.D.O. Anderson, "Nonlinear regulator theory and an inverse optimal control problem," *IEEE Trans. Automat. Contr.*, vol. AC-18, pp. 460-464, 1973.

[37] H. Nijmeijer and A.J. van der Schaft, *Nonlinear Dynamical Control Systems*, New York: Springer-Verlag, 1990.

[38] J. Palis, jr., and W. de Melo, *Geometric Theory of Dynamical Systems*, Springer, New York, 1982.

[39] A.J. van der Schaft, "On nonlinear observers", *IEEE Trans. Autom. Contr.*, vol. AC-30, pp. 1254-1256, 1985.

[40] A.J. van der Schaft, "On a state space approach to nonlinear $\mathcal{H}_\infty$ control," *Syst. Contr. Lett.*, vol. 16, pp. 1-8, 1991.

[41] A.J. van der Schaft, "On the Hamilton-Jacobi equation of nonlinear $\mathcal{H}_\infty$ optimal control, " in *Proc. 1st EEC*, Grenoble, July 1991, Hermes, Paris, pp. 649-654.

[42] A.J. van der Schaft, "Relations between $(\mathcal{H}_\infty-)$ optimal control of a nonlinear system and its linearization," in *Proc. 30th CDC*, Brighton, UK, 1991, pp. 1807-1808.

[43] A.J. van der Schaft, "L_2-gain analysis of nonlinear systems and nonlinear state feedback $\mathcal{H}_\infty$ control," *IEEE Trans. Autom. Contr.*, vol. AC-37, pp. 770-784, 1992.

[44] A.J. van der Schaft, "Nonlinear $\mathcal{H}_\infty$ control and Hamilton-Jacobi inequalities," *Proc. IFAC NOLCOS '92* (ed. M. Fliess), Bordeaux, 1992, pp. 130-135.

[45] A.J. van der Schaft, "Complements to nonlinear $\mathcal{H}_\infty$ optimal control by state feedback", *IMA J. Math. Contr. Inf.*, vol. 9, pp. 245-254, 1992.

[46] C. Scherer, "$\mathcal{H}_\infty$-control by state feedback: An iterative algorithm and characterization of high-gain occurrence," *Syst. Contr. Lett.*, vol. 12, pp. 383-391, 1989.

[47] C. Scherer, *The Riccati inequality and state space $\mathcal{H}_\infty$ control theory*, Doctoral Dissertation, University of Würzburg, Germany, 1991.

[48] A.A. Stoorvogel, *The $\mathcal{H}_\infty$ control problem: a state space approach*, Prentice Hall, Englewood Cliffs, 1992.

[49] G. Tadmor, "Worst-case design in the time domain: the maximum principle and the standard $\mathcal{H}_\infty$ problem," *Math. Contr. Sign. Syst.*, 3, pp. 301-324, 1990

[50] M. Takegaki and S. Arimoto, "A new feedback method for dynamic control of manipulators," *Trans. ASME, J. Dyn. Systems, Meas. Control* 103, pp. 119-125, 1981.

[51] J.C. Willems, "The generation of Lyapunov functions for input-output stable systems," *SIAM J. Contr.*, vol. 9, pp. 105-133, 1971.

[52] J.C. Willems, "Least squares stationary optimal control and the algebraic Riccati equation," *IEEE Trans. Automat. Contr.*, vol. AC-16, pp. 621-634, 1971.

[53] J.C. Willems, "Dissipative dynamical systems, Part I: General theory," *Arch. Rat. Mech. Anal.*, vol. 45, pp. 321-351, 1972.

[54] S. Weiland, "Theory of approximation and disturbance attenuation for linear systems," Ph.D. Thesis, Rijksuniversiteit Groningen, 1991.

[55] G. Zames, "Feedback and optimal sensitivity: model reference transformations, multiplicative seminorms, and approximate inverses," *IEEE Trans. Automat. Contr.*, vol. AC-26, pp. 301-320, 1981.

7. Learning Control and Related Problems in Infinite-Dimensional Systems

Y. Yamamoto*

Abstract

The basic features of a special type of learning control scheme, currently known as *repetitive control* are reviewed. It is seen that this control scheme also induces varied interesting theoretical problems—particularly those related to infinite-dimensional systems. They include such problems as the internal model principle, minimal representation of transfer functions, fractional representations, stability characterizations, correspondence of internal and external stability, etc. This article intends to give a comprehensive overview of the repetitive control scheme as well as the discussion of these related theoretical problems for infinite-dimensional systems.

1 What is Learning Control?

The word *learning* almost always intrigues us. It reflects our very fundamental desire of having an intelligent artificial object that can perform complicated tasks instead of ourselves.

Control itself was in fact a product of such a desire. ¿From the very beginning, control was destined to have some intelligent power to be able to automatically adjust itself to the operating environment. It is not hard to imagine that the concept of feedback (in connection with the sensitivity reduction against disturbances) had a striking conceptual appeal to those working in the field. The secret here is that the control system can adjust itself (or, more appealingly, "learn to react") to the fluctuation, disturbances, etc. that could be an obstacle to performing its task.

However, we generally no longer regard an ordinary control system as something realizing learning. What is then expected of "learning" in the

*Division of Applied Systems Science, Faculty of Engineering, Kyoto University, Kyoto 606, Japan. Supported in part by the Tateishi Science and Technology Foundation.

present day context? What are the characteristics of learning? As an example, let us take a look at the field of neural networks where learning has been one of the central issues [50]. Roughly speaking, given a multi-layer neural network, one applies the so-called error backpropagation learning mechanism as follows: Given learning data prepared in the form of input/output pairs, we apply an input to the network, observe the corresponding output, take the error from the desired output, and then adjust the interconnecting coefficients among neuron-like devices so as to reduce this error signal. Repeating the same procedure for a large number of times, we expect the network to acquire the input/output relations encoded in the learning data; if the learning data are appropriate, it is also expected that the network acquires the essential knowledge packed in the data.

The essence of the procedure above rests upon i) prepare a relatively loosely organized memory device (e.g., multi-layer neural network), ii) map the problem of acquisition of the input/output relation to an optimization in a parameter space (adjustment of interconnecting coefficients), and then iii) solve this problem in a certain iterative procedure (large number of backpropagation). In this last step, taking the error signal and feeding it back to the system (network) iteratively seem to be essential.

However, the same methodology can be observed also in various control aspects: Servomechanism is based on a similar idea although one usually does not call it a learning system. Learning control has appeared in varied forms in control theory, and in one way or another, they utilize a relatively less organized memory, and try to reorganize the pertinent parameters by some iterative methods.

The kind of learning control scheme we are going to deal with in this article also makes use of this methodology. As an example, consider the motion control problem of a robotic arm. Even today, its path control often relies on human teaching, and this requires the assistance of skilled technicians. It would be highly desirable if the robot itself could "learn" how to control itself for a given desired motion trajectory. One way of handling this is to give such a trajectory as a reference command, measure the error signal, and then implement a mechanism to reduce this error signal. This requires a large number of repetitions, but robots can perform the same task many times, so the repetition here is not a major problem. In general, many mechanical systems can, or are designed to, perform the same type of movements, as is often the case with revolving motions. Thus there are many such devices that can be embraced in this framework.

A relatively new servo control scheme called *repetitive control* is based on this principle. It yields an iterative (repetitive) learning mechanism for a system to acquire an open-loop control input when the reference command is given in the form of periodic signals.

Our objective is two-fold: we first introduce this scheme, and discuss its basic properties: its learning capabilities, basic construction, stability/stabilizability conditions, etc. This discussion further motivates more detailed and advanced study of infinite-dimensional systems, and it is seen that the scheme is closely related to many interesting aspects of infinite-dimensional systems: stability, internal model principle, minimal representation of transfer functions, fractional representations, state space realizations, eigenfunction completeness, internal and external stability, etc. Although this article is by no means intended to be a complete survey of these topics, it is hoped that it gives a comprehensive account on these aspects of infinite-dimensional systems with repetitive control as a standing motivating example.

For simplicity of exposition, we confine ourselves to the single-input-/single-output case, and all functions are scalar-valued. The notation is fairly standard: $L^2[0, \infty)$ and $L^2_{loc}[0, \infty)$ are the spaces of square, and locally square, integrable functions, respectively. $\mathbf{C}_\sigma := \{s \in \mathbf{C}; \operatorname{Re} s > \sigma\}$ where $\sigma \in \mathbf{R}$. $\mathcal{L}[f]$ or $\hat{f}$ denotes the Laplace transform. For a function or a distribution f, its *support* supp f is the smallest closed set outside of which f is zero. The following notation for Hardy spaces will also be used:

- $H^\infty(\mathbf{C}_\sigma) :=$ bounded holomorphic functions on $\mathbf{C}_\sigma$.

- $H^\infty := H^\infty(\mathbf{C}_0)$, with standard norm

$$\|\psi\|_\infty := \sup_{s \in \mathbf{C}_0} |\psi(s)| = \sup_{-\infty < \omega < \infty} |\psi(j\omega)|. \tag{1}$$

- $H^\infty_- := \bigcup_{\sigma < 0} H^\infty(\mathbf{C}_\sigma)$.

2   Repetitive Control

We have seen that learning mechanisms are often accompanied with repeating actions of the same sort [1]. There are many practical situations where repeating actions are not only possible but also desirable. Motion control of robots, control of NC machines, hard-disc drive control or many mechanical systems having revolving mechanisms inside, etc. are such examples. The central idea is that instead of "teaching" a machine, e.g., robot, exactly how to behave, we give a reference command (e.g., a desired trajectory) to it, and by incorporating a suitable correction mechanism, we let the machine to acquire the desired control action automatically by itself. In order that this scheme works, it is necessary that the same trial actions be repeatable on the machine. This methodology is employed for a number of applications, and has proven to be successful: [1], [41], [45], [42], [11], [2], just to name a few.

---

[1] We do not deal with unsupervised learning as studied e.g., in [56].

This is the guiding principle of the relatively new control scheme, now called *repetitive control* (e.g., [33], [45], [42], [29], [11]): repeat the same type of action, and by the repetition the system acquires the desired control action. In the case of robots, there is a question of nonlinearity, but one can also reset the initial state at each step so that each trial action can be separated [1]. On the other hand, repetitive control forms a complete closed-loop system, and is closer to the classical servo control scheme, and induces various interesting technical or theoretical questions which are the themes of this article. To see its basic idea more clearly, let us go back to some historical accounts related to this scheme.

Repetitive control was first introduced in the control of proton synchrotron magnetic power supply [33]. In order to obtain the desired proton acceleration pattern, it was necessary to control the current supply in a specific curve (for a rough sketch, see Fig. 2.1 [43]). The highest current level is quite high (about $3 \times 10^3 A$), but the precision requirement is still very high: relative precision of 10^{-4}.

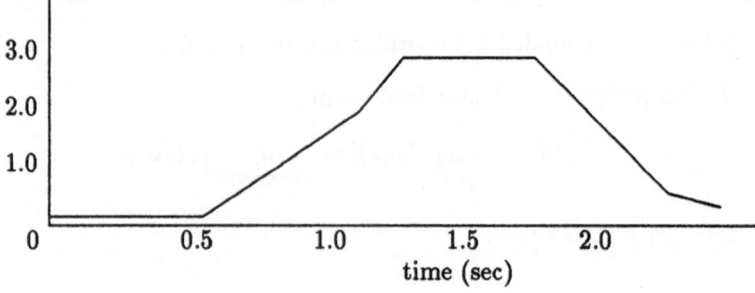

Figure 2.1: Current Supply Curve

The first thing one might try is to attempt an open-loop type control: identify the plant dynamics, find the inverse system, and then solve for the desired input pattern via this inverse system. However, this requires very high precision in the identification; as a result, it was difficult to attain the desired precision 10^{-4}.

If the system cannot be known to this precision and if the performance requirement is this high, is it possible at all to accomplish this task? Although seemingly hopeless, the very nature of the synchrotron was the key to this problem: Since protons turn around in the synchrotron, this reference command repeats by itself, i.e., it is periodic. Once this synchrotron is activated, it is kept under operation for at least two weeks. Therefore, if one can implement a mechanism that has the ability of self-correction so that in the "steady state" the system tracks this periodic signal, then for all practical purposes, the objective will be accomplished. Therefore, Inoue, Nakano and others [33], came up with the idea of i) feeding the reference input as above periodically to the ring magnet, ii) storing the error signal of complete one period, and then iii) feeding it back to the system (with a suitable compensator). This is in complete correspondence with the learning mechanism discussed in the Introduction.

In the context of servomechanism design, this amounts to designing a compensator with the property

- the closed-loop system asymptotically tracks the reference periodic signal, and

- this property must hold for small variations of plant parameters (robust tracking property).

The shape of this periodic function is rather complex so that it is simpler to require that this servo system track every periodic signal of a fixed period L.

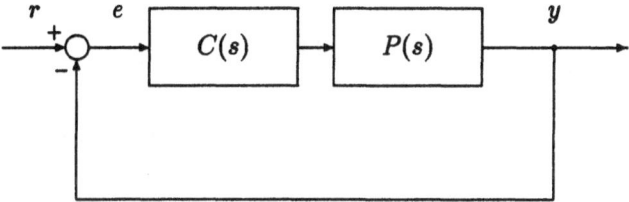

Figure 2.2: Unity Feedback System

Let us elaborate further. Let $P(s)$ be a continuous-time plant, and let $C(s)$ be a compensator, $C(s)P(s)$ the loop transfer function, and we form the unity feedback system as in Fig. 2.2. What sort of condition should $C(s)$ satisfy to meet the requirements above? If the reference signals are generated by a finite-dimensional system, the now classical *internal model principle* [23], [4] tells us that $C(s)$ must have the ability of producing all the reference signals demanded on the system. The typical case is $C(s) = 1/s$ for the tracking to step functions. Another is $C(s) = 1/(s^2 + \omega^2)$ which allows for tracking to the sinusoid $\sin \omega t$. But what kind of $C(s)$ should be taken to track every periodic signal of a fixed period L?

Recall that every periodic function of period L can be expanded in the Fourier series:

$$r(t) = \sum_{n=-\infty}^{\infty} r_n \exp(2n\pi jt/L).$$

To track $\exp(2n\pi jt/L)$ and $\exp(-2n\pi jt/L)$, we must incorporate

$$\frac{1}{s^2 + 4n^2\pi^2/L^2}$$

into $C(s)$. Since this must hold for every n, how about

$$C(s) = s \prod_{n=0}^{\infty} \frac{1}{s^2 + 4n^2\pi^2/L^2} \ ?$$

Well, not quite, because the product here is infinite and one has to be a little more careful. In fact, we readily see that the infinite product here is identically zero. To remedy this, we can take instead

$$C(s) = 1 \bigg/ \left(s \prod_{n=1}^{\infty} \left(1 + \frac{L^2 s^2}{4n^2\pi^2} \right) \right) = 1 \bigg/ \left(\frac{1}{2\pi} \left(e^{Ls/2} - e^{-Ls/2} \right) \right). \quad (2)$$

The second equality readily follows from the well known identity

$$\sinh \pi s = \pi s \prod_{n=1}^{\infty} \left(1 + \frac{s^2}{n^2} \right).$$

Multiplying the denominator by $2\pi e^{Ls/2}$, we obtain the transfer function $C(s) = 1/(e^{Ls} - 1)$ which can be realized by the delay element e^{-Ls} as in Fig. 2.3.

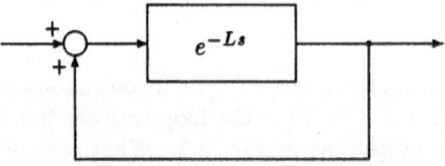

Figure 2.3: Repetitive Compensator

In fact, it is readily seen that the delay element stores the function of the past one period, so that the system in Fig. 2.3 can work as a "periodic signal generator." Now the classical internal model principle [23] says that if we include the model of the exogenous signal generator in the loop and stabilize the unity feedback system, then the closed-loop system asymptotically tracks the exogenous signals. Although this is proved for finite-dimensional systems, it is expected that a closed-loop system with this *repetitive compensator* works in a similar way as follows:

- any periodic signal of period L can be generated by this $C(s)$ with suitable initial function;

- with internal stability, the output tracks any periodic exogenous signal.

More intuitively, given a plant and a periodic reference signal, we feed it into the plant, store the entire error function of one period, and then feed it back to the plant. If the closed-loop configuration is appropriate, the error will tend to zero. This is the restatement of the idea of the authors of [33], and in fact the original synchrotron was successfully controlled by this new method. As mentioned earlier, variety of applications can be embraced in this framework, and some of them are in actual use: [41], [45], [42], [11], [2].

But does it really work in this way? Here the compensator $1/(e^{Ls} - 1)$ contains a delay, and has infinitely many poles $2n\pi j/L$, $n = 0, \pm 1, \ldots$ on the imaginary axis. Although the idea is quite intuitive, it is desirable to give a more firm ground on the treatment of servo problems of this kind. The explanations above relied much upon Fourier series, but a more general method would give a better perspective. We can ask many theoretical questions: If this scheme works, is this the simplest construction that satisfies the requirement (necessity question)? What is the more natural way of handling such a class of systems? Can the internal stability of the overall feedback system be checked in terms of poles of the transfer function? When does it work?

In the next section, we will start by looking at the question of stability/stabilizability question of the repetitive control scheme, and then see how it leads to more general theoretical problems of a certain class of infinite-dimensional systems.

3 Repetitive Control–Continued

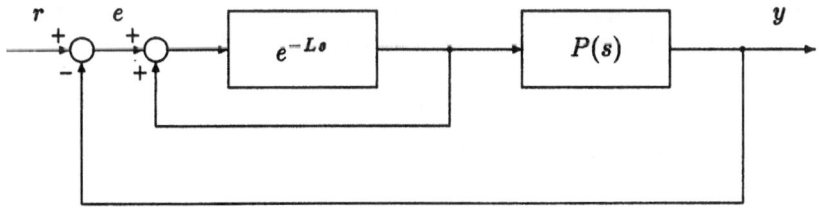

Figure 3.1: Repetitive Control System

As illustrated in the previous section, a typical repetitive control system takes the construction shown in Fig. 3.1. Let us note again that if we replace the repetitive compensator $1/(e^{Ls} - 1)$ by the integrator $1/s$, then it is exactly the same as the classical servo control system tracking to step reference signals. In this section, we give further analysis on the tracking and stability property of repetitive control systems.

3.1 Tracking and Stability Conditions

Suppose $P(s)$ does not vanish at any poles $s = \pm 2n\pi j/L$ $n = 0, 1, \ldots$ of the repetitive compensator. Let

$$W_{er}(s) := \frac{1}{1 + P(s)/(e^{Ls} - 1)} = \frac{e^{Ls} - 1}{e^{Ls} - 1 + P(s)}$$

be the transfer function from r to e in Fig. 3.1. Then $s = \pm 2n\pi j/L$ become the transmission zeros of $W_{er}(s)$. This means that a sinusoidal mode $\sin(2n\pi t/L)$ becomes unobservable from e, so that if the closed-loop system is stable, then the system asymptotically tracks this sinusoid. Since this is true for every n, any finite Fourier series (having period L) can be tracked. Assuming continuity in the limit, it seems plausible that the repetitive control system asymptotically tracks any periodic signal of a fixed period L.

This argument relies upon Fourier series, and it is certainly more desirable to have a general theory that can embrace the repetitive controller as a special case, not depending on a specialized structure such as that of Fourier series. In fact, there are several approaches to the internal model principle for distributed systems [21], [19], [7], [15], etc. However, let us note for the moment that since this repetitive compensator contains infinitely many poles $\pm 2n\pi j/L$, $n = 0, 1, \ldots$ on the imaginary axis, it is often excluded from the existing frameworks.

Let us start with some simple stability analysis. Suppose for simplicity that $P(s)$ is stable. An easy loop transformation converts Fig. 3.1 to Fig. 3.2. (By the stability of $P(s)$, we have neglected the initial value responses.) A straightforward application of the small gain theorem (e.g., [16]) yields that the converted system is L^2 input/output stable if

$$\|1 - P\|_\infty < 1, \tag{3}$$

where $\| \cdot \|_\infty$ denotes the standard H^∞ norm (1).

Several problems arise:

- Condition (3) is only a sufficient condition. How close is it to necessity?

- It assures only the L^2 input/output stability, i.e., when we apply L^2 inputs with zero initial state we get L^2 outputs. Can exponential

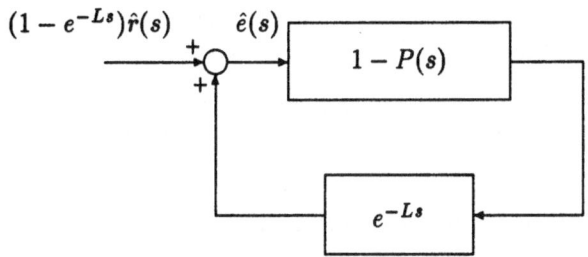

Figure 3.2: Equivalent System

stability be concluded from this? This is particularly important for the consideration of unknown initial states.

• When can this condition (3) be satisfied?

First let us note that the delay e^{-Ls} introduces a large amount of phase shift, so that condition (3) is actually very close to necessity, especially in the high frequency range.

The second point requires more attention. In the next subsection we will make use of results on delay-differential equations [28] to show that this condition actually guarantees exponential stability. In later developments, we will adopt a more abstract framework, and discuss relationships between internal and external stability in a general setting.

However, the third point causes a serious problem. It is readily seen that condition (3) can *never* be satisfied for a *strictly proper* $P(s)$. It is satisfied only for a plant with relative degree 0 (and the direct transmission term 1). Otherwise, not only does this condition fail, but stabilization is in fact impossible (e.g., [44]). Since this means that the plant should have a direct transmission term from the input to output, this could hardly occur in reality. Is this inevitable? If so, does this mean that repetitive control is meaningless?

Let us look at Fig. 3.1 again. There is a direct feedback path from the delay e^{-Ls} into itself with gain 1, if $P(s)$ is strictly proper. Such a system is known to be a delay system of *neutral* type [28], and it is known (e.g., [44], [51]) that such a system is hard to stabilize. More specifically, if the feedback gain of the delay into itself is k, then *there is an infinite chain of poles that asymptote the line* $\{s \in \mathbf{C}; \operatorname{Re} s = \log k\}$, and *this is determined solely by this gain* k [28]. In the configuration of Fig. 3.1, this gain can be affected only through the direct transmission path of $P(s)$, so that $P(s)$ should have relative degree zero in order that Fig. 3.1 be exponentially stabilized.

This is certainly an ironical fact. Is it impossible to realize repetitive control? We should ask the following questions:

- Is the construction of Fig. 3.1, in particular the implementation of the repetitive compensator $1/(e^{Ls} - 1)$, mandatory for the objective of repetitive control?

- If so, is there any way to make a reasonable compromise?

The answer to the first question is positive, i.e., if we want to realize the repetitive control in the strict sense, we must incorporate $1/(e^{Ls} - 1)$ as an internal model. This is a generalization of the necessity of the internal model principle, and will be proved in Section 4 in a more general setting. Recall that in the argument using the Fourier series in (2), we have multiplied the infinite product by $e^{Ls/2}$ to get this form $1/(e^{Ls} - 1)$, so there is some nonuniqueness in the infinite product representations; this situation will be clarified in Sections 4 and 5. In any case, since $1/(e^{Ls} - 1)$ must be incorporated, the stability condition (3) cannot be relaxed as it stands.

The situation is, however, not totally impossible. Let us recall the requirement of repetitive control. If its objective is fulfilled, the closed-loop system Fig. 3.1 asymptotically tracks *any* periodic signal of period L. Then even a discontinuous periodic function should be tracked as an output of the continuous-time plant $P(s)$, and this is absurd unless the relative degree of $P(s)$ is zero. So this restriction comes from the apparently unrealistic overspecification of tracking in a very high frequency band. One way of handling this is to introduce a low-pass filter in front of the delay term, thereby replacing the delay element e^{-Ls} by $f(s)e^{-Ls}$ for some strictly proper stable rational filter $f(s)$ (for a related but different discussion, see also [32]). This relaxes the tracking requirement in the high frequency range, thereby relaxing the stability condition. In fact this procedure makes the total system *retarded* rather than neutral. As in the same way in (3), the small gain condition becomes

$$\|f(1 - P)\|_\infty < 1. \tag{4}$$

Clearly the high frequency band condition is relaxed here compared to (3), and this condition can be satisfied with strictly proper $P(s)$. Although this *modified repetitive control* system has less ability in tracking in the high frequency band, it has the advantage of wider applicability. In reality, the low-pass characteristic of the filter does not become a drawback, since we do not need tracking in a very high frequency range anyway. This modified repetitive control scheme is used in most of applications, and has proven to be successful [41], [42], [45], [43].

3.2 Modified Repetitive Control—State Space Analysis

To guarantee exponential stability we give some state space analysis for the modified repetitive control system. Since the treatment here employs functional differential equations, and exponential stability will be discussed in later sections from the external viewpoint, readers who are not interested in such technical details can skip this subsection without much loss.

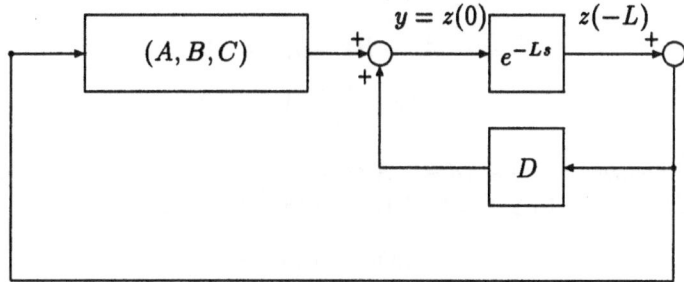

Figure 3.3: Modified Repetitive Control System

Let

$$\dot{x}(t) \;=\; Ax(t) + Bu(t) \tag{5}$$
$$y(t) \;=\; Cx(t) + Du(t) \tag{6}$$

be a minimal realization of $f(s)(1 - P(s))$. Note that we do not exclude the case of strict repetitive controller $f(s) = 1$. Replace $1 - P(s)$ by $f(s)(1 - P(s))$ in Fig. 3.2, and rewrite the diagram as Fig. 3.3. Then this modified repetitive control system is represented by the following well-known M_2 functional differential equation model [51]:

$$\frac{d}{dt}\begin{bmatrix} x_t \\ z_t \end{bmatrix} \;=\; \begin{bmatrix} Ax_t + Bz_t(-L) \\ \frac{\partial}{\partial \theta}z_t(\theta) \end{bmatrix}, \tag{7}$$
$$y(t) \;=\; z_t(0).$$

Here x_t is the state of $f(s)(1 - P(s))$, and $z_t(\theta)$, $-L \le \theta \le 0$ is the function stored in the delay at time t. The pair $(x_t, z_t(\cdot))$ belongs to the so-called M_2 space $\mathbf{R}^n \times L^2[-L, 0]$. We also set the input to be zero since it is irrelevant to stability. Let A denote the right-hand side operator in (7). Its domain $D(A)$ must be specified as

$$D(A) := \{(x, z(\cdot)) \in M_2; z \in W_2^1[-L, 0], z(0) = Cx + Dz(-L)\}$$

where $W_2^1[0, h]$ denotes the first-order Sobolev space. The boundary condition $z(0) = Cx + Dz(-L)$ specifies how the output Cx and feedthrough term $Dz(-L)$ is fed into the delay in Fig. 3.3. It is known [28] that

- the exponential stability of this system is determined by the location of the spectrum, and

- every point in the spectrum is an eigenvalue

Note that $\lambda \in \mathbf{C}$ is an eigenvalue of A if and only if

$$\lambda x - Ax - Bz(-L) \; = \; 0 \qquad (8)$$

$$\lambda z - \frac{d}{d\theta}z(\theta) \; = \; 0 \qquad (9)$$

for some nonzero $(x, z) \in D(A)$. A straightforward calculation along with $z(0) = Cx + Dz(-L)$ yields that this holds if and only if

$$e^{\lambda L} - f(\lambda)(1 - P(\lambda)) = 0. \qquad (10)$$

If condition (4) is satisfied, it is easily seen [29] that the least upper bound of the real parts of all such λ is negative. This implies the exponential stability of the modified repetitive control system.

3.3   Synthesis

Let us give an indication on the trade-off between the tracking accuracy and stability margin. As the low-pass filter $f(j\omega)$ decays faster as $\omega \to \infty$, the stability condition (4) becomes easier to satisfy. On the other hand, instead of the repetitive compensator $1/(e^{Ls} - 1)$, we have $fe^{-Ls}/(1 - f(s)e^{-Ls})$, so that at $\omega = 2n\pi/L$ the tracking error is proportional to $1 - f(2n\pi j/L)$. In practice, according to the nature of reference signals, a suitable bandwidth should be specified which in turn leads to a reasonable choice of $f(s)$.

Let us write $P(s) = C(s)P_0(s)$, separating the controller $C(s)$. Once such $f(s)$ is chosen, the small gain condition (4), $\|f(1 - CP_0)\|_\infty < 1$, reduces precisely to the so-called model matching problem in the H^∞ control problem, and this can be solved using standard techniques [22], [18]. For some synthesis examples, see [29].

4   More on The Internal Model Principle

We have seen that repetitive control falls into the category of delay-differential systems, and stability analysis was given based on the results in such a framework. We have also shown that there are certain limitations in stabilizing repetitive control systems, and to relax them the modified repetitive control scheme was introduced. These arguments were based upon an analogy to the classical internal model principle, and the necessity of the repetitive compensator $1/(e^{Ls} - 1)$ was shown as the internal model for a periodic signal generator. This necessity relied upon the somewhat intuitive argument using Fourier series and infinite product representations. Strictly speaking, therefore, it is not yet quite proved that incorporating this compensator is the *minimal* construction for realizing repetitive control. In what follows, we place this problem in a more general framework, study the internal model principle in this setting, and also attempt to exhibit what kind of theoretical problems this can lead to. This includes fractional representations in the time domain, their Laplace transforms and growth order estimates, close relationships with infinite product representations, state space representations, eigenfunction completeness, relation between internal and external stability, etc.

We start by reformulating our problem in a class of complex analytic functions.

4.1 Minimality of Representations

Examining the argument in Section 2 again, we find that the following points are left open:

- For a given infinite sequence $\{\lambda_1, \ldots, \lambda_n, \ldots\}$ of poles, with corresponding modes $\{e^{\lambda_1 t}, \ldots, e^{\lambda_n t}, \ldots\}$, what is the natural class of transfer functions obtained by forming the corresponding infinite product as in (2)? Is the representation unique? If not, what is the simplest form?

- Under the same condition, what is the "minimal" state space representation consisting of these "modes" $\{e^{\lambda_1 t}, \ldots, e^{\lambda_n t}, \ldots\}$? If a state space realization is associated to them, when is the state space "spanned" by them, i.e., eigenfunction complete?

While the first question concerns minimality in the frequency domain, the second is concerned with that of realizations. Introduced and motivated by the repetitive control scheme, what we would now like to investigate is some interesting interplay between them, bridged by various complex analytic structures.

Let us again take the repetitive compensator $1/(e^{Ls} - 1)$. It is the ratio of *entire*, i.e., analytic on the whole plane, functions where the denominator vanishes at $2n\pi j/L$, $n = 0, \pm 1, \ldots$ As we have already seen, these zeros combined together give rise to the internal model for the periodic reference signals. *Is this the only function that has this property?*

The answer is negative. The well known Weierstrass factorization theorem [49] tells us that there is much freedom due to the choice of convergence factors to make infinite product representations convergent. However, a more restricted class of entire functions of *finite order* makes the factorization unique. The result is known as the Hadamard factorization theorem [5], and we give the special case of order one which concerns us here. An entire function $f(s)$ is said to be of *order* at most 1 if for every ϵ, there exists $r > 0$ such that

$$|f(s)| \leq e^{|s|^{1+\epsilon}}, \quad |s| > r. \tag{11}$$

It is said to be of *exponential type* if there exist $C, K > 0$ such that

$$|f(s)| \leq Ce^{K|s|} \tag{12}$$

for all $s \in \mathbb{C}$. Clearly a function of exponential type is of order at most one, but not conversely [5]. Let $f(s)$ be of order at most one with nonzero zeros $\{\lambda_1, \ldots, \lambda_n, \ldots\}$, counted according to multiplicities. Then $f(s)$ admits the infinite product representation

$$f(s) = s^m e^{as+b} \prod_{n=1}^{\infty} \left(1 - \frac{s}{\lambda_n}\right) \quad \text{or}$$

$$f(s) = s^m e^{as+b} \prod_{n=1}^{\infty} \left(1 - \frac{s}{\lambda_n}\right) \exp\left(\frac{s}{\lambda_n}\right) \tag{13}$$

depending on whether or not $\sum_{n=1}^{\infty} 1/|\lambda_n| < \infty$. This is a special case of the Hadamard factorization theorem [5]. An important point to be noted here is that the rate of zeros $\lambda_n \to \infty$ controls the growth order of $f(s)$. For example, if f has only finitely many zeros, then f is a polynomial; if $\sum_{n=1}^{\infty} 1/|\lambda_n| < \infty$, the growth is slower than exponential type. We will encounter a more elaborate result, which is necessary for proving the internal model principle (Theorem 4.3). In any case, except the factor e^{as} which roughly corresponds to a delay, the zeros $\{\lambda_1, \ldots, \lambda_n, \ldots\}$ uniquely determine an entire function of order at most one.

Is this the best we can say? Actually, functions like $e^{Ls} - 1$ are not only of order one, but also of exponential type, and a more detailed characterization is possible. To this end, we will establish a link with Laplace transforms, and consider fractional representations over a convolution algebra.

4.2   Fractional Representations

Fractional representations give a compact and powerful tool for handling tracking and the internal model principle [4], [21], etc. For infinite-dimensional systems, there are various fractional representation approaches. Let us list some of them. Let $\mathcal{A}$ denote the Banach algebra consisting of L^1 functions and delays with ℓ^1 coefficients. $\mathcal{A}_-$ is the subset consisting of those with some exponential decay rate. The Callier-Desoer algebra ([6], [7]) $\hat{B}$ consists of fractions of $\hat{A}_-$ with denominator being bounded away from 0 at ∞ in C_0. Another possibility related to $\mathcal{A}$ is its quotient field since it is an integral domain. Inouye [34] considered transfer functions of bounded type; this class is the quotient field of H^∞. Another choice is the class of pseudorational impulse responses [58]. While the first three are closer to the fractional representations over stable rational transfer functions, the pseudorational class is closer to fractions of polynomials. It is not possible to say which algebra is most suitable without specifying a particular problem. For some comparisons, see [40] for an excellent survey on the subject and related topics.

Let us now introduce the class of *pseudorational* impulse responses or transfer functions. This class is closer to the fractions of polynomials, and is suitable for realization. Another advantage is that it can handle the case of infinitely many unstable poles, as is the case with the repetitive compensator $e^{Ls} - 1$. Also, this class has a very close connection with entire functions of exponential type and infinite product representations.

Consider the input/output relations described by the convolution equation

$$q * y = p * u \tag{14}$$

where u and y denote the input and output functions, and instead of polynomials in the differential operator, we consider q and p in the class of distributions with compact support [52], [54]. We require that q and p have support in $(-\infty, 0]$ [2]. This allows finite-order differentiations, delays (as represented by convolution with the Dirac delta distribution δ_a), finite-time integral operator, etc.

If q is invertible as a distribution, we may rewrite (14) as

$$y = q^{-1} * p * u \tag{15}$$

where $q^{-1} * p$ plays the role of the impulse response. For example, the delay-differential equation

$$\dot{y} = y(t) + y(t-1) + u(t) \tag{16}$$

can be expressed as

$$y = (\delta' * \delta_{-1} - \delta_{-1} - \delta)^{-1} * \delta_{-1} * u. \tag{17}$$

Here δ denotes the delta function, δ' its derivative and δ_a its shift as follows:

$$(\delta' * y)(t) = y'(t), \qquad (\delta_a * y)(t) = y(t-a).$$

---

[2] This is in full analogy with the discrete-time case where transfer functions are expressed as the ratio of polynomials in the forward shift operator z.

We call such impulse responses *pseudorational* as a direct generalization of the rational case. (For the precise technical definition, see [58].) To discuss their transfer functions, let us introduce the Laplace transform for such distributions. Since they have support in $(-\infty, 0]$, it is clearly natural to consider the two-sided Laplace transform:

$$\mathcal{L}[q](s) = \hat{q}(s) := \langle q, e^{-st} \rangle = \int_{-\infty}^{\infty} q(t)e^{-st}dt. \qquad (18)$$

Typical examples are

$$\mathcal{L}[\delta'](s) = s, \qquad \mathcal{L}[\delta_a](s) = e^{-as}.$$

A complex function $G(s)$ is *pseudorational* if it admits such a factorization $G(s) = \hat{p}(s)/\hat{q}(s)$ for distributions p, q with conditions above. Since q has compact support, definition (18) gives rise to an entire function [52]. Hence the transfer function $\hat{p}(s)/\hat{q}(s)$ is a meromorphic function. Actually, we can say much more. The celebrated Paley-Wiener theorem assures that these Laplace transforms are entire functions of exponential type:

Theorem 4.1 ([46], [52], [54], [49]) *A complex function $\psi(s)$ is the Laplace transform of a distribution with compact support in $(-\infty, 0]$ if and only if it satisfies the estimate*

$$|\psi(s)| \leq \begin{cases} (1 + |s|)^m \exp(a \operatorname{Re} s), & \operatorname{Re} s \geq 0 \\ (1 + |s|)^m, & \operatorname{Re} s \leq 0 \end{cases} \qquad (19)$$

for some integer $m \geq 0$ and positive a.

Furthermore, if this estimate is satisfied, then the support of ψ is contained in $[-a, 0]$. For the ease of reference, we will call this class of entire functions *Paley-Wiener* class.

The exponential growth comes from the compact support property. However, $\psi(s)$ can grow only with the polynomial order along the imaginary axis. This is due to the fact that $\psi(s)$ is the Laplace transform of a distribution on the real line. For example, e^{js} can grow exponentially along the imaginary axis, but its time domain counterpart (inverse Laplace transform) would be δ_{-j}, which is not a function (distribution) on the real line. Actually, the maximal growth one can allow for $\psi(s)$ on the $j\omega$ axis such that its inverse Laplace transform can be justified as a time domain concept is the so-called *infra-exponential growth* (i.e., slower growth than any exponentials); otherwise its inverse Laplace transform will have support outside the real line (as is the case with e^{js}). This fact was pointed out by Carleman [9], and its realization is given by Sato hyperfunctions; see, e.g., [36].

4.3   Internal Model Principle

We have introduced the notion of pseudorational transfer functions. Let us now outline what is involved in the repetitive control and the internal model principle. For simplicity, consider the unity feedback system depicted in Fig. 4.1. Assume that the reference signal generator $1/\hat{d}(s)$ is totally

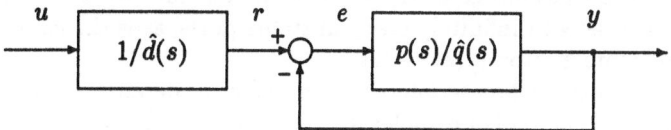

Figure 4.1: Unity Feedback System

unstable. This condition is to be elaborated in the sequel, but for the time being, let us understand that this means that $\hat{d}(s)$ do not have any zeros in the open left half plane.

The transfer function $W_{eu}(s)$ from u to e is given by

$$W_{eu}(s) = \frac{\hat{q}(s)}{\hat{d}(s)(\hat{q}(s) + \hat{p}(s))}. \tag{20}$$

Let us understand stability in the sense that the transfer function belongs to H_-^{∞}, assume the closed-loop stability. Then $\hat{q}(s) + \hat{p}(s)$ should not have any unstable zeros. Furthermore, $W_{eu} \in H_-^{\infty}$ implies that all the unstable zeros of $\hat{d}(s)$ must be canceled by $\hat{q}(s)$. In other words, $\hat{q}(s)/\hat{d}(s)$ must be an entire function. Actually, we can say more. Let d be the inverse Laplace transform of $\hat{d}(s)$, and assume

$$r(d) := \sup\{t \in \operatorname{supp} d\}. \tag{21}$$

This means that $1/\hat{d}(s)$ does not have any redundant delay part (see Section 6.1). We can now state the desired internal model principle in this setting:

Theorem 4.2 ([61]) *Under the hypotheses above, $\hat{q}(s)/\hat{d}(s)$ is an entire function in the Paley-Wiener class (19). In other words, $\hat{d}(s)$ divides $\hat{q}(s)$ in the Paley-Wiener class, so that the denominator $\hat{q}(s)$ of the forward loop should contain $\hat{d}(s)$ as an internal model. Conversely, if $\hat{q}(s)$ contains $\hat{d}(s)$ as an internal model in this sense, and if the closed-loop system is stable, then the system asymptotically tracks any signal generated by $1/\hat{d}(s)$*

Choosing $\hat{d}(s) = e^{Ls} - 1$ now completes the picture for repetitive control, except some questions on stability and state space realizations. That is, the internal model $1/(e^{Ls} - 1)$ is necessary, so that the limitation on stabilizability given in Section 3 is also an inevitable consequence. Note also that condition (21) forces $\hat{d}(s)$ to be in this form $e^{Ls} - 1$ rather than $e^{Ls/2} - e^{-Ls/2}$ as obtained first in (2). The state space interpretation of this condition $\hat{q}(s)/\hat{d}(s)$ will be discussed in the next section.

This type of result has been studied and given in many different settings: [21], [7], [15], [19], etc. However, as announced in the beginning of this section, an advantage of the pseudorational framework is that it can handle the case with infinitely many unstable poles, as is the case with the repetitive compensator $1/(e^{Ls} - 1)$.

For the experts who are interested in technical details, we here give an indication on what kind of complex analytic problems Theorem 4.2 involves, and see what (21) amounts to mean. It shows some interesting interplay between growth estimates and algebraic divisibility conditions. For those who are not, the rest of the section can be skipped without much loss.

We first have to show that $\hat{q}(s)/\hat{d}(s)$ is of exponential type. It is easily seen to be of order at most one. This follows from the Hadamard factorization (13). In fact, if $\hat{q}(s)$ admits the factorization (13) and if $\hat{q}(s)/\hat{d}(s)$ is entire, then division by $\hat{d}(s)$ results only in reducing some factors in (13), so that by the Hadamard factorization theorem, it is again of order at most one. However, showing that it is indeed of exponential type is considerably trickier. For example, in the factorization

$$e^s - 1 = 2\pi s e^{s/2} \left(\prod_{n=1}^{\infty} \left(1 - \frac{s}{2n\pi j}\right) e^{s/2n\pi j} \right) \left(\prod_{n=-1}^{-\infty} \left(1 + \frac{s}{2n\pi j}\right) e^{-s/2n\pi j} \right)$$

the two products on the right are of order one, but not of exponential type, while their product $e^{Ls} - 1$ is. The reason is that $\sum_{n=1}^{\infty} 1/n$ is divergent but $\sum_{n=-m, n\neq 0}^{m} j/n$ is uniformly bounded (e.g., [5]). As mentioned earlier, the growth order is governed by the growth order of zeros, and functions of exponential type in general belong to the marginal case where a quantity like $\sum 1/\lambda_n$ is only conditionally convergent. The following miraculous result is due to Lindelöf:

Theorem 4.3 ([38], [5]) *Let $\psi(s)$ be an entire function of order at most one with zeros $\lambda_1, \lambda_2, \ldots$, excluding the zeros at the origin. Then it is of exponential type if and only if*

1. *the number of zeros $n(r)$ inside the circle $\{|s| \leq r\}$ satisfies $n(r) = O(r)$; and*

2. *$S(r) := \sum_{|\lambda_n| \leq r} 1/\lambda_n$ is bounded in r.*

According to this theorem, it is easily seen that the division of $\hat{q}(s)$ by $\hat{d}(s)$ increases neither $n(r)$ nor $S(r)$. Thus it is of exponential type. It can also be proved that it has only polynomial growth along the imaginary axis so that it is the Laplace transform of a distribution with compact support (Paley-Wiener theorem), but we omit its proof here ([61]). Let ψ be this inverse Laplace transform. It remains to show that ψ has support contained in $(-\infty, 0]$. This is where (21) comes into play. The well known Titchmarsh theorem (local version, [17]) on nonvanishing of convolution shows that if two distributions have support in $(-\infty, 0]$ both not identically zero in a neighborhood of 0, then their convolution is also nonzero around 0. This yields $r(\psi_1 * \psi_2) = r(\psi_1) + r(\psi_2)$, so that

$$r(q) = r(\psi * d) = r(\psi) + r(d) \qquad (22)$$

along with $r(d) = 0$ yields $r(\psi) = r(q) \leq 0$ and hence $\operatorname{supp} \psi \subset (-\infty, 0]$. This condition (21) looks only a technical assumption, but in fact it is not. As we will see in the next section, the canonical realization associated to $1/\hat{d}(s)$ is *eigenfunction complete* if and only if (21) is satisfied (Theorem 5.1). In other words, the signals generated by $1/\hat{d}(s)$ is completely determined by its poles, so that it is a minimal representation that contains all such poles.

5   Further Analysis Related to Realization

We have reached the point where we need to discuss properties related to realizations. However, it is well known [3] that there are infinitely many nonisomorphic realizations for distributed parameter systems. This shows that there is much freedom in associating a state space model to a given transfer function. Actually, even the problem of the opposite direction can present a nontrivial problem. For example, given a system in the state space form, say,

$$\dot{x} = Ax + Bu, \quad y = Cx,$$

it is not always straightforward to associate the notion of transfer functions to this system, and this is related to the problem of admissibility. Much has been investigated recently (e.g., [57]), and one of the nice classes where such a correspondence is well established is the so-called Prichard-Salamon class of systems [51], [47], [13], [14]; under some assumptions of admissible stabilizability/detectability, the transfer functions of a Prichard-Salamon system belongs to the Callier-Desoer algebra [14]. However, we will not go into details here; interested readers are referred to [40], [13] for excellent surveys of the material.

¿From the realization viewpoint, the most standard way of constructing a model is perhaps that of adopting some standard state space (e.g., L^2, H^2 etc.) and employ the shift semigroup in there as a canonical model for state evolution. This methodology is taken by [24] for discrete-time systems. Although strictly in an input/output setting, Georgiou and Smith also makes use of the characterization of shift invariant subspaces [27]. The characterization of right shift invariant subspaces of H^2 space is known as the Beurling-Lax theorem [31]: given a right shift invariant subspace L of H^2, it is of the form $L = BH^2$ for some inner function B. It also has some deep connections with Sarason's theorem, and frequency-domain H^∞ control theory for distributed systems ([20]; see also [64]).

5.1   Fuhrmann Type Realization

The realization that will be introduced here is also based on shift semigroups. Let $g(t) = q^{-1} * p$ be a pseudorational impulse response, and consider the following subspace of $L^2_{loc}[0, \infty)$:

$$X^q := \{x \in L^2_{loc}[0, \infty); \operatorname{supp}(q * x) \subset (-\infty, 0]\}. \qquad (23)$$

This actually represents the space of all input-free outputs by $1/\hat{q}(s)$, and gives a time-domain analog of the Fuhrmann realization for finite-dimensional systems [25]. It is easy to see that this space is closed in $L^2_{loc}[0, \infty)$ and is invariant under the left shift operators:

$$(\sigma_t x)(\tau) := x(\tau + t). \qquad (24)$$

As is well known, these shift operators constitute a strongly continuous semigroup, with the infinitesimal generator $A = d/dt$. Furthermore, X^q is essentially a Hilbert space; its topology is determined by bounded-time data (see Theorem 5.1 below). Then the following functional differential equation is naturally derived as a standard model [58]:

$$\begin{aligned} \frac{dx_t(\tau)}{dt} &= \frac{\partial x_t(\tau)}{d\tau} + g(\tau)u(t) \\ y(t) &= x_t(0), \end{aligned} \qquad (25)$$

$x_t(\cdot) \in X^q$. Let us denote this system by Σ^q. The following facts are known for this model:

Theorem 5.1 ([58, 59])

1. X^q is isomorphic to a Hilbert space; its topology is determined by the L^2 norm $\|x\|_{[0,T]}$ on the interval $[0, T]$ for some $T > 0$.

2. $X^d \subset X^q$ if and only if $q = d * \alpha$ for some distribution α with compact support in $(-\infty, 0]$, and this means $X^q = X^d \oplus X^\alpha$.

3. The spectrum $\sigma(A)$ of the infinitesimal generator A is precisely the zeros of $\hat{q}(s)$. Furthermore, they are all eigenvalues. The eigenvectors are given by $\{e^{\lambda t}; \hat{q}(\lambda) = 0\}$.

4. The space spanned by these eigenvectors is dense in X^q if and only if $r(q) = 0$ (cf. (21)).

Let us return to the internal model principle Theorem 4.2. Condition $\hat{d}(s)|\hat{q}(s)$ in the Paley-Wiener class means that $q = d * \alpha$ for some distribution with compact support in $(-\infty, 0]$, and according to 2. above, this means that all signals generated by X^d can be generated by X^q. The role of (21) in Theorem 4.2 is now also clear from these facts above. It requires that X^q be "spanned" essentially by the eigenvectors; this in turn assures that the corresponding state space is minimally generated by the infinite product representation (13), because each factor in such a representation corresponds to an eigenvalue and vice versa (cf. discussion in Section 4.1).

What if condition (21) is not satisfied? Since supp $d \subset (-\infty, 0]$, this means $d = \delta_{-a} * d_1$ for some $a > 0$ and $r(d_1) = 0$. Then by fact 2. above, $X^d = X^{\delta_{-a}} \oplus X^{d_1}$. While X^{d_1} is eigenfunction complete as stated above, $X^{\delta_{-a}}$ corresponds to the pure delay equation $y(t) = u(t - a)$. Clearly this part is irrelevant to asymptotic tracking, and this is precisely why we required condition (21) in Theorem 4.2.

Let us close this section by giving a short remark on fact 3. above. The differential equation

$$(\lambda I - A)\, x = \left(\lambda - \frac{d}{dt} \right) x(t) = 0$$

always admits a nonzero solution $e^{\lambda t}$ in $L^2_{loc}[0, \infty)$. However, λ is an eigenvalue of A only when this gives an eigenvector in X^q, i.e., when $e^{\lambda t} \in X^q$. Let us take the Laplace transform $\mathcal{L}[q * e^{\lambda t}] = \hat{q}(s)/(s - \lambda)$. This function again belongs to the Paley-Wiener class (so that $\mathrm{supp}(q * e^{\lambda t}) \subset (-\infty, 0]$) if and only if $\hat{q}(\lambda) = 0$. This is how the zeros of $\hat{q}(s)$ give rise to the eigenvalues.

6   Internal and External Stability

So far we have not discussed stability in terms of transfer functions. However, to complete the story for the internal model principle given in Section 4, we need to have some criteria on stability in terms of transfer functions or impulse responses.

How can we deduce stability from conditions on external behavior? What kind of representation is suitable for transfer functions for this purpose?

Let us recall that for infinite-dimensional systems, the location of spectrum does not, in general, determine stability [63]. The counterexample is given as an infinite direct sum of finite-dimensional subsystems, so that

attempting to deduce stability from the local information based on each
pole and eigenfunction expansion seems to be difficult to work out.

Let us illustrate another difficulty with the example by Logemann [39].

$$G(s) = \frac{1}{s+1} \cdot \frac{e^{hs}}{s(e^{hs} - 1) + ae^{hs}}, \ a > 0. \tag{26}$$

This is the transfer function of a neutral delay-differential system, and as
is the case with repetitive control, it has a chain of poles approaching the
imaginary axis. Hence the standard realization is not exponentially stable
by the stability criterion for neutral systems [28]. Nevertheless, $G(s) \in H^\infty$.
The reason is that the rate that these poles asymptote the imaginary axis is
of polynomial order, so that it is possible to cancel this effect by multiplying
a suitable "convergence factor" (in this case $1/(s + 1)$). This will improve
the behavior on the imaginary axis ("roll off at infinity"), but definitely
not the stability property. Of course, we should require that spectrum be
strictly separated from the imaginary axis, but as this example shows, it
cannot be checked from the condition $G(s) \in H^\infty$.

All these arise from the fact that (exponential) stability is a topological property. In order to conclude stability of a realization, we should
impose some condition on the topology of the state space. This problem of
establishing connection with internal stability and external stability (e.g.,
$G \in H^\infty$) attracted recent research interest, and a great deal becomes
known. One direction is to assume stabilizability and detectability (which
are obvious necessary conditions for stability) and proving stability based
on conditions on spectrum or transfer functions. The answer is generally affirmative, and much has been done in this direction: [35], [8], [12],
[48], [14], etc. Generally speaking, once we assume suitable notions of sta-
bilizability/detectability in some class of systems (e.g., Prichard-Salamon
class), then $G \in H^\infty$ implies exponential stability. (The example (26) is
not stabilizable.)

However, again due to infinite-dimensionality, stabilizability/detectabil-
ity can be a problem. For example, it is known that approximate reachability does not necessarily imply stabilizability (e.g., [55]). Because of such
a gap, one may not sometimes want to assume a priori stabilizability. On
the other hand, we can still prove internal stability based on an external
condition by placing conditions on the type of realizations. We give one
such case for the pseudorational class in what follows.

6.1 Stability Analysis Based on Infinite Products

Let us first state the following theorem:

Theorem 6.1 *Let $G(s) = \hat{p}(s)/\hat{q}(s)$ be a pseudorational transfer function. The realization Σ^q given by (25) is exponentially stable if* $\sup\{\mathrm{Re}\,\lambda;\ \hat{q}(\lambda) = 0\} < 0.$

Roughly speaking, exponential stability is related to the estimate of the fundamental solution, and this is in turn related to the question of growth order estimate of its Laplace transform $1/\hat{q}(s)$ on a line $\{\mathrm{Re}\,s = -\alpha\}$ for some $\alpha > 0$ (actual reasoning is much more involved). So the question here is basically whether we can establish a boundedness estimate on $\{\mathrm{Re}\,s = -\alpha\}$, given the condition $\sup\{\mathrm{Re}\,\lambda;\ \hat{q}(\lambda) = 0\} < 0$. By shifting coordinates, we can move this line $\{\mathrm{Re}\,s = -\alpha\}$ to the imaginary axis.

Showing this is nontrivial, but let us illustrate the method. As noted earlier, the growth rate of zeros of $\hat{q}(s)$ gives rise to a very precise estimate of $\hat{q}(s)$ or $1/\hat{q}(s)$, and theorems like 4.3 are the key to the issue here. However, the readers who are not interested in technical details can skip the rest of this subsection.

Invoke the Hadamard factorization

$$\hat{q}(s) = s^m e^{as+b} \prod_{n=1}^{\infty} \left(1 - \frac{s}{\lambda_n}\right) \exp\left(\frac{s}{\lambda_n}\right) \tag{27}$$

for the generic case of order one, and suppose $\mathrm{Re}\,\lambda_n < -c < 0$. Then by Lindelöf's Theorem 4.3, $S(r) := \sum_{|\lambda_n| \leq r} 1/\lambda_n$ is a bounded function in r. Since

$$\mathrm{Re}\left(\sum \frac{1}{\lambda_n}\right) = \sum \mathrm{Re}\,\frac{1}{\lambda_n} = \sum \frac{\mathrm{Re}\,\lambda_n}{|\lambda_n|^2},$$

and these terms are all negative, we have the absolute convergence

$$\sum_{n=1}^{\infty} \frac{|\mathrm{Re}\,\lambda_n|}{|\lambda_n|^2} < \infty. \tag{28}$$

Now since G is pseudorational, q^{-1} is a distribution, so a standard theory [52] implies that there exists $\sigma > 0$ such that $G(s) \in H^{\infty}(\mathbf{C}_\sigma)$. The question is then to move this estimate to the imaginary axis. We must give a uniform estimate for

$$\left|\frac{1/\hat{q}(j\omega)}{1/\hat{q}(\sigma + j\omega)}\right|.$$

Write $s = \sigma + j\omega$, and bring in the factorization (27) to get

$$\frac{1/\hat{q}(s-\sigma)}{1/\hat{q}(s)} = e^{\alpha\sigma} \prod_{n=1}^{\infty} \frac{(1 - \frac{s}{\lambda_n})}{(1 - \frac{s-\sigma}{\lambda_n})} \exp\left(\frac{\sigma}{\lambda_n}\right).$$

Denote this product by $F(s)$. We need to give a uniform estimate on the line $\{\text{Re } s = \sigma\}$. We have

$$
\begin{aligned}
|F(s)| &= \exp(\text{Re Log } F(s)) \\
&= \exp\left(\sum_{n=1}^{\infty} \text{Re } \frac{\sigma}{\lambda_n} - \sum_{n=1}^{\infty} \text{Re}\{\text{Log}(1 - \frac{s-\sigma}{\lambda_n}) - \text{Log}(1 - \frac{s}{\lambda_n})\}\right) \\
&= \exp\left(\sum_{n=1}^{\infty} \frac{\sigma \text{ Re } \lambda_n}{|\lambda_n|^2} - \sum_{n=1}^{\infty} \text{Re}\{\text{Log}(1 + \frac{\sigma}{\lambda_n - s})\}\right).
\end{aligned}
$$

where we take the principal branch for Log. By (28), the first sum on the right-hand side is clearly finite. A more involved but similar argument ([60]) along with the expansion

$$
\text{Log}(1 + \frac{\sigma}{\lambda_n - s}) = \sum_{k=1}^{\infty}(-1)^{k-1}\frac{1}{k}(\frac{\sigma}{\lambda_n - s})^k
$$

shows that the second term is also uniformly bounded, and hence the desired condition $1/\hat{q}(s) \in H^{\infty}$ (or $1/\hat{q}(s) \in H^{\infty}_{-}$) follows.

6.2 Stability and Topology

It is well known that for a given (irrational) transfer function there exist infinitely many nonisomorphic, but approximately reachable and observable, realizations (e.g., [3]). Although we have repeatedly emphasized that stability is a topological property, it is still desirable if stability is preserved among all such realizations. In fact, the results obtained by assuming stabilizability/detectability partially answer the question (cf. discussion in the beginning of this section). Then what if we do not assume a priori stabilizability/detectability? Fuhrmann [24, 26] gave an example of a discrete time system in which even the notion of spectrum is not preserved among all such realizations. Since the "A" operator is bounded there, stability is determined by the spectral radius, and it is not preserved either. Recently an easy continuous-time example was given by Zwart et al. [65]. The idea here is to find an impulse response, and take a state space to be the subspace in $L^2[0, \infty)$ with the standard norm, generated by the outputs by this impulse response. One can impose a different norm on this space with different weighting

$$
\|x\|^2 := \int_0^{\infty} |x(t)|e^{-2t} dt \tag{29}
$$

with its completion. While in $L^2[0, \infty)$ the shift semigroup σ_t (24) satisfies $\|\sigma_t\| = 1$, with this weighting $\|\sigma_t\| = e^t$. So they exhibit completely different stability properties. Furthermore, with suitable choice of an impulse response, both systems can be made approximately reachable and observable. Hence preservation of exponential stability among all approximately reachable and observable realizations seems difficult.

One reason this does not occur for finite-dimensional systems is that there topology is determined on bounded time data. But this is also true for the pseudorational class as Theorem 5.1 shows. However, no affirmative result of sufficient generality seems known at present.

7 Some Related Questions

We have reviewed some fundamental issues in repetitive control and also seen how they motivate and yield various problems in infinite-dimensional systems. We have concentrated on some issues arising in the interplay between time and frequency domain concepts: Paley-Wiener theorem, Lindelöf theorem, Hadamard factorization have played particularly important roles.

However, various related problems were left not discussed. Let us conclude by making some remarks on them and on possible future directions.

7.1 More on Fractional Representations

We have seen that the pseudorational class of transfer functions offers a nice framework for a certain class of systems. This class includes, for example, neutral or retarded delay systems, some wave equations, systems with distributed delays, etc. However, as we have seen, both the numerator and denominator have to satisfy the Paley-Wiener estimate (Theorem 4.1), and this places restriction on the rate of growth along the imaginary axis (at most of polynomial order). It would be of great interest to see if a more general factorization, for example, ratios of H^∞ functions (e.g., [34]) can lead to a tractable time domain treatment as discussed here. However, as pointed out in Section 4, in order to have some relationship with Laplace transforms some restriction must be imposed on the growth order of transfer functions. Perhaps, the precise detail is yet to be seen in future developments.

7.2 Robustness and Related Issues

We have not discussed robustness issues for repetitive control systems. In fact, if we further assume some sort of regularity assumption on the transfer function, then the usual small gain condition (e.g., [10]) implies internal exponential stability also [62]. This means that for a restricted class of systems, H^∞ transfer functions yield not only L^2 input/output stability [16] but internal stability also (see, e.g., [37] for related issues).

This is applicable to modified repetitive control systems. Consider the modified repetitive control system with plant perturbation Δ given in Fig. 7.1. Assume for simplicity that $P(s)$ and $C(s)$ are stable. As established

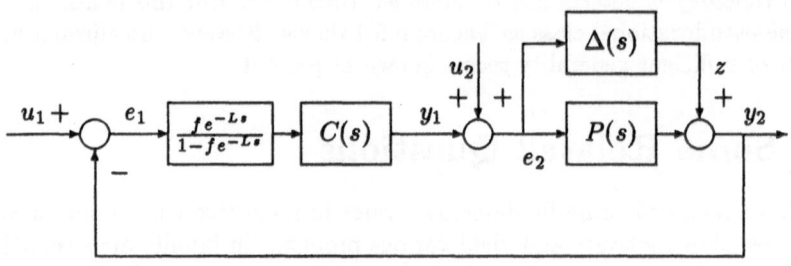

Figure 7.1: Modified Repetitive Control System

in Section 3,

$$\|f(s)(I - PC)\|_\infty := \gamma < 1,$$

assures stability. Suppose that the perturbation $\Delta(s)$ satisfies

$$|\Delta(j\omega)| < |r(j\omega)| \quad \text{for all } \omega.$$

for some H^∞ function $r(s)$. Then the modified repetitive control system above remains stable if

$$\|r(s)f(s)C(s)\|_\infty < 1 - \gamma. \tag{30}$$

The detail can be found in [62].

7.3 Digital Repetitive Control and Ripples

In implementing repetitive control into a real plant, it is quite natural to resort to digital control (e.g., [11], [53]). Then with digital memory devices the repetitive compensator becomes $1/(z^N - 1)$, which is no longer infinite-dimensional, but merely a finite-dimensional discrete-time system. Ironically, there would be no strong restriction on the relative degree of the plant for stabilizability, as encountered in the continuous-time systems. If we need digital repetitive controllers only, and if no problems arise, what is the use of all the fabuluous theory discussed so far?

   Actually, the tracking property that can be assured by a digital repetitive control concerns its behavior only at sampled instants, and its intersample behavior is generally not guaranteed. If its continuous-time counterpart does not satisfy the stabilizability condition as a delay system, then there can be some problems in the digital repetitive control. In general, we can imagine that precise tracking at sampled instants may be accomplished at the expense of intersample behavior, since it is in general not taken into account in the digital design. This is in fact shown in [30] by an example that shows a very large ripple in a certain tracking problem. Characterizing precisely what kind of systems present such behavior is a theme for future study.

Acknowledgments

A large part of the research related to repetitive control is a joint work with Professor S. Hara; his collaboration is gratefully acknowledged. The author would also like to thank Professor M. Nakano and the Society for Instrument and Control Engineers for allowing the use of Fig. 2.1.

He also wishes to thank Professors R. F. Curtain, T. T. Georgiou, A. Tannenbaum, H. Zwart for their helpful discussions and comments and to Professor M. Ikeda for helpful discussions and valuable comments on the final draft.

References

[1] S. Arimoto, S. Kawamura and F. Miyazaki, "Bettering operation of robots by learning," *J. Robotic System*, **1**: 123-149, 1984.

[2] D. Balala, "Učiaci sa servosystém priemyselného robota (Learning servosystem of industrial robots)," *Zborník Vedeckých Prác*: 451-457 1985.

[3] J. S. Baras, R. W. Brockett and P. A. Fuhrmann, "State-space models for infinite-dimensional systems," *IEEE Trans. Autom. Contr.*, **AC-19**: 693-700, 1974.

[4] G. Bengtsson, "Output regulation and internal models—a frequency domain approach," *Automatica*, **13**: 333-345, 1977.

[5] R. P. Boas Jr., *Entire Functions*, Academic Press, 1954.

[6] F. M. Callier and C. A. Desoer, "An algebra of transfer functions for distributed linear time invariant systems," *IEEE Trans. Circuit and Systems*, **CAS-25**: 651-662, 1978 (Correction in vol. 26, p. 360).

[7] F. M. Callier and C. A. Desoer, "Stabilization, tracking and disturbance rejection in multivariable convolution systems," *Annales de la Societé Scientifique de Bruxelles*, **94**: 7-51, 1980.

[8] F. M. Callier and C. A. Desoer, "Distributed system transfer functions of exponential order," *Int. J. Control* **43**: 1353-1373, 1986.

[9] T. Carleman, *L'Integrale de Fourier et Questions qui s'y Rattachent*, Institut Mittag-Leffler, Uppsala, 1944.

[10] M. J. Chen and C. A. Desoer, "Necessary and sufficient conditions for robust stability of linear distributed feedback systems," *Int. J. Control*, **35**: 255-267, 1982.

[11] K.-K. Chew and M. Tomizuka, "Digital control of repetitive errors in disk drive systems," *IEEE Control Systems Magazine*, **10-1**: 16-20, 1990.

[12] R. F. Curtain, "Equivalence of input-output stability and exponential stability for infinite dimensional systems," *Math. Systems Theory*, **21**: 19-48, 1988.

[13] R. F. Curtain, "A synthesis of time and frequency domain methods for the control of infinite-dimensional systems: a system theoretic approach," *SIAM Frontiers in Applied Mathematics*, 1989.

[14] R. F. Curtain, H. Logemann, S. Townley and H. Zwart, "Well-posedness, stabilizability and admissibility for Prichard-Salamon systems," *Report 260, Institut für Dynamische Systeme, Universität Bremen*, 1992.

[15] C. A. Desoer and C. L. Gustafson, "Design of multivariable feedback systems with simple unstable plant," *IEEE Trans. Autom. Control*, **AC-29**: 901-908, 1984.

[16] C. A. Desoer and M. Vidyasagar, *Feedback Systems: Input-Output Properties*, Academic Press, 1975.

[17] W. F. Donoghue, *Distributions and Fourier Transforms*, Academic Press, 1969.

[18] J. C. Doyle, K. Glover, P. P. Khargonekar and B. A. Francis, "State-space solutions to standard $\mathcal{H}_\infty$ and $\mathcal{H}_2$ control problems," *IEEE Trans. Autom. Control*, **AC-34**: 831-847, 1989.

[19] P. M. G. Ferreira and F. M. Callier, "The tracking and disturbance rejection problem for distributed parameter systems," *IEEE Trans. Autom. Control*, **AC-27**: 478-481, 1982.

[20] C. Foias, A. Tannenbaum and G. Zames, "Some explicit formulae for the singular values of certain Hankel operators with factorizable symbol," *SIAM J. Math. Anal.*, **19**: 1081-1089, 1988.

[21] B. A. Francis, "The multivariable servomechanism problem from the input-output viewpoint," *IEEE Trans. Autom. Control*, **AC-22**: 322-328, 1977.

[22] B. A. Francis, *A Course in H_∞ Control Theory*, Lecture Notes in Control and Information Sci., **88**, Springer, 1987.

[23] B. A. Francis and W. M. Wonham, "The internal model principle for linear multivariable regulators," *J. Appl. Math. and Optimiz.*, **2**: 170-194, 1975.

[24] P. A. Fuhrmann, "On realization of linear systems and applications to some questions of stability" *Math. Systems Theory*, **9**: 132-141, 1974.

[25] P. A. Fuhrmann, "Algebraic system theory: an analyst's point of view," *J. Franklin Inst.*, **301**: 521-540 1976.

[26] P. A. Fuhrmannn, *Linear Systems and Operators in Hilbert Space*, McGraw-Hill, 1981.

[27] T. T. Georgiou and M. C. Smith, "Graphs, causality and stabilizability: linear, shift-invariant systems on $\mathcal{L}_2[0, \infty)$," *Proc. 31st CDC*: 1024-1029, 1992.

[28] J. K. Hale, *Theory of Functional Differential Equations*, Springer, 1977.

[29] S. Hara, Y. Yamamoto, T. Omata and M. Nakano, "Repetitive control system: a new type servo system for periodic exogenous signals," *IEEE Trans. Autom. Control*, **AC-33**: 659-668, 1988.

[30] S. Hara, M. Tetsuka and R. Kondo, "Ripple attenuation in digital repetitive control systems," *Proc. 29th CDC*: 1679-1684, 1990.

[31] K. Hoffman, *Banach Spaces of Analytic Functions*, Prentice Hall, 1962.

[32] M. Ikeda and M. Takano, "Repetitive control for systems with nonzero relative degree," *Proc. 29th CDC*: 1667-1672, 1990.

[33] T. Inoue, M. Nakano, T. Kubo, S. Matsumoto and H. Baba, "High accuracy control of a proton synchrotron magnet power supply," *Proc. IFAC 8th World Congress*, **XX**: 216-221, 1981.

[34] Y. Inouye, "Parametrization of compensators for linear systems with transfer functions of bounded type," *Proc. 27th CDC*: 2083-2088, 1988.

[35] C. A. Jacobson and C. N. Nett, "Linear state-space systems in infinite-dimensional space: the role and characterization of joint stabilizability/detectability," *IEEE Trans. Autom. Control*, **33**: 541-549, 1988.

[36] A. Kaneko, *Introduction to Hyperfunctions*, Reidel, 1988.

[37] P. P. Khargonekar and K. Poola, "Robust stabilization of distributed systems," *Automatica*, **22**: 77-84, 1986.

[38] F. Lindelöf, "Sur les fonctions entieres d'ordre entier," *Ann. Sci. Ecole Norm. Sup.*, **(3) 22**: 369-395, 1905.

[39] H. Logemann, "On the transfer matrix of a neutral system: Characterizations of exponential stability in input-output terms," *Syst. Contr. Letters*, **9**: 393-400, 1987.

[40] H. Logemann, "Stabilization and regulation of infinite-dimensional systems using coprime factorizations," *Report 266, Institut für Dynamische Systeme, Universität Bremen*, to appear in *Proc. 10th Int. Conf. on Analysis and Optimization of Systems: State and Frequency Domain Approach for Infinite-Dimensional Systems*, 1992.

[41] T. Mita and E. Kato, "Iterative control and its application to motion control of robot arm—a direct approach to servo problems," *Proc. 24th CDC*: 1393-1398, 1985.

[42] N. Nakano and S. Hara, "Microprocessor-based repetitive control," *Microprocessor-Based Control Systems*, Reidel, 1986.

[43] N. Nakano, T. Inoue, Y. Yamamoto and S. Hara, *Repetitive Control*, SICE Publications, 1989 (in Japanese).

[44] D. A. O'Connor and T. J. Tarn, "On stabilization by state feedback for neutral differential difference equations," *IEEE Trans. Autom. Control*, **28**: 615-618, 1983.

[45] T. Omata, S. Hara and M. Nakano, "Nonlinear repetitive control with application to trajectory control of manipulators," *J. Robotic System*, **4**: 631-652, 1987.

[46] R. E. A. C. Paley and N. Wiener, *Fourier Transforms in the Complex Domain*, Amer. Math. Soc. Colloquium Publ., **19**, 1934.

[47] A. J. Prichard and D. Salamon, "The linear quadratic control problem for infinite-dimensional systems with unbounded input and output operators," *SIAM J. Control & Optimiz.*, **25**: 121-144, 1987.

[48] R. Rebarber, "Conditions for the equivalence of internal and external stability for distributed parameter systems," *Proc. 30th CDC*: 3008-3013, 1991.

[49] W. Rudin, *Real and Complex Analysis*, McGraw-Hill, 1966.

[50] D. E. Rumelhart, J. L. McClelland and PDP Research Group, "Learning internal representations by error propagation," *Parallel Distributed Processing Vol.1*: 318-362, The MIT Press, 1986.

[51] D. Salamon, *Control and Observation of Neutral Systems*, Pitman, 1984.

[52] L. Schwartz, *Théorie des Distributions*, 2me Edition, Hermann, 1966.

[53] M. Tomizuka, T.-S. Tsao and K.-K. Chew, "Analysis and synthesis of discrete-time repetitive controllers," *Trans. ASME, J. Dynamic Syst. Measurement and Control*, **11**: 353-358, 1989.

[54] F. Treves, *Topological Vector Spaces, Distributions and Kernels*, Academic Press, 1967.

[55] R. Triggiani, "On the stabilizability problem in Banach space," *J. Math. Anal. Appl.*, **52**: 383-403, 1975.

[56] Ya Z. Tsypkin, "Self-learning—What is it?" *IEEE Trans. Autom. Control*, **AC-13**: 608-612, 1968.

[57] G. Weiss, "Admissibility of unbounded control operators," *SIAM J. Control & Optimiz.*, **27**: 527-545, 1989.

[58] Y. Yamamoto, "Pseudo-rational input/output maps and their realizations: a fractional representation approach to infinite-dimensional systems," *SIAM J. Control & Optimiz.*, **26**: 1415-1430, 1988.

[59] Y. Yamamoto, "Reachability of a class of infinite-dimensional linear systems: an external approach with applications to general neutral systems," *SIAM J. Control & Optimiz.*, **27**: 217-234, 1989.

[60] Y. Yamamoto, "Equivalence of internal and external stability for a class of distributed systems," *Math. Control, Signals and Systems*, **4**: 391-409, 1991.

[61] Y. Yamamoto and S. Hara, "Relationships between internal and external stability with applications to a servo problem," *IEEE Trans. Autom. Control*, **AC-33**: 1044-1052, 1988.

[62] Y. Yamamoto and S. Hara, "Internal and external stability and robust stability condition for a class of infinite-dimensional systems," *Automatica*, **28**: 81-93, 1992.

[63] J. Zabczyk, "A note on C_0-semigroups," *Bull. l'Acad. Pol. de Sc. Serie Math.*, **23**: 895-898, 1975.

[64] K. Zhou and P. P. Khargonekar, "On the weighted sensitivity minimization problem for delay systems," *Syst. Control Lett.*, **8**: 307-312, 1987.

[65] H. Zwart, Y. Yamamoto and Y. Gotoh, "On the Stability Uniformity of Infinite-Dimensional Systems," to appear in *Proc. 10th Int. Conf. on Analysis and Optimization of Systems: State and Frequency Domain Approach for Infinite-Dimensional Systems*, 1992.

8. An Algebraic Approach to Linear and Nonlinear Control

M. Fliess* S.T. Glad †

1 Introduction

The analysis and design of control systems has been greatly influenced by the mathematical tools being used. Maxwell introduced linear differential equations in the 1860's. Nyquist, Bode and others started the systematic use of tranfer functions, utilizing complex analysis in the 1930's. Kalman brought forward state space analysis around 1960. For nonlinear systems, differential geometric concepts have been of great value recently. We will argue here that algebraic methods can be very useful for both linear and nonlinear systems. To give some motivation we will begin by looking at a few examples.

1.1 Some examples of dynamical systems

1.1.1 Example Consider a unit mass in a gravitaional field. Let the height coordinate be y and the vertical velocity v. Assume that the gravitational acceleration is unity. Then the system is described by the equations

$$\frac{1}{2}v^2 + y = 0, \quad \dot{y} - v = 0 \tag{1}$$

(where the dot denotes time derivative). The first equation says that the sum of kinetic and potential energy is constant (for simplicity zero) and the second one is the definition of velocity. By differentiating the first equation and substituting from the second one we obtain

$$v(\dot{v} + 1) = 0 \tag{2}$$

*Laboratoire des Signaux et Systemes, C.N.R.S.-E.S.E., Plateau du Moulon, 91190 Gif-sur-Yvette, France. Email: fliess@frese51.bitnet. Work partially supported by the G.R. "Automatique" of the French "Centre National de la Recherche Scientifique".

†Department of Electrical Engineering, Linköping University, S-58183 Linköping, Sweden. Email: torkel@isy.liu.se. Work partially supported by the Swedish Research Council for Engineering Sciences and by the G.R. "Automatique" of the French "Centre National de la Recherche Scientifique".

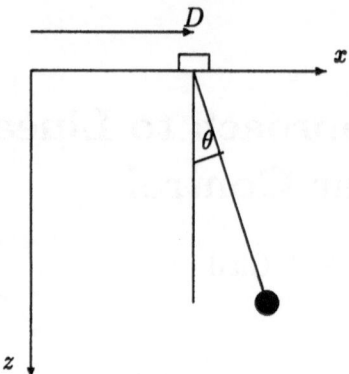

Figure 1.1: A two dimensional crane

We see that the model really describes two different physical situations: either $v = 0$, i. e. a particle with a fixed position, or else $\dot{v} = -1$, i. e. a particle falling with a constant acceleration.

1.1.2 Example Consider the following system in standard state space form.

$$\begin{aligned}
\dot{x}_1 &= x_2^2 \\
\dot{x}_2 &= u \\
y &= x_1
\end{aligned} \tag{3}$$

By differentiations and some simple algebra we can get the following three equations

$$\ddot{y}^2 - 4u^2\dot{y} = 0, \quad 2ux_2 - \ddot{y} = 0, \quad x_1 - y = 0 \tag{4}$$

The first equation gives an input-output relation while the second and third ones show that the system is observable. In fact we have explicit equations from which to calculate x_1 and x_2, given the input, output and their derivatives.

1.1.3 Example Consider a crane consisting of a trolley moving on a horizontal support, with the load suspended in a rope, figure 1.1. The equations describing the system are

$$\begin{aligned}
m\ddot{x} &= -T\sin\theta \\
m\ddot{z} &= -T\cos\theta + mg \\
x &= R\sin\theta + D \\
z &= R\cos\theta
\end{aligned} \tag{5}$$

where x and z are the coordinates of the load and D is the horisontal position of the trolley. R is the length of the rope, T the tension of the rope and m the mass of the load.

Eliminating T and θ we get a purely algebraic relation between the variables

$$\begin{array}{rcl} (\ddot{z} - g)(x - D) &=& \ddot{x}z \\ (x - D)^2 + z^2 &=& R^2 \end{array} \tag{6}$$

Differentiating the geometric constraints gives

$$R\ddot{\theta} + 2\dot{R}\dot{\theta} + \ddot{D}\cos\theta + g\sin\theta = 0$$

With $x_1 = \theta$, $x_2 = \dot{\theta}$ we get the description

$$\begin{array}{rcl} \dot{x}_1 &=& x_2 \\ \dot{x}_2 &=& -2\frac{\dot{R}}{R}x_2 - \frac{\ddot{D}}{R}\cos x_1 - \frac{g}{R}\sin x_1 \end{array} \tag{7}$$

A more detailed discussion of this example can be found in [17].

1.2 Algebraic methods

The simple examples above show that there are several interesting aspects of dynamic systems, and in particular control systems, that can be investigated using algebraic manipulations, together with differentiations. We also see that some of the descriptions are different from the usual state space ones. We will give a survey of an algebraic approach dealing with these matters..

Although algebraic methods have been present since a long time in control, mainly in linear system theory under the influence of Kalman (see [35]), algebra does not constitute by any means a mainstay of our discipline. What is algebra? Shafarevich in his marvellous introduction [48] to the subject has asserted, following Weyl [53] that it is related to the process of "coordinatization", which means much more than counting and measuring. Anyhow it offers a wealth of concepts and techniques which not only help solving classical problems but also permit to recast many aspects of control, which were taken for granted.

Our two basic ingredients are modules in the linear case and differential fields in the general nonlinear situation. Modules serve in a more general manner than in Kalman [36, 37], as they encompass all the variables and permit to treat on an equal footing the time-varying case.

Differential fields on the other hand are much less classical. They stem from differential algebra which was created between the two World Wars by the American mathematician J. F. Ritt, [46] for extending methods and results from algebra which was then blooming under the influence of E. Noether and her pupils (see van der Waerden's treatise in two volumes, [52]).

In the same spirit as Willems, [54]–[58], with the geometric notion of trajectories, we define systems irrespective of any denomination of the variables. When a dynamics is given, i. e. when a set of input channels is selected, we arrive at a state variable representation for which the following conclusions hold:

- Linear dynamics can always be given a classical Kalman realization.

- In the nonlinear case two phenomena *sui generis* might occur.

  - Equations might be implicit with respect to the derivatives of the state and/or with respect to the output.

  - Derivatives up to finite order of the input might be present in the equations.

Are these facts, which are a direct consequence of our formalism, *artefacts* or confirmed by some realistic physical models? The implicit character is indeed verified in several examples, e. g. 1.1.1, 1.1.2 and 1.1.3 above. In examples 1.1.2 and 1.1.3 there is a choice between ordinary state space descriptions and non-classical, implicit ones. In example 1.1.2 we saw that the nonclassical one gave important information about observability.

The presence of input derivatives is more subtle since it can be asserted that the derivatives of highest order should be chosen as the true control. Notice however that the highest order is by no means unique. The crane example (1.1.3) shows even more (see section 5.3) and [17]): The algebro-differential equations, which are obtained from the basic principles of mechanics, do not contain derivatives of the input; those derivatives appear when writing the state variable representation. Once the presence of input derivatives is admitted, state transformations become dependent on the input and a finite number of its derivatives. Such transformations which were until now rejected by most theoreticians permit one to write down new canonical forms, [12] which might be useful in various situations.

2   Linear systems

2.1   A short review of module theory

2.1.1 Elementary linear algebra deals with vector spaces over fields. Module theory is more general as it covers linear structures over rings. It enjoys a rich body of results which are useful in many chapters of mathematics. As we only treat finite-dimensional linear systems, which may be time varying, we only need modules over principal ideal rings.

2.1.2 A *differential field*, [39], k is a commutative field which, here, is equipped with a single derivation $\frac{d}{dt} = $ " $\cdot$ " such that

$$\forall a \in k, \quad \frac{da}{dt} = \dot{a} \in k \tag{8}$$

$$\forall a, b \in k, \quad \frac{d}{dt}(a+b) = \dot{a} + \dot{b} \tag{9}$$

$$\frac{d}{dt}(ab) = \dot{a}b + a\dot{b} \tag{10}$$

2.1.3 A *constant* is an element $c \in k$ such that $\dot{c} = 0$. A *field of constants* is a differential field which only contains constants.

2.1.4 Examples.

1. The fields $\mathbb{Q}, \mathbb{R}, \mathbb{C}$ of rational, real and complex numbers, as well as any finite field are trivial examples of fields of constants.

2. The field $R(t)$ of real rational functions in the indeterminate t is a differential field with respect to $\frac{d}{dt}$.

3. A field of meromorphic funtions in the single variable t over an open connected domain of $\mathbb{R}$ or $\mathbb{C}$ is a differential field with respect to $\frac{d}{dt}$.

2.1.5 Denote by $k\left[\frac{d}{dt}\right]$ the set of linear differential operators of the form

$$\sum_{\text{finite}} a_\alpha \frac{d^\alpha}{dt^\alpha}, \quad a_\alpha \in k$$

It can be endowed in an obvious way with a ring structure. This ring in general is noncommutative as demonstrated by the following example:

$$a\frac{d}{dt}\left(\frac{d}{dt}\right) = a\frac{d^2}{dt^2}, \quad \frac{d}{dt}\left(a\frac{d}{dt}\right) = \dot{a}\frac{d}{dt} + a\frac{d^2}{dt^2}$$

Therefore, $\frac{d}{dt}\left(a\frac{d}{dt}\right)$ and $a\frac{d}{dt}\left(\frac{d}{dt}\right)$ are equal if and only if a is a constant: $k\left[\frac{d}{dt}\right]$ is commutative if and only if k is a field of constants.

2.1.6 $k\left[\frac{d}{dt}\right]$, even in the general case, is a *principal ideal ring* and satisfies the left and right *Ore properties* which permit to define the quotient skew field $k\left(\frac{d}{dt}\right)$. see e. g. [2].

2.1.7 Let M be a left $k\left[\frac{d}{dt}\right]$-module. An element $m \in M$ is said to be *torsion* if, and only if, there exists $\pi \in k\left[\frac{d}{dt}\right]$, $\pi \neq 0$, such that $\pi m = 0$. In other words m satisfies a linear differential equation with coefficients in k. A module such that all its elements are torsion is said to be *torsion*.

2.1.8 We now have the following
Proposition Let M be finitely generated left $k\left[\frac{d}{dt}\right]$-module. The following two conditions are equivalent.

1. M is torsion.

2. The dimension of M as a k-vector space is finite.

2.1.9 The set of all torsion elements of a module M is a submodule T, called the *torsion submodule* of M

2.1.10 The next definition is in fact a characterization which holds for finitely generated left $k\left[\frac{d}{dt}\right]$-modules: Such a module M is said to be *free* if and only if its torsion module is trivial, i. e. is equal to $\{0\}$. Then, there exists a finite set $b = (b_1, \ldots, b_\mu)$ called a *basis*, such that any $m \in M$ can be written

$$m = \gamma_1 m_1 + \cdots + \gamma_\mu m_\mu, \quad \gamma_1, \ldots, \gamma_\mu \in k\left[\frac{d}{dt}\right]$$

in a unique way. Two bases posess the same number of elements, which is called the *rank* of M.

2.1.11 Theorem Any finitely generated left $k\left[\frac{d}{dt}\right]$-module M can be decomposed into a direct sum

$$M = T \oplus \Phi$$

where T is the torsion submodule and Φ is a free module.
    Notice that $\Phi = M/T$ is unique up to isomorphism.

2.2   What is a linear system?

2.2.1 A *linear system* is a finitely generated left $k\left[\frac{d}{dt}\right]$-module Λ.

2.2.2 Example Take, as in Willems [54, 55, 56], a system Σ corresponding to a finite set $w = (w_1, \ldots, w_q)$ of quantities which are related by a finite set of homogeneous linear differential equations over k. Let

$$E_\alpha\left(w_i^{(\nu_j)}\right) = \sum_{\text{finite}} a_{\alpha, i, j} w_i^{(\nu_j)} = 0, \quad (a_{\alpha, i, j} \in k) \tag{11}$$

Introduce the free left $k\left[\frac{d}{dt}\right]$-module $\mathcal{F}$ spanned by $\bar{w} = (\bar{w}_1, \ldots, \bar{w}_q)$. Let $\Xi \subset \mathcal{F}$ be the submodule spanned by

$$e_\alpha = E_\alpha\left(\bar{w}_i^{(\nu_j)}\right) = \sum_{\text{finite}} a_{\alpha, i, j} \bar{w}_i^{(\nu_j)}$$

The quotient module $\Lambda = \mathcal{F}/\Xi$ is the module corresponding to Σ. Note that the components of the residue $w = (w_1, \ldots, w_q)$ of $\bar{w}$, i. e., of the image of $\bar{w}$ in Λ, satisfy (11).

2.2.3 A *linear dynamics* $\mathcal{D}$ is a linear system $\mathcal{D}$ where we distinguish a finite set $u = (u_1, \ldots, u_m)$ such that the quotient module $\mathcal{D}/[u]$ is torsion ($[u]$ denotes the module spanned by the components of u, which play the role of *input variables*), see [13]. The input is said to be *independent* if and only if $[u]$ is a free module. An *output* $y = (y_1, \ldots, y_p)$ is a finite set of elements in $\mathcal{D}$.

2.2.4 Example Consider the single-input, single-output system

$$a\left(\frac{d}{dt}\right) y = b\left(\frac{d}{dt}\right) u, \quad a, b \in k\left[\frac{d}{dt}\right], \quad a \neq 0$$

Like in 2.2.2 take the free left $k\left[\frac{d}{dt}\right]$-module $\mathcal{F} = (\bar{u}, \bar{y})$ spanned by $\bar{u}, \bar{y}$ and denote by $\Xi \subset \mathcal{F}$ the submodule spanned by $a\left(\frac{d}{dt}\right)\bar{y} - b\left(\frac{d}{dt}\right)\bar{u}$. The quotient module $\mathcal{D} = \mathcal{F}/\Xi$ is the system module. Let u, y be the residues of $\bar{u}, \bar{y}$ in $\mathcal{D}$ and $\underline{y}$ the residue of y in $\mathcal{D}/[u]$. Then $\underline{y}$ which satisfies $a\left(\frac{d}{dt}\right)\underline{y} = 0$, is torsion.

2.3 State variable representation

2.3.1 Take a dynamics $\mathcal{D}$ with input $u = (u_1, \ldots, u_m)$ and a possible output $y = (y_1, \ldots, y_p)$. According to 2.1.8, the finitely generated torsion module $\mathcal{D}/[u]$ is finite-dimensional as a k-vector space. Take a basis $\xi = (\xi_1, \ldots, \xi_n)$ of it. The derivatives $\dot{\xi} = (\dot{\xi}_1, \ldots, \dot{\xi}_n)$ depend k-linearly on ξ. This yields

$$\frac{d}{dt}\begin{pmatrix} \xi_1 \\ \vdots \\ \xi_n \end{pmatrix} = F \begin{pmatrix} \xi_1 \\ \vdots \\ \xi_n \end{pmatrix} \tag{12}$$

where F is an $n \times n$ matrix over k. The components of the vector $\tilde{y} = (\tilde{y}_1, \ldots, \tilde{y}_p)$ in $\mathcal{D}/[u]$ of y also depends k-linearly on ξ:

$$\begin{pmatrix} \tilde{y}_1 \\ \vdots \\ \tilde{y}_p \end{pmatrix} = H \begin{pmatrix} \xi_1 \\ \vdots \\ \xi_n \end{pmatrix} \tag{13}$$

where H is a $p \times n$ matrix over k.

2.3.2 Now pull back (12) and (13) to $\mathcal{D}$. Denote by $\eta = (\eta_1, \ldots, \eta_n)$ an

n-tuple of elements in $\mathcal{D}$ such that its residue in $\mathcal{D}/[u]$ is ξ. We obtain

$$
\frac{d}{dt}\begin{pmatrix} \eta_1 \\ \vdots \\ \eta_n \end{pmatrix} = F \begin{pmatrix} \eta_1 \\ \vdots \\ \eta_n \end{pmatrix} + \sum_{\alpha=0}^{\nu} \mathcal{G}_\alpha \frac{d^\alpha}{dt^\alpha} \begin{pmatrix} u_1 \\ \vdots \\ u_m \end{pmatrix}
$$

$$
\begin{pmatrix} y_1 \\ \vdots \\ y_p \end{pmatrix} = H \begin{pmatrix} \eta_1 \\ \vdots \\ \eta_n \end{pmatrix} + \sum_{\beta=0}^{\mu} \mathcal{H}_\beta \frac{d^\beta}{dt^\beta} \begin{pmatrix} u_1 \\ \vdots \\ u_m \end{pmatrix}
\tag{14}
$$

η is called a *(generalized) state*. Two such states η and $\bar{\eta} = (\bar{\eta}_1, \ldots, \bar{\eta}_n)$, the residues of which in $\mathcal{D}/[u]$ are bases of the latter vector space, are related by an input-dependent relation

$$
\begin{pmatrix} \bar{\eta}_1 \\ \vdots \\ \bar{\eta}_n \end{pmatrix} = P \begin{pmatrix} \eta_1 \\ \vdots \\ \eta_n \end{pmatrix} + \sum_{\text{finite}} Q_j \frac{d^j}{dt^j} \begin{pmatrix} u_1 \\ \vdots \\ u_m \end{pmatrix}
\tag{15}
$$

where

- P is an invertible $n \times n$ matrix over k.
- the Q_j:s are $n \times m$ matrices over k.

2.3.3 Assume in (14) that $\mathcal{G}_\nu \neq 0$, $\nu \geq 1$. Let

$$
\begin{pmatrix} \eta_1 \\ \vdots \\ \eta_n \end{pmatrix} = \begin{pmatrix} \eta_1^+ \\ \vdots \\ \eta_n^+ \end{pmatrix} + \mathcal{G}_\nu \frac{d^{\nu-1}}{dt^{\nu-1}} \begin{pmatrix} u_1 \\ \vdots \\ u_m \end{pmatrix}
\tag{16}
$$

It yields

$$
\frac{d}{dt}\begin{pmatrix} \eta_1^+ \\ \vdots \\ \eta_n^+ \end{pmatrix} = F \begin{pmatrix} \eta_1^+ \\ \vdots \\ \eta_n^+ \end{pmatrix} + \sum_{\alpha=0}^{\nu-1} \mathcal{G}_\alpha^+ \frac{d^\alpha}{dt^\alpha} \begin{pmatrix} u_1 \\ \vdots \\ u_m \end{pmatrix}
\tag{17}
$$

where the maximum order of derivations of u now is strictly less than ν. Repeating the elimination procedure yields

$$
\frac{d}{dt}\begin{pmatrix} x_1 \\ \vdots \\ x_n \end{pmatrix} = F \begin{pmatrix} x_1 \\ \vdots \\ x_n \end{pmatrix} + G \begin{pmatrix} u_1 \\ \vdots \\ u_m \end{pmatrix}
$$

$$
\begin{pmatrix} y_1 \\ \vdots \\ y_p \end{pmatrix} = H \begin{pmatrix} x_1 \\ \vdots \\ x_n \end{pmatrix} + \sum_{\gamma=0}^{\epsilon} J_\gamma \frac{d^\gamma}{dt^\gamma} \begin{pmatrix} u_1 \\ \vdots \\ u_m \end{pmatrix}
\tag{18}
$$

where no derivatives of the input appear in the dynamics, which is the classic form introduced by Kalman. Such a state variable representation is called a *Kalman representation* or *realization*. The corresponding state is called a *Kalman state*. Two such Kalman states x and $\bar{x} = (\bar{x}_1, \ldots, \bar{x}_n)$ are related by the usual input-independent transformation

$$\begin{pmatrix} \bar{x}_1 \\ \vdots \\ \bar{x}_n \end{pmatrix} = P \begin{pmatrix} x_1 \\ \vdots \\ x_n \end{pmatrix} \tag{19}$$

since the presence of the input would reintroduce the presence of input derivatives in the state-variable representation.

The vector valued differential polynomial

$$\sum_{\gamma=0}^{\epsilon} J_\gamma \frac{d^\gamma}{dt^\gamma} \tag{20}$$

is uniquely defined and is called the *impulsive polynomial*. Its degree is called the *index*.

2.3.4 Let us summarize the previous section:
Theorem Any linear dynamics $\mathcal{D}$ with input u and output y can be given a Kalman state variable representation (18). The dimension of the Kalman state is equal to the dimension of $\mathcal{D}/[u]$, viewed as a k-vector space.

2.3.5 Application Consider the input output system.

$$A(\frac{d}{dt}) \begin{pmatrix} y_1 \\ \vdots \\ y_p \end{pmatrix} = \begin{pmatrix} u_1 \\ \vdots \\ u_p \end{pmatrix} \tag{21}$$

where A is a $p \times p$ matrix over $k[\frac{d}{dt}]$, which is invertible as a matrix over the quotient field $k(\frac{d}{dt})$. The invertibility condition ensures the existence and uniqueness of the solution of (21). It is a well known problem in numerical analysis to determine the minimum order of differentiability of u which is necessary for integrating (21). This integer is of course the index. The realization (14) demonstrates moreover that derivatives of u do not intervene in the integration procedure (see [19] for the details).

2.4   Controllability and observability

2.4.1 A linear system is aid to be *controllable* if and only if the module Λ is free. A linear dynamics $\mathcal{D}$ with input u is said to be *controllable* if and only if the associated linear system is so. We thus see that this new definition of controllability is independent of any denomination of the variables.

2.4.2 Take the representation (18) where we might assume for the sake of simplicity that $k = \mathbb{R}$. The construction of (18) demonstrates a one-to-one correspondence between the Kalman uncontrollable space and the torsion submodule of $\mathcal{D}$. This leads to the

Theorem A Kalman state-variable representation is controllable if and only if its associated module is free.

2.4.3 The controllability concept can be compared with Willems' concepts, [57, 58]. Consider the framework of 2.2.2 Let $k = \mathbb{R}$. The choice of the functional analytic nature of the trajectories can be manifold. The class of C^∞ functions however appears to be simple and natural. Denote by $C^\infty(t_1, t_2)$ the $R[\frac{d}{dt}]$-module of C^∞ functions on the interval (t_1, t_2) where $-\infty \le t_1 < t_2 \le +\infty$. A trajectory of Σ on (t_1, t_2) is a mapping

$$\tau(t_1, t_2) : \mathcal{D} \to C^\infty(t_1, t_2)$$

in the category of $R[\frac{d}{dt}]$-modules. Thus $\tau(t_1, t_2)$ assigns to any element of $\mathcal{D}$ a funtion in $C^\infty(t_1, t_2)$. The set of all trajectories of Σ on (t_1, t_2) is the set $\mathrm{Hom}(\mathcal{D}, C^\infty(t_1, t_2))$ of homomorphisms in the category of $R[\frac{d}{dt}]$-modules.

2.4.4 Set $-\infty \le t_1 \le t_2 < t_3 \le t_4 \le +\infty$. A trajectory

$$\tau(t_2, t_3) : \mathcal{D} \to C^\infty(t_2, t_3)$$

is said to be a *restriction* of a trajectory

$$\tau(t_1, t_4) : \mathcal{D} \to C^\infty(t_1, t_4)$$

if and only if, for any $\xi \in \mathcal{D}$, the functions $\tau(t_1, t_4)\xi$ and $\tau(t_2, t_3)\xi$ coincide on (t_2, t_3).

2.4.5 Set $-\infty \le t_1' < t_2' < t_3' < t_4' \le +\infty$. Trajectories

$$\tau(t_1', t_2') : \mathcal{D} \to C^\infty(t_1', t_2')$$

and

$$\tau(t_3', t_4') : \mathcal{D} \to C^\infty(t_3', t_4')$$

are said to be *compatible* if and only if there exists a trajectory

$$\tau(t_1', t_4') : \mathcal{D} \to C^\infty(t_1', t_4')$$

such that $\tau(t_1', t_2')$ and $\tau(t_3', t_4')$ are restrictions of $\tau(t_1', t_4')$. $\tau(t_1', t_2')$ is the *past trajectory* and $\tau(t_3', t_4')$ the *future trajectory*.

2.4.6 Definition The linear system Σ is controllable à la Willems, [15], if and only if past and future trajectories are always compatible.

2.4.7 Theorem The linear system is controllable à la Willems if, and only if, its module $\mathcal{D}$ is free.

2.4.8 Assume that $\mathcal{D}$ is free. Let $(b_1, \ldots, b_m)$ be a basis of $\mathcal{D}$. Any trajectory

$$\tau(t_1, t_2) : \mathcal{D} \to C^{\infty}(t_1, t_2)$$

is of course completely determined by the knowledge of

$$\tau(t_1, t_2)b = (\tau(t_1, t_2)b_1, \ldots, \tau(t_1, t_2)b_m)$$

Well known proprties of C^{∞} functions make it possible to define for any $i = 1, \ldots, m$, a function $f_i(t)$ on $C^{\infty}(t_1, t_4)$ such that

$$f_i(t) = \left\{ \begin{array}{ll} \tau(t_1', t_2')b_i(t), & t_1' < t < t_2' \\ \tau(t_3', t_4')b_i(t), & t_3' < t < t_4' \end{array} \right.$$

By setting $\tau(t_1', t_4')b_i = f_i$, we see that $\tau(t_1', t_2')$ and $\tau(t_3', t_4')$ are compatible.

2.4.9 Now assume that Σ is controllable à la Willems, but that $\mathcal{D}$ is not free. Then following paragraph 2.1.11 we restrict the trajectories to the torsion submodule T.

T is finite dimensional when considered as an $\mathbb{R}$-vector space. Take $\xi \in T$. The sequence $\xi, \dot{\xi}, \ldots, \xi^{(\nu)}, \ldots$ therefore spans a finite dimensional vector space. In other words it means that ξ satisfies a homogeneous linear differential equation:

$$\xi^{(n)} + c_1 \xi^{(n-1)} + \cdots + c_n \xi = 0 \quad (c_1, \ldots, c_n \in R) \tag{22}$$

Any trajectory $\tau(t_1, t_2)\xi$ is an element of $C^{\infty}(t_1, t_2)$ which satisfies (22). Assume that the past trajectory $\tau(t_1', t_2')\xi$ and the future trajectory $\tau(t_3', t_4')\xi$ do not lie on the same integral curve of (22). As different solutions of (22) do not intersect, $\tau(t_1', t_2')\xi$ and $\tau(t_3', t_4')\xi$ are incompatible: The system module $\mathcal{D}$ must be free.

2.4.10 Remark We therefore see that the decomposition $\mathcal{D} = \Phi \oplus T$ in paragraph 2.1.11 parallells the one given by Nieuwenhuis and Willems [41] of a system as a direct sum of a controllable and an autonomous subsystem.

2.4.11 An element $\lambda \in \Lambda$ is said to be *observable with respect to a set* $\zeta = \{\zeta_i, \quad i \in I\}$ *of elements in* Λ if and only if λ belongs to the module $[\zeta]$ spanned by the ζ_i. It means that λ is a finite k-linear combination of the ζ_i and of their derivatives.

2.4.12 Take a linear dynamics $\mathcal{D}$ with input $u = (u_1, \ldots, u_m)$ and output $y = (y_1, \ldots, y_p)$. An element $\delta \in \mathcal{D}$ is said to be *observable* if and only if it is observable with respect to u and y, i. e. if and only if it belongs to the module $[u, y]$. The dynamics $\mathcal{D}$ is said to be *observable* if and only if any subset of $\mathcal{D}$ is observable. It is equivalent to saying that the two modules $\mathcal{D}$ and $[u, y]$ coincide.

2.4.13 Remark We easily see that a dynamics is observable if and only if any component of a generalized state in a representation of type (14) is so. Our definition of observability thus corresponds to the notion of *(re)-constructibility* as discussed in several textbooks, [34], [1].

2.4.14 The notion of constructibility being equivalent to that of observability in the classical sense, we have proved the
Proposition A dynamics $\mathcal{D}$ with input u and output y is observable if and only if it is observable in the classical sense.

2.4.15 The *minimal* state variable representation or realisation of a system with input u and output y is the representation of the system $[u, y]$ with input u and output y. We thus have the following characterization (see also [57], [58])
Proposition A state variable representation is minimal if and only if it is observable.

2.4.16 Example The minimal Kalman realization of the single-input, single- output system $\dot{y} = \dot{u}$ is

$$\begin{aligned} \dot{x} &= 0 \\ y &= x + u \end{aligned} \tag{23}$$

It is observable but not controllable.

2.4.17 Remark The apparent contradiction with the well known result of Kalman [37] on the equivalence between minimality of a realization and its property of being both controllable and observable can be explained as follows: Kalman's setting applies to transfer matrices whereas the transfer function of $\dot{y} = \dot{u}$ is 1.

2.4.18 Application. *Hidden modes* or *decoupling zeros* express a lack of controllability or observability. We refer to [47] for a survey of the vast litterature on this subject.

The ground field k is of course a field of constants. Assume that Λ is not controllable, i. e. not free. The derivation $\frac{d}{dt}$ acts as a k-linear endomorphism on the torsion submodule T. Its eigenvalues in an algebraic closure $\bar{k}$ of k are the *input decoupling zeros*.

Assume that the dynamics $\mathcal{D}$ is unobservable, i.e. that $[u, y] \subset \mathcal{D}$ (strict inclusion). The derivation $\frac{d}{dt}$ acts as before as a k-linear endomorphism on the quotient module $\mathcal{D}/[u, y]$, which is torsion. Its eigenvalues in $\bar{k}$ of k are the *output decoupling zeros*. The details are given in [14].

3   Nonlinear systems

3.1   A short view of differential algebra

3.1.1 A *differential ring*, [39], R is a commutative ring which here is equipped with a single derivation $\frac{d}{dt} = \dot{\ }$ such that

$$\forall a \in R, \quad \frac{da}{dt} = \dot{a} \in R \tag{24}$$

$$\forall a, b \in R, \quad \frac{d}{dt}(a+b) = \dot{a} + \dot{b} \tag{25}$$

$$\frac{d}{dt}(ab) = \dot{a}b + a\dot{b} \tag{26}$$

A *differential field* is a differential ring which is a field (compare with 2.1.2).

3.1.2 To any differential ring R we may associate a ring of *differential polynomials* $R\{z_1, \ldots, z_\nu\}$ in ν *differential indeterminates*, which is the ring of polynomials in the z_i:s and all their formal derivatives $\dot{z}_i, \ddot{z}_i, \ldots, z_i^{(\alpha)}, \ldots$. $R\{z_1, \ldots, z_\nu\}$ carries an obvious structure of differential ring.

3.1.3 Let k be a given differential ground field. A system of *algebraic differential equations* $p_i = 0$, $i \in I$ in ν variables is determined by a subset $\{p_i | i \in I\}$ of $k\{z_1, \ldots, z_\nu\}$. As is usual in i. e. nondifferential commutative algebra and algebraic geometry, a geometric interpretation is related to such a system. *Differential algebraic geometry* considers the set of solutions in some sufficiently big differential overfield $\mathcal{K}$ of k, which is called *universal* and which is the differential analogue of the algebraic closure.

3.1.4 Differential systems might of course possess the same *variety* of solutions. Like in the classic algebraic geometry, this phenomenon is taken into account by ideal theory. A *differential ideal* $\mathcal{I} \subseteq k\{z_1, \ldots, z_\nu\}$ is an ideal which is closed with respect to derivation, i. e., $\frac{d}{dt}\mathcal{I} = \{\dot{p} | p \in \mathcal{I}\} \subseteq \mathcal{I}$. Two differential ideals define the same variety if and only if they have the same *radical ideal*. (Here the radiacal of $\mathcal{I}$ consists of all p, such that some power of p is in $\mathcal{I}$.())

3.1.5 We might therefore assume that a differential ideal is *perfect*, i. e. coincides with its own radical. For such ideals the celebrated Lasker-Noether and Hilbert basis theorems can be stated:

- Any perfect differential ideal is a finite intersection of prime differential ideals. (A prime differential ideal has the property that p or q belongs to the ideal whenever pq does.())

- Any perfect differential ideal may be generated by a finite set of elements in $k\{z_1, \ldots, z_\nu\}$.

3.1.6 Example In example 1.1.1 we have a non-prime ideal: $v(\dot{v} + 1)$ belongs to the ideal but neither v nor $\dot{v} + 1$ does (it is not possible to get any of them out of the original polynomials using differentiations, additions and multiplications). The Lasker-Noether decomposition in this case corresponds to a splitting of the physical model into two cases: the stationary particle and the particle with constant acceleration.

3.1.7 Field theory plays a crucial role in our approach. Recall that it was invented in the second half of the nineteenth century in order to avoid lengthy manipulations of algebraic equations. We first start with classic non differential field extensions and proceed to the differential analogues.

3.1.8 Take a (non-differential) field extension E/F, i.e., two fields F, E such that $F \subseteq E$. Only two solutions are possible

1. An element of E is said to be *algebraic* over F if and only if it satisfies an algebraic equation over F. For instance $\sqrt{2}$ satisfies $x^2 - 2 = 0$ and is algebraic over $\mathbb{Q}$. The extension E/F is *algebraic* if and only if any element of E is algebraic over F.

2. An element $a \in E$ is said to be *transcendental* over F if and only if it is not algebraic over F. This implies the nonexistence of a single variable polynomial $p(x)$ over F, $p \not\equiv 0$, such that $p(a) = 0$. The extension E/F is said to be *transcendental* if and only if it is not algebraic, i. e. if and only if there exists at least one element of E which is transcendental over F. The extension $\mathbb{R}/\mathbb{Q}$, for example, is transcendental since e and π are transcendental over $\mathbb{Q}$.

3.1.9 There exist for transcendental extensions concepts which are the nonlinear analogues of dimension and basis for a vector space. A set $\{\xi_i \,|\, i \in I\}$ of elements in E is said to be *F-algebraically dependent* if and only if there exists at least one nonzero polynomial $p(x_1, \ldots, x_\nu)$ over F such that $p(\xi_{i_1}, \ldots, \xi_{i_\nu}) = 0$. A set which is not *F*-algebraically dependent is said to be *F-algebraically independent*. It is possible to define *F*-algebraically independent sets which are maximal with respect to inclusion. Such a set is called a *transcendence basis* of E/F. Two such sets have the same cardinality, i. e. the same number of elements, which is the transcendence degree of E/F. It is denoted tr $d^\circ E/F$.

3.1.10 Examples

1. The extension E/F is algebraic if and only if tr $d^\circ E/F = 0$.

2. Take $E = F(s_1, \ldots, s_n)$ i. e. the field of rational functions in n indeterminates $s_1, \ldots, s_n$ with coefficients in F. This models multidimensional rational transfer functions. Since $s_1, \ldots, s_n$ are F-algebraically independent, they constitute a transcendence basis of

$F(s_1, \ldots, s_n)/F$. Moreover $\operatorname{tr} \operatorname{d}^o F(s_1, \ldots, s_n)/F = n$. Such extensions are called *purely transcendental* extensions.

3. The general picture is as follows: most of the time it is impossible to find a transcendence basis $x = \{x_i \mid i \in I\}$ such that E coincides with the field $F(x)$ of rational functions. Nevertheless the extension $E/F(x)$ has the most important property of being algebraic.

3.1.11 Take a *tower* of three fields F, E, D, i. e. such that $F \subseteq E \subseteq D$. Then

$$\operatorname{tr} \operatorname{d}^o D/F = \operatorname{tr} \operatorname{d}^o D/E + \operatorname{tr} \operatorname{d}^o E/F$$

3.1.12 A *differential (field) extension* L/K is given by two differential fields K, L such that
i) $K \subseteq L$
ii) The derivation of K is the restriction to K of the derivation of L.

In analogy with the non-differential case two solutions are possible:

1. An element $\xi \in L$ is said to be *K-differentially algebraic* if and only if it satisfies an algebraic differential equation $p(\xi, \dot{\xi}, \ldots, \xi^{(d)}) = 0$ where p is a polynomial over K in ξ and its derivatives. For instance e^t satisfies $\dot{y} - y = 0$ and is differentially algebraic over $\mathbb{Q} < y >$ (the algebraic differential equations in one indeterminate, with rational coefficients). The extension L/K is said to be *differentially algebraic* if and only if any element of L is K-differentially algebraic.

2. An element of L is said to be *K-differentially transcendental* if and only if it is not *K-differentially algebraic* i. e. if and only if it does not satisfy any algebraic differential equation.

A set $\{\xi_i \mid i \in I\}$ of elements in L is said to be *K-differentially algebraically dependent* if and only if the set $\{\xi_i^{(\nu_i)} \mid i \in I, \nu_i = 0, 1, 2, \ldots\}$ of derivatives of any order is K-algebraically dependent. In other words the ξ_i satisfy some algebraic differential equation. A set which is not K-differentially algebraically dependent is said to be *K-differentially algebraically independent*. A K-differentially algebraically independent set which is maximal with respect to inclusion is called a *differential transcendence basis* of L/K. Two such bases have the same cardinality which is the *differential transcendence degree* of L/K. It is denoted $\operatorname{diff} \operatorname{tr} \operatorname{d}^o L/K$.

3.1.13 Examples

1. The extension L/K is differentially algebraic if and only if $\operatorname{diff} \operatorname{tr} \operatorname{d}^o L/K = 0$.

2. Take $L = K < u_1, \ldots, u_m >$ i. e. the differential overfield of K generated by m K differentially independent elements. Any element of

$K < u_1, \ldots, u_m >$ is a rational expression over K in the u_i and a finite number of their derivatives. The u_i constitute a differential transcendence basis of $K < u_1, \ldots, u_m > /K$ and diff tr $d^o K < u_1, \ldots, u_m > /K = m$. The extension $K < u_1, \ldots, u_m > /K$ is said to be *purely differentially transcendental*.

3. Let $\xi = \{\xi_i \mid i \in I\}$ be a differential transcendence basis of L/K. In general, L and $K < \xi >$ do not coincide. But the extension $L/K < \xi >$ is differentially algebraic.

3.1.14 Take a tower K, L, M of differential fields, i. e. such that $K \subseteq L \subseteq M$. Then

$$\text{diff tr } d^o M/K = \text{diff tr } d^o M/L + \text{diff tr } d^o L/K$$

3.1.15 The following most important result holds for finitely generated differential extensions.:
Theorem A finitely generated differential extension is differentially algebraic if and only if its (non-differential) transcendence degree is finite.

In more down-to-earth but less precise language this transcendence degree is the number of initial conditions which are needed for computing the solutions.

3.1.16 Consider again a finitely generated differentially algebraic extension L/K. Assume moreover that K is not a field of constants. Then the theorem of the *differential primitive element*, which generalizes a well known result in the non-differential case, states the existence of an element $\pi \in L$ such that $L = K < \pi >$. The proof shows that "almost all" elements in L, but not in K, are differentially primitive.

3.1.17 The classic differential calculus in analysis and differential geometry possesses a nice counterpart in commutative algebra (see e. g. [40]) which has been extended to differential fields by Johnson [33]. It permits the translation of properties of differential extensions into the language of linear algebra.

3.1.18 Associate to a differential extension L/K the left $L[\frac{d}{dt}]$ module $\Omega_{L/K}$ spanned by the so-called (Khler) differentials $d_{L/K}a$, $a \in L$. The mapping $d_{L/K} : L \to \Omega_{L/K}$ enjoys the following properties:

$$\forall a, b \in L, \quad d_{L/K}(a+b) = d_{L/K}a + d_{L/K}b \qquad (27)$$

$$d_{L/K}(ab) = d_{L/K}(a)\, b + a\, d_{L/K}b \qquad (28)$$

$$d_{L/K}(\dot{a}) = \frac{d}{dt}d_{L/K}a \qquad (29)$$

$$\forall c \in K, \quad d_{L/K}c = 0 \qquad (30)$$

The last property means that elements of K behave like constants with respect to $d_{L/K}$.

3.1.19 The three following properties will be utilized:

- A set $(z_1, \ldots, z_m)$ is a differential transcendence basis of L/K if and only if $(d_{L/K} z_1, \ldots, d_{L/K} z_m)$ is a maximal set of $L[\frac{d}{dt}]$-linearly independent elements in $\Omega_{L/K}$. Thus diff tr d$^o L/K = $ rk $\Omega_{L/K}$.

- The extension L/K is differentially algebraic if, and only if, $\Omega_{L/K}$ is torsion. Then a set $(x_1, \ldots, x_n)$ is a transcendence basis of L/K, if and only if, $(d_{L/K} x_1, \ldots, d_{L/K} x_n)$ is a basis of $\Omega_{L/K}$ as an L-vector space. Thus tr d$^o L/K = \dim \Omega_{L/K}$.

- The extension L/K is algebraic if and only if $\Omega_{L/K}$ is trivial, i. e. $\Omega_{L/K} = \{0\}$.

3.2 What is a nonlinear system?

3.2.1 Following Willems [54]–[58], we define a system by the set of trajectories of a finite number of variables $\{w_1, \ldots, w_q\}$, where we do not distinguish between input, state, output and other variables. When such a set is differentially algebraic, i. e. when the differential equations are algebraic, we can take advantage of the formalism of section 3.1. Paragraphs 3.1.3–3.1.5 demonstrate that the trajectories might be defined by a perfect differential ideal. Take a prime differential ideal $\mathcal{P}$ in the decomposition described in paragraph 3.1.5. The quotient differential ring $R = k\{w_1, \ldots, w_q\}/\mathcal{P}$ is integral, i. e. without zero divisors. The quotient differential field K of R will play a crucial role.

3.2.2 Let k be a given differential ground field. A *system*, [11], is a finitely generated differential extension K/k. Let $u = (u_1, \ldots, u_m)$ be a finite set of differential quantities, i. e. of elements which can be formally derivated to any order, which play the role of input variables. Denote by $k < u >$ the differential field generated by k and by the components of u. A *dynamics* is a finitely generated differential algebraic extension $\mathcal{D}/k < u >$. An *output* $y = (y_1, \ldots, y_p)$ is a finite set of elements in $\mathcal{D}$.

3.2.3 Remark Take a linear system, i. e. a finitely generated left $k[\frac{d}{dt}]$-module Λ. The symmetric tensor product of Λ, as a k-vector space, can be endowed a structure of integral differential ring R. The differential quotient field K of R gives the field theoretic description of 3.2.2 of our linear system.

3.3 State variable representation

3.3.1 Take a dynamics $\mathcal{D}/k < u >$. We know from paragraph 3.1.15 that the (non differential) transcendence degree of $\mathcal{D}/k < u >$ is finite, say n. Choose a transcendence basis $x = (x_1, \ldots, x_n)$. The components of $\dot{x} = (\dot{x}_1, \ldots, \dot{x}_n)$ and of $y = (y_1, \ldots, y_p)$ are k-algebraically dependent on the components of x. It yields

$$
\begin{aligned}
A_i(\dot{x}_i, x, u, \dot{u}, \ldots, u^{(\alpha_i)}) &= 0, \quad i = 1, \ldots, n \\
B_j(y_j, x, u, \dot{u}, \ldots, u^{(\beta_j)}) &= 0, \quad j = 1, \ldots, p
\end{aligned}
\tag{31}
$$

where the A_i and the B_j are polynomials over k in their arguments. Take two *(generalized) states* x and $\bar{x} = (\bar{x}_1, \ldots, \bar{x}_n)$ i. e. two transcendence bases of $\mathcal{D}/k < u >$. Any component of x is $k < u >$-algebraically dependent on the components of $\bar{x}$ and vice versa. It yields:

$$
\begin{aligned}
p_i(x_i, \bar{x}, u, \dot{u}, \ldots, u^{(\pi_i)}) &= 0 \\
\bar{p}_i(\bar{x}_i, x, u, \dot{u}, \ldots, u^{(\bar{\pi}_i)}) &= 0
\end{aligned}
\quad i = 1, \ldots, n
\tag{32}
$$

3.3.2 There are antinomies with the now prevailing viewpoint which is usually expressed within the differential geometric language (see, e. g. [32], [42]). Let us list and discuss them:

1. The equations (31) are implicit with respect to $\dot{x}$ (and y). This fact is confirmed by many realistic case studies, such as nonlinear electric circuits (see [16]). The vanishing of the $\frac{\partial A_i}{\partial \dot{x}_i}$ is related to *impasse points* which might give rise to strange qualitative behaviour, such as discontinuities (see e. g. [29]).

2. The presence of input derivatives in (31) and the corresponding input dependence of the state transformations (32) were utilized in several more practically oriented works for obtaining various "canonical" forms. The fact that input derivatives might appear, as in example 1.1.3, (7), is more subtle to interpret. In any case it means a possibility of choosing between different orders of derivation of the control variable.

3.3.3 Remark It is sometimes possible to lower the order of derivation of u in the dynamics (31). This problem was first treated by Freedman and Willems [21] and by Glad [23] and fully solved by Delaleau [4] and by Delaleau and Respondek [5].

3.3.4 Take a dynamics $\mathcal{D}/k < u >$. From 3.1.16, we know the possibility of choosing a differential primitive element ξ for this finitely generated differential extension. It yields the *generalized controller canonical form*

$$
\dot{x}_1 = x_2
$$
$$
\vdots
$$
$$
\dot{x}_i = x_{i+1}
\tag{33}
$$
$$
\vdots
$$
$$
c(\dot{x}_n, x_1, \ldots, x_n, u, \dot{u}, \ldots, u^{(j)}) = 0
$$

where $x_1 = \xi$, $x_2 = \dot{\xi},..,x_n = \xi^{(n-1)}$ and c is a polynomial over k in its arguments. This form, which does not hold for constant linear systems has been utilized in [20] for discontinuous controls.

3.4 Tangent linear systems, controllability and observability

3.4.1 Take a system K/k. Following 3.1.18, we define the *tangent (or variational) linear system* as the left $K[\frac{d}{dt}]$-module $\Omega_{K/k}$ of Khler differentials. To a dynamics $\mathcal{D}/k < u >$ is associated the *tangent (or variational) linear dynamics* $\Omega_{\mathcal{D}/k}$ with input $d_{\mathcal{D}/k}u = (d_{\mathcal{D}/k}u_1, \ldots, d_{\mathcal{D}/k}u_m)$. The *tangent (or variational) output* associated to $y = (y_1, \ldots, y_p)$ is $d_{\mathcal{D}/k}y = (d_{\mathcal{D}/k}y_1, \ldots, d_{\mathcal{D}/k}y_p)$.

3.4.2 Sussmann and Jurdjevic [51] have introduced in the differential geometric setting the notion of strong accessibility for a dynamics of type $\dot{x} = f(x, u)$. Sontag [49], [50] and Coron [3] have demonstrated that strong accessibility is equivalent to the controllability of the tangent system along almost any trajectory.

3.4.3 We are therefore lead to the following definition: A system K/k is said to be *controllable* (or *strongly accessible*) if and only if the tangent linear system is controllable, i. e. if and only if the module $\Omega_{K/k}$ is free. Notice that, as in the linear case, controllability is independent of any distinction between the variables.

3.4.4 Take a system K/k and choose a subset $z = \{z_i \mid i \in I\}$ of K. An element $\zeta \in K$ is said to be *observable with respect to* z, if and only if, ζ is (non-differentially) algebraic over the differential field $k < z >$ generated by the z_i. Take a dynamics $\mathcal{D}/k < u >$, with input $u = (u_1, \ldots, u_m)$ and output $y = (y_1, \ldots, y_p)$. An element $\zeta \in \mathcal{D}$ is said to be *observable* if and only if it is observable with respect to u and y, i. e. if and only if it is algebraic over $k < u, y >$. The dynamics $\mathcal{D}/k < u >$ is said to be *observable* if and only if the extension $\mathcal{D}/k < u, y >$ is algebraic, i. e. if and only if any element of $\mathcal{D}$ is observable. A more detailded discussion is given in [9], [10], [24], [27]. See also [45]

3.4.5 Remark The intuitive meaning is the following: the observable quantity ζ can be expressed as a root of a (non-differential) algebraic equation,i. e., without integrating a differential equation and knowing initial conditions.

3.4.6 Take a dynamics $\mathcal{D}/k < u >$ with output y. Consider the tangent linear dynamics $\Omega_{\mathcal{D}/k}$ with input $d_{\mathcal{D}/k}u$ and output $d_{\mathcal{D}/k}y$. We know from 3.1.19 that the nonlinear dynamics is observable if and only if its tangent linear one is so. Notice moreover that it can be shown that the classical matrix rank condition for linear systems gives back the well known rank condition due to Hermann and Krener, [30] which holds for dynamics of the type $\dot{x} = f(x, u)$.

3.4.7 Remark It is perhaps worth wile to stress that the classic notions of strong accessibility and observability for nonlinear systems can be seen, from some epistemological standpoint, as linear concepts, since they are characterized by their linear counterparts.

3.5 Flatness

3.5.1 Assume from now on, as is often done in algebraic geometry, that all the quantities we are considering belong to a "big" enough differential overfield of k, which may be *universal*, [39].

3.5.2 Two systems K/k and $\tilde{K}/k$ are said to be *equivalent*, or *equivalent by endogeneous feedback* if and only if any element of K is (non-differentially) algebraic over $\tilde{K}$ and vice versa. Two dynamics $\mathcal{D}/k < u >$ and $\tilde{\mathcal{D}}/k < \tilde{u} >$ are *equivalent*, or *equivalent by endogeneous feedback* if and only if the two corresponding systems $\mathcal{D}/k$ and $\tilde{\mathcal{D}}/k$ are so.

3.5.3 Property If the two systems K/k and $\tilde{K}/k$ are equivalent, then diff tr d$^o K/k$ = diff tr d$^o \tilde{K}/k$.
Proof. Let L be the differential field generated by K and $\tilde{K}$: the extensions L/K and $L/\tilde{K}$ are algebraic. Then

$$\text{diff tr d}^o K/k = \text{diff tr d}^o L/k = \text{diff tr d}^o \tilde{K}/k$$

3.5.4 Remark The last property means that two equivalent systems possess the same number of independent input channels.

3.5.5 Consider again two equivalent dynamics $\mathcal{D}/k < u >$ and $\tilde{\mathcal{D}}/k < \tilde{u} >$ with generalized states $x = (x_1, \ldots, x_n)$ and $\tilde{x} = (\tilde{x}_1, \ldots, \tilde{x}_{\tilde{n}})$. The algebraicity of any element of $\mathcal{D}$ over $\tilde{\mathcal{D}}$ (and $\tilde{\mathcal{D}}$ over $\mathcal{D}$) yields

$$\begin{array}{rcll} \phi_i(u_i, \tilde{x}, \tilde{u}, \dot{\tilde{u}}, \ldots, \tilde{u}^{(\nu_i)}) & = & 0, & i = 1, \ldots, m \\ \sigma_\alpha(x_\alpha, \tilde{x}, \tilde{u}, \dot{\tilde{u}}, \ldots, \tilde{u}^{(\mu_\alpha)}) & = & 0, & \alpha = 1, \ldots, n \\ \tilde{\phi}_i(\tilde{u}_i, x, u, \dot{u}, \ldots, u^{(\tilde{\nu}_i)}) & = & 0, & i = 1, \ldots, m \\ \tilde{\sigma}_{\tilde{\alpha}}(\tilde{x}_{\tilde{\alpha}}, x, u, \dot{u}, \ldots, u^{(\tilde{\mu}_{\tilde{\alpha}})}) & = & 0, & \tilde{\alpha} = 1, \ldots, \tilde{n} \end{array} \qquad (34)$$

where the ϕ_i, σ_α, $\tilde{\phi}_i$ and $\tilde{\sigma}_{\tilde{\alpha}}$ are polynomials over k. These correspondences define a peculiar type of dynamic feedback called *endogeneous*, since it does not necessitate the introduction of variables which are transcendental over $\mathcal{D}$ and/or $\tilde{\mathcal{D}}$. If we know x and $\tilde{x}$, we can calculate u from $\tilde{u}$ and vice versa without integrating a differential equation. Endogeneous feedbacks, like quasi-static ones (see [6]) are therefore related to invertible "integral-free" transformations (*umkehrbar integrallose Transformationen* in German) introduced a long time ago by Hilbert [31].

3.5.6 A system K/k is said to be *(differentially) flat*, [18], if and only if it is equivalent to a purely differentially transcendental extension. A dynamics $\mathcal{D}/k < u >$ is said to be *(differentially) flat* if and only if the corresponding system is so.

A differential transcendence basis $y = (y_1, \ldots, y_m)$ of L/k such that $L = k < y >$ is called a *linearizing* (or *flat*) *output*. It plays a somewhat analogous role to the *flat coordinates* in the differential geometric approach to the Frobenius theorem.

4 Constructive methods

One of the appealing aspects of the algebraic methods, from the point of view of applications, is the possibility of constructing explicitly many of the objects of interest. Much of this is described in principle already in the work by Ritt, [46]. Since calculations at that time had to be done by hand, they were seldom practical however. With the availability of computer algebra systems like Macsyma, Reduce, Maple, Mathematica, Axiom,..., the situation has altered, and quite a few things can be calculated. Below a short overview will be given. For more details the reader should consult [7], [8], [26], [43], [44], [22].

4.1 Differential polynomials

4.1.1 As mentioned in 3.1.2 and 3.1.3 the concrete representation of nonlinear systems is in terms of differential equations or differential polynomials. A differential polynomial is simply a polynomial in a number of variables and their derivatives. In examples 1.1.1, 1.1.2 and 1.1.3, for example, the left hand sides of (1), (4) are differential polynomials. In (3) and (6) we get the corresponding differential polynomials by transferring all terms to the left hand side. Not all differential equations correspond to differential polynomials: in (5) and (7) there are sine and cosine terms for instance. Comparing (5), (7) and (6) we see however that descriptions that are not in terms of differential polynomials can sometimes be converted to differential polynomial ones.

4.2 Ranking

4.2.1 When doing calculations one usually wants to end up with differential polynomials that are as simple as possible. Of course simple might mean different things in different contexts. The concept of ranking is one way of expressing simplicity which turns out to be useful.

It starts with the observation that differential equations are more complex if they involve higher derivatives. For a single variable y it is then natural to form the ranking

$$y < \dot{y} < \ddot{y} < y^{(3)} < y^{(4)} < \cdots \qquad (35)$$

Here $<$ means "is ranked lower than". (Mathematically speaking the ranking is a total ordering.)

4.2.2 If two variables u and y are involved, two types of rankings seem natural. One could say that polynomials involving, say u only, are simpler than those involving y. Then the ranking would be

$$u < \dot{u} < \ddot{u} < \cdots < y < \dot{y} < \ddot{y} < \cdots \qquad (36)$$

One could also say that higher order derivatives are always considered more complicated, no matter what the variables are. Then a natural ranking could be

$$u < y < \dot{u} < \dot{y} < \ddot{u} < \ddot{y} < \cdots \qquad (37)$$

4.2.3 Requirements for rankings It turns out that any type of ranking can be used in the algorithms we are going to present, provided two conditions are satisfied. Let $u^{(\mu)}$ and $y^{(\nu)}$ be arbitrary derivatives of arbitrary variables. Then the ranking should be such that, for arbitrary positive σ

$$y^{(\nu)} \quad < \quad y^{(\nu+\sigma)} \qquad (38)$$

$$u^{(\mu)} \quad < \quad y^{(\nu)} \Rightarrow u^{(\mu+\sigma)} < y^{(\nu+\sigma)} \qquad (39)$$

Among three variables u, y and v there are a number of possible rankings, e.g.

$$u < \dot{u} < \ddot{u} < \cdots < y < \dot{y} < \ddot{y} < \cdots < v < \dot{v} < \ddot{v} < \cdots \qquad (40)$$

$$u < v < y < \dot{u} < \dot{v} < \dot{y} < \ddot{u} < \ddot{v} < \ddot{y} < \cdots \qquad (41)$$

$$u < \dot{u} < \ddot{u} < \cdots < v < y < \dot{v} < \dot{y} < \ddot{v} < \ddot{y} < \cdots \qquad (42)$$

4.2.4 Often in algorithms one generates sequences of derivatives where each element has strictly lower rank the the preceding one. Examples of such sequences are

$$v^{(17)}, \; v^{(12)}, \; v^{(3)}, \; y^{(4)}, \; y^{(2)}, \; u^{(55)}, \; u^{(23)}, \; u^{(6)}$$

for the ranking (40) and

$$y^{(11)}, u^{(9)}, v^{(8)}, y^{(7)}, v^{(7)}, u^{(6)}, y^{(5)}, v^{(3)}, u^{(2)}$$

for the ranking (41). It is an important fact that such sequences can not be infinitely long. This is obvious for the ranking (41) since there are only finitely many derivatives that *can* be lower than the first one. It is somewhat less obvious for (40) since there are infinitely many derivatives that are ranked lower than say $v^{(77)}$.

4.2.5 In fact we have the
Proposition A sequence of derivatives, each one ranked lower than the preceding one, can only have finite length.
Proof Let $y_1, \ldots, y_p$ denote all the variables whose derivatives appear anywhere in the sequence. For each y_j let σ_j denote the order of the first appearing derivative. There can then be only σ_j lower derivatives of y_j in the sequence. The total number of elements is thus boumded by $\sigma_1 + \cdots + \sigma_p + p$.

4.2.6 The ranking of variables automatically carries over to a ranking of differential polynomials. When comparing two polynomials the procedure is as follows.

1. Find the highest ranking derivative in each polynomial. This is called the *leader* of the polynomial.

2. If the polynomials have different leaders, the one with the highest ranking leader is ranked highest.

3. If the polynomials have the same leader, the one with the highest power of the leader is ranked highest.

4.2.7 Examples

1. Let the ranking be (40) and consider

$$A: \quad u\ddot{y}v + 1, \quad \text{leader} = v$$

$$B: \quad \ddot{u} + \left(y^{(5)}\right)^2, \quad \text{leader} = y^{(5)}$$

   Here B is ranked lower.

2. Now consider the ranking (41) and the same polynomials.

$$A: \quad u\ddot{y}v + 1, \quad \text{leader} = \ddot{y}$$

$$B: \quad \ddot{u} + \left(y^{(5)}\right)^2, \quad \text{leader} = y^{(5)}$$

   Now A is ranked lower.

3. Consider again the ranking (41) and the polynomials

$$A: \quad v^2\dot{y} + \ddot{u}^2, \quad \text{leader} = \ddot{u}$$

$$B: \quad \dot{v} + y + \ddot{u} + \ddot{u}^5, \quad \text{leader} = \ddot{u}$$

   Here A is ranked lower since its leader is raised to a lower power (2) than in B (5).

4. Consider again the ranking (41) and the polynomials

$$A: \; v^2 \dot{y} + \ddot{u}^2, \quad \text{leader} = \ddot{u}$$

$$B: \; v^3 + \dot{y} + \ddot{u} + \ddot{u}^2, \quad \text{leader} = \ddot{u}$$

Here the polynomials have the same leader with the same highest power. They have thus the same ranking.

¿From the last example we see that different differential polynomials can have the same ranking. Mathematically this means that the ranking is only a pre-ordering for differential polynomials.

4.2.8 Sometimes it is of interest to compare two differential polynomials in the following way. Suppose the differential polynomial A has leader $y^{(\nu)}$. The differential polynomial B is said to be

1. *partially reduced* with respect to A, if it contains no higher derivative of y than $y^{(\nu)}$,

2. *reduced* with respect to A, if it is partially reduced, and is a polynomial of lower degree than A in $y^{(\nu)}$.

4.2.9 Examples

1. Let the ranking be (40) and consider the same differential polynomials as in example 4.2.7.

$$A: \; u\ddot{y}v + 1, \quad \text{leader} = v$$

$$B: \; \ddot{u} + \left(y^{(5)} \right)^2, \quad \text{leader} = y^{(5)}$$

Here B is reduced with respect to A (it contains no derivative of v and is a degree zero polynomial in v). However A is also reduced with respect to B (it does not contain $y^{(5)}$ or higher derivatives of y).

2. Let the ranking be (41) and consider

$$A: \; u\left(\ddot{y} \right)^3 v + 1, \quad \text{leader} = \ddot{y}$$

$$B: \; \left(\ddot{y} \right)^2 + y, \quad \text{leader} = \ddot{y}$$

Here B is reduced with respect to A but not vice versa.

4.2.10 ¿From example 1 in paragraph 4.2.9 we see that it is possible for differential polynomials to be mutually reduced with respect to each other. A set of differential polynomials, all of which are reduced with respect to each other, is called an *autoreduced set*.

4.2.11 Example In example 1.1.2 (4) defines an autoreduced set under the ranking $u^{(\cdot)} < y^{(\cdot)} < x_2^{(\cdot)} < x_1^{(\cdot)}$. The leaders are $\ddot{y}$, x_2 and x_1 respectively.

4.2.12 Example Consider

$$a_1(x_1,\ldots,x_n,u)\dot{x}_1 \quad - \quad b_1(x_1,\ldots,x_n,u) \tag{43}$$

$$\vdots$$

$$a_n(x_1,\ldots,x_n,u)\dot{x}_n \quad - \quad b_n(x_1,\ldots,x_n,u) \tag{44}$$

$$a_{n+1}(x_1,\ldots,x_n,u)y_1 \quad - \quad b_{n+1}(x_1,\ldots,x_n,u) \tag{45}$$

$$\vdots$$

$$a_{n+p}(x_1,\ldots,x_n,u)y_p \quad - \quad b_{n+p}(x_1,\ldots,x_n,u) \tag{46}$$

$$\tag{47}$$

where a_i and b_i are *ordinary polynomials*, i.e. they involve no derivatives of the variables. This is an autoreduced set under the ranking

$$u < \dot{u} < \ddot{u} < \cdots$$

$$< x_1 < \cdots < x_n < y_1 < \cdots < y_p < \dot{x}_1 < \cdots < \dot{x}_n < \dot{y}_1 < \cdots < \dot{y}_p < \cdots$$

We see that the equations one get, when putting the differential polynomials (43) – (46) equal to zero, are equivalent to the state space description

$$\dot{x}_1 = \frac{b_1(x_1,\ldots,x_n,u)}{a_1(x_1,\ldots,x_n,u)} \tag{48}$$

$$\vdots$$

$$\dot{x}_n = \frac{b_n(x_1,\ldots,x_n,u)}{a_n(x_1,\ldots,x_n,u)} \tag{49}$$

$$y_1 = \frac{b_{n+1}(x_1,\ldots,x_n,u)}{a_{n+1}(x_1,\ldots,x_n,u)} \tag{50}$$

$$\vdots$$

$$y_p = \frac{b_{n+p}(x_1,\ldots,x_n,u)}{a_{n+p}(x_1,\ldots,x_n,u)} \tag{51}$$

$$\tag{52}$$

(if the denominators are nonzero). State space descriptions can thus be considered special cases of autoreduced sets.

4.2.13 Let
$$\mathbf{A} = A_1,\ldots,A_r; \quad \mathbf{B} = B_1,\ldots,B_s$$

be two autoreduced sets of differential polynomials. Let them be ordered in increasing rank so that

$$A_1 < \cdots < A_r, \quad B_1 < \cdots < B_s$$

A and **B** can now be ranked using a pairwise comparison of the differential polynomials.

1. Let j be the lowest index such that A_j is ranked differently from B_j (assuming such an index exists). Then $A_j < B_j$ implies that $\mathbf{A} < \mathbf{B}$ while $B_j < A_j$ implies $\mathbf{B} < \mathbf{A}$.

2. If A_i and B_i have equal rank for $= 1, \ldots, \min(r, s)$, then **A** is ranked lower if $r > s$ while **B** is ranked lower if $s > r$.

4.2.14 Examples

1. Let the ranking be given by (40) and consider the autoreduced sets

$$\mathbf{A} : \dot{u}^2 + u, \quad \ddot{y}^5 + uy, \quad \dot{v} + y + u$$

$$\mathbf{B} : u^3\dot{u}^2 - 1, \quad \dot{y} + y, \quad \ddot{v} + uy$$

The first polynomials in each set are equally ranked. The second polynomial in **B** is ranked lower than the corresponding one in **A** and thus **B** is ranked lower than **A**.

2. Let the ranking be given by (40) and consider

$$\mathbf{A} : \dot{u} + u$$

$$\mathbf{B} : u^3\dot{u} - 1, \quad \dot{y} + y$$

Here **B** is ranked lower since it contains more differential polynomials.

4.2.15 Let Σ be a set of differential polynomials. If **A** is an autoreduced set in Σ, such that no lower autoreduced set can be formed in Σ, then **A** is called a *characteristic set* of Σ.

4.3   Applications of characteristic sets

4.3.1 Characteristic sets can be used to make the ideas put forward in section 3.1 concrete and algorithmic. Consider the situation described in 3.1.3, where a system is described by a set of differential equations in the physical variables $z_1, \ldots, z_\nu$. Assume that the corresponding differential ideal is prime. Choose a ranking of the variables, e.g.

$$z_1 < \dot{z}_1 < \cdots < z_2 < \dot{z}_2 < \cdots < z_3 < \dot{z}_3 < \cdots$$

A characteristic set will then typically have the following form

$$A_1(z_1,\ldots,z_m,z_{m+1}), A_2(z_1,\ldots,z_m,z_{m+1},z_{m+2}),\ldots,A_p(z_1,\ldots,z_\nu) \quad (53)$$

where $m + p = \nu$. We can immediately read off the differential transcendence degree, defined in 3.1.12: it is m, and z_1 through z_m form a differential transcendence basis. This is because z_1 through z_m can not be differentially algebraically dependent (if they were, there would be a differential polynomial in only those variables, and it would be possible to form a lower autoreduced set than (53)). On the other hand, (53) shows that the other variables are differentially algebraically dependent on z_1 through z_m. This shows that z_1 through z_m can play the role of inputs to the system and that all other variables are defined by differential equations, once the inputs are known. Of course we could choose a different ranking of the variables and get a different characteristic set with a different natural choice of inputs. However the general theory of 3.1.12 shows that the *number* of inputs would remain the same.

4.3.2 A special case of (53) in the previous example is given by (4) in example 1.1.2. Here u, y, x_2 and x_1 correspond to z_1 through z_4. The differential transcendence degree is one, with u as basis element.

4.3.3 In 3.1.15 it was mentioned that the non-differential transcendence degree of an extension is finite and related to the number of initial conditions needed. By Kolchin [38] this number is called the *order* of the system. In [24] it was shown that
Proposition The order of a system is equal to the sum of the orders of the leaders in a characteristic set.

4.3.4 Examples

1. As we saw in 4.2.12 a state space description corresponds to an autoreduced set which can be shown to be a characteristic set of a certain prime differential ideal. It follows (since the only derivatives are the first order ones of the x_i) that in this case Kolchin's definition of order coincides with the usual one for dynamic systems.

2. Consider again (4) of 1.1.2. We see that the order of the system is 2, which agrees with the state space description.

4.3.5 In 3.1.16 a differentially primitive element was mentioned. Looking again at (4) of 1.1.2 we see that y plays the role of a primitive element, since the other variables are rational expressions in y and its derivatives over the differential field generated by the input.

4.4   Remainders

4.4.1 Example Consider example 1.1.1 where we have the differential polynomials.

$$A_1 = \frac{1}{2}v^2 + y, \quad A_2 = \dot{y} - v \tag{54}$$

We note that this is an autoreduced set for the ranking

$$y < v < \dot{y} < \dot{v} < \cdots$$

Now consider the differential polynomial

$$F = \ddot{y}$$

which describes the acceleration of the particle. We get

$$v(F - A_2') - A_1' + A_2 = -v$$

so we can write

$$vF + v \in [A_1, A_2] \quad \text{or} \quad vF = -v + [A_1, A_2]$$

where $[A_1, A_2]$ denotes the differential ideal generated by A_1 and A_2, i.e. everything that can be generated from those differential polynomials using addition, multiplication with other polynomials and differentiation. We see that v is "what remains of F when A_1 and A_2 are taken into account". Therefore v will be called the *remainder* of F with respect to the autoreduced set (54). We will now generalize this notion.

4.4.2 Suppose we have an autoreduced set

$$\mathbf{A}: \quad A_1, \ldots, A_r$$

and a differential polynomial F. Then, as a first step in generalizing the example of the previous section, we might try to find a differential polynomial R which is partially reduced with respect to $\mathbf{A}$ and such that

$$QF = R + [\mathbf{A}]$$

As in the previous paragraph $[\mathbf{A}]$ denotes all the differential polynomials that can be generated from $\mathbf{A}$ using addition, multiplication with other polynomials and differentiation. What should Q be? In the example we had $Q = v$ where v was the partial derivative of A_1 with respect to v. Let us define the *separant* S_G of a differential polynomial G as

$$S_G = \frac{\partial G}{\partial u_G}, \quad u_G = \text{leader of } G$$

4.4.3 The importance of the separant lies among other things in the following property
Proposition Let the differential polynomial A have the leader $y^{(\nu)}$. Then the k:th derivative of A has the form

$$A^{(k)} = S_A y^{(\nu+k)} + B$$

where the differential polynomial B contains only variables and derivatives of variables which are lower ranked than $y^{(\nu+k)}$.
Proof Differentiating A using the chain rule gives an expression of the form

$$A' = S_A y^{(\nu+1)} + B_1$$

where neither S_A nor B_1 contains $y^{(\nu+1)}$. Since B_1 contains one differentiation of derivatives that were lower ranked than the leader, it must be lower ranked than $y^{(\nu+1)}$, from property (39). Differentiating again gives

$$A'' = S_A y^{(\nu+2)} + B_2$$

where $B_2 = S'_A y^{(\nu+1)} + B'_1$. Again from property (39), B_2 must be lower ranked than $y^{(\nu+2)}$. Proceeding in this fashion one obtains the proof of the proposition.

4.4.4 If F is not partially reduced with respect to $\mathbf{A}$, then there is some derivative in F which can be obtained by differentiating one of the leaders of $\mathbf{A}$. Let us call the highest ranked such derivative the *highest unreduced derivative* of F with respect to $\mathbf{A}$.

4.4.5 We can now state the
Proposition Suppose F is not partially reduced with respect to $\mathbf{A}$. Then there exists a differential polynomial $\tilde{F}$, no higher than F and with a lower highest unreduced derivative, such that

$$S_{A_j}^\sigma F = \tilde{F} + [A]$$

for some A_j in $\mathbf{A}$ and some integer σ.
Proof Let the highest unreduced derivative of F be $y^{(\mu)}$. Since F is not partially reduced, there is some A_j in $\mathbf{A}$, with leader $y^{(\nu)}$ such that $y^{(\mu)} = y^{(\nu+d)}$ with d strictly positive. Write F as a polynomial in $y^{(\mu)}$

$$F = \sum_{k=0}^{e} J_k (y^{(\mu)})^k$$

The d:th derivative of A_j is of the form

$$A_j^{(d)} = S_{A_j} y^{(\nu+d)} + B = S_{A_j} y^{(\mu)} + B$$

Define F_1 as

$$F_1 = S_{A_j}F - J_e(y^{(\mu)})^{e-1}A_j^{(d)} = S_{A_j}\sum_{k=0}^{e-1}J_k(y^{(\mu)})^k - J_e(y^{(\mu)})^{e-1}B$$

If $e = 1$, then F_1 is free of $y^{(\mu)}$. If not, we repeat the process with F replaced by F_1. Since the degree in $y^{(\mu)}$ is decreased each time, we finally end up with a differential polynomial free of $y^{(\mu)}$, which we denote $\tilde{F}$. Since the highest ranked variable in $A_j^{(d)}$ is $y^{(\mu)}$ which has rank lower than or equal to the leader of F, we see that the rank of F is not increased. Since $\tilde{F}$ is free of $y^{(\mu)}$, it has a lower highest unreduced derivative than F

4.4.6 The construction in the proof of Proposition 4.4.5 suggests the following algorithm.

Algorithm 1

Input data: a differential polynomial F and an autoreduced set $\mathbf{A}$.

1. Put $P = F$

2. If P is partially reduced with respect to $\mathbf{A}$ then stop.

3. Let $y^{(\mu)}$ be the highest unreduced derivative of P.
   Let A_j be the element in $\mathbf{A}$ whose leader is a derivative of y.
   Let d be such that $y^{(\nu+d)} = y^{(\mu)}$.

   (a) Put $\tilde{P} = P$

   (b) Let $J_e(y^{(\mu)})^e$ be the higest degree term with $y^{(\mu)}$ in $\tilde{P}$.

   (c) Replace $\tilde{P}$ by
   $$S_{A_j}\tilde{P} - J_e(y^{(\mu)})^{e-1}A_j^{(d)}$$

   (d) If $\tilde{P}$ still contains $y^{(\mu)}$ then go to step 3b

4. Put $P = \tilde{P}$ and go to step 2.

4.4.7 The essential properties of the algorithm are summarized in the **Proposition** Algorithm 1 will terminate after a finite number of steps with a differential polynomial P satisfying

$$\prod_{A_j \in \mathbf{A}} S_{A_j}^{\sigma_j}F = P + [\mathbf{A}]$$

for some nonnegative integers σ_j. P is partially reduced with respect to $\mathbf{A}$ and no higher than F.

Proof The highest unreduced derivative of P will be lower for each new P that is generated by the algorithm. From Proposition 4.2.5 it follows that only a finite number of P:s can be generated. The remainder of the proposition follows directly from Proposition 4.4.5.

4.4.8 Definition The polynomial P that is produced by Algorithm 1 is denoted the *partial remainder* of F with respect to **A**. We will sometimes use the notation

$$P = \text{prem}_{\mathbf{A}}(F)$$

4.4.9 Having developed an algorithm to compute the partial remainder, it is natural to proceed and compute a differential polynomial that is not only partially reduced but reduced with respect to an autoreduced set. To begin with we define the *initial* I_G of the differential polynomial G. This is the coefficient of the highest power of the leader.

4.4.10 Example For the ranking (40)

$$G = u\ddot{y}^3 + y^2\ddot{y}^3 + uy\ddot{y}^2 + \dot{y}^2 + uy$$

has the leader $\ddot{y}$. Consequently the initial is

$$I_G = u + y^2$$

4.4.11 We can now define the counterpart of Algorithm 1.

Algorithm 2
Input data: A differential polynomial A and a differential polynomial P, partially reduced with respect to A.

1. Put $R = P$

2. Let v be the leader of A.
   Write A in the form $A = I_A v^a + B$ where B contains lower degree terms in v.

3. If R is reduced with respect to A then stop.

4. Write R in the form $R = Jv^e + S$ where S contains lower degree terms in v.

5. Replace R with $I_A R - v^{e-a} J A$

6. Go to step 3

The 5:th step of the algorithm will remove the degree v^e-terms. Since the degree is decreased at each step, the algorithm will finally stop at step 3.

4.4.12 Definition If A and P are differential polynomials, P partially reduced with respect to A, then the R resulting from Algorithm 2 is called the *remainder* of P with respect to A and is denoted

$$R = \text{rem}_A(P)$$

4.4.13 This definition is easily extended: Let F be a differential polynomial and $\mathbf{A}$ an autoreduced set. Let $\mathbf{A}$ consist of the differential polynomials $A_1, \ldots, A_p$, ordered in increasing rank. Then the *remainder* of F with respect to $\mathbf{A}$ is denoted $\operatorname{rem}_{\mathbf{A}}(F)$ and is given by

$$\operatorname{rem}_{\mathbf{A}}(F) = \operatorname{rem}_{A_1}\left(\operatorname{rem}_{A_2}(\ldots \operatorname{rem}_{A_p}(\operatorname{prem}_{\mathbf{A}}(F))\ldots)\right)$$

4.4.14 ¿From the development of the algorithm for computing the remainder the following property easily follows. Let F have the remainder R with respect to the autoreduced set $\mathbf{A}$. Then

$$\prod_{A_j \in \mathbf{A}} S_{A_j}^{\sigma_j} I_{A_j}^{\nu_j} F = R + [\mathbf{A}]$$

for some nonnegative integers σ_j and ν_j. The remainder R is reduced with respect to $\mathbf{A}$ and no higher than F.

4.4.15 Example Suppose we have differential polynomials in the three variables u, y and v with the ranking (40). Let

$$F = v^{(4)}$$

$$\mathbf{A}: \quad \dot{y}^2 - 1, \quad \ddot{v} - uy$$

F is not partially reduced since it contains the second derivative of $\ddot{v}$, which is the leader of A_2. Applying Algorithm 1 gives

$$F - \ddot{A}_2 = \ddot{u}y + 2\dot{u}\dot{y} + u\ddot{y}$$

This differential polynomial is not partially reduced either since it contains the derivative of the leader of A_1. One more step of the algorithm gives

$$2\dot{y}(\ddot{u}y + 2\dot{u}\dot{y} + u\ddot{y}) - u \cdot 2\dot{y}\ddot{y} = 2\ddot{u}y\dot{y} + 4\dot{u}\dot{y}^2 = P$$

P is partially reduced with respect to $\mathbf{A}$ and therefore constitutes the partial remainder of F. Using definition 4.4.13 we get

$$\operatorname{rem}_{A_2}(P) = P$$

since P is already reduced with respect to A_2. Finally we get

$$\operatorname{rem}_{\mathbf{A}}(F) = \operatorname{rem}_{A_1}(P) = P - 4\dot{u}A_1 = 2\ddot{u}y\dot{y} + 4\dot{u}$$

4.4.16 Example Let $\mathbf{A}$ be as in the previous example, but let $F = u^{(3)}$. In this case we immediately get

$$\operatorname{rem}_{\mathbf{A}}(F) = F$$

since F is already reduced with respect to $\mathbf{A}$.

4.4.17 Example Let $F = y^{(3)}$ and

$$\mathbf{A}: \quad v\dot{y} - 1, \quad y\dot{v}^2 - 2$$

Then one gets successively

$$P := vF - \ddot{A}_1 = -\ddot{v}\dot{y} - 2\dot{v}\ddot{y}$$

$$P := 2y\dot{v}P + \dot{y}\dot{A}_2 = -4y\dot{v}^2\ddot{y} + \dot{y}^2\dot{v}^2$$

$$P := v\dot{y}^2\dot{v}^2 + 4\dot{v}^3\dot{y}y$$

This is the partial remainder. Continuing gives

$$R := P - 4\dot{y}\dot{v}A_2 = v\dot{y}^2\dot{v}^2 + 8\dot{v}\dot{y}$$

$$R := yR - \dot{y}^2 vA_2 = 8\dot{v}\dot{y}y + 2\dot{y}^2 v$$

$$R := R - 2\dot{y}A_1 = 8\dot{v}\dot{y}y + 2\dot{y}$$

$$R := vR - (2 + 8y\dot{v})A_1 = 2 + 8y\dot{v}$$

This is the remainder.

4.4.18 Example Consider again example 4.4.1. With

$$\mathbf{A}: \quad \frac{1}{2}v^2 + y, \quad \dot{y} - v$$

we can write the result of the calculation as

$$\mathrm{rem}_{\mathbf{A}}(F) = -v$$

4.5   Basic building blocks of Ritt's algorithm

4.5.1 With the tool's of the previous sections it is possible to formulate an algorithm that computes the Lasker-Noether decomposition mentioned in 3.1.5. It also gives characteristic sets for each component. The algorithm is outlined in Ritt's work, [46] and also in [39]. Ritt's algorithm starts with a finite set of differential polynomials

$$\Phi = \{\phi_1, \ldots, \phi_N\}$$

The aim is to construct a finite number of auctoreduced sets

$$\mathbf{A_i}, \quad i = 1, \ldots, n_A$$

each $\mathbf{A_i}$ being a characteristic set of a prime differential ideal Π_i, such that

$$\sqrt{[\Phi]} = \Pi_1 \cap \cdots \cap \Pi_{n_A}$$

where $\sqrt{[\Phi]}$ is the radical ideal of $[\Phi]$ (see 3.1.4).

4.5.2 We assume that we have a function Aul() that computes a characteristic set for a *finite* set of differential polynomials. One building block then uses Aul and the remainder calculations alternately until no new differential polynomials are generated.

4.5.3 Algorithm R Input: A finite set of differential polynomials

$$\Phi = \{\phi_1, \dots, \phi_N\}$$

1. Compute $\mathbf{A} := \mathrm{Aul}(\Phi)$
2. If $\{\mathrm{rem}_{\mathbf{A}}(\phi) \neq 0, \ \phi \in \Phi\} = \emptyset$ go to step 5.
3. Compute some $R := \mathrm{rem}_{\mathbf{A}}(\phi), \ \phi \in \Phi, \ R \neq 0$
4. Put $\Phi := \Phi \cup \{R\}$ and go to step 1.
5. Stop with the ouput $\mathbf{A}$

The algorithm thus computes a function

$$\mathbf{A} = R(\Phi)$$

The ouput $\mathbf{A}$ is an autoreduced set with the property

$$\mathbf{A} \subset [\Phi] \subset [\mathbf{A}] : H_{\mathbf{A}}^{\infty} \tag{55}$$

Here H_A denotes the product of all separants and initials of the A_i and $[\mathbf{A}] : H_{\mathbf{A}}^{\infty}$ is the set

$$\{F \mid H_A^n F \in [\mathbf{A}], \text{ some nonnegative integer } n\}$$

i. e. the set of all differential polynomials whose remainder with respect to $\mathbf{A}$ is zero.

4.5.4 To call this an algorithm is perhaps not completely appropriate since the details of steps 2 and 3 remain open. Of course both steps would probably be done together in a computer implementation. If there are several nonzero remainders, the choice of one might effect the speed of the algorithm greatly. It is possible to augment Φ with several or all nonzero remainders at each step. This would however lead to many unnecessary remainder calculations. Also it is worth noting that all new generated polynomials belong to $[\Phi]$, where Φ is the original set of polynomials. Therefore step 3 actually has to be checked only against those ϕ's.

4.5.5 It would be nice to know if the rightmost set of (55) is prime. This is the task of Ritt's factorization algorithm.

Algorithm Irred Input: An autoreduced set $\mathbf{A} = A_1, \dots, A_p$

1. Replace the leader of each A_i with a new variable z_i

2. Replace each variable and derivative of a variable except the leaders with a new variables w_1, w_2 etc. This can be done in any convenient fashion. Denote the result $\mathbf{B} = B_1, \ldots, B_p$

3. Regard $\mathbf{B}$ as ordinary (non differential polynomials) in $\mathcal{F}[z, w]$ and introduce a ranking so that $z_1, \ldots, z_p$ are still leaders and $\mathbf{B}$ an autoreduced set.

4. Do steps 5,6,7 below for $j = 1, \ldots, p$

5. Let $w = \tau$ and $z_1 = \eta_1, \ldots, z_{j-1} = \eta_{j-1}$ be a generic solution to the prime differential ideal generated by $B_1, \ldots, B_{j-1}$.

6. If B_j is irreducible over $\mathcal{F}(\tau, \eta_1, \ldots, \eta_{j-1})$ then proceed, else stop with output: No.

7. If $j = p$ then stop with output: Yes.

4.5.6 The importance of Ritt's algorithm lies in the following fact.
Proposition If the answer of the algorithm Irred is "Yes", then

$$[\mathbf{A}] : H_A^\infty$$

is a prime differential ideal with $\mathbf{A}$ as a characteristic set.
Proof See [46].

4.5.7 Remark Ritt's factorization test can be done in such a way that, when the answer is "No", two differential polynomials F and G are produced. They have the property that

- F and G are reduced with respect to $\mathbf{A}$
- $FG \in [\mathbf{A}]$

This is described in [46].

4.5.8 Remark If $\mathbf{A} = R(\Phi)$ computed by algorithm R has $H_{\mathbf{A}} \in \mathcal{F}$ then the algorithm *Irred* immediately shows that

$$[\Phi] = [\mathbf{A}]$$

is prime.

4.5.9 Sometimes it is necessary to split a set of differential polynomials. The basic fact is the
Proposition
$$\sqrt{[F_1, \ldots, F_r, GH]} =$$
$$= \sqrt{[F_1, \ldots, F_r, G]} \cap \sqrt{[F_1, \ldots, F_r, H]}$$

4.5.10 Another important fact is the

Proposition Let Φ be a set of differential polynomials, let $\mathbf{A} = R(\Phi)$ and let *Irred* applied to $\mathbf{A}$ give the answer "Yes". Then

$$\sqrt{[\Phi]} = ([\mathbf{A}] : H_A^\infty) \cap \sqrt{[\Phi, H_A]}$$

Proof It is clear that the left hand side is included in the right hand side. Let F belong to the right hand side. Then, for some integer $m \geq 0$, $F^m = F_1 + F_2$ with $F_1 \in [\Phi]$ and $F_2 \in [H_A]$. Also $H_A^r F \in [\mathbf{A}] \subset \sqrt{[\Phi]}$. It follows that $H_A^{(\nu)} F \in \sqrt{[\Phi]}$ for any $\nu \geq 0$. Consequently $F_2 F \in \sqrt{[\Phi]}$, so that $F^{m+1} \in \sqrt{[\Phi]}$.

4.6 Ritt's algorithm

4.6.1 We can now state the basic form of Ritt's algorithm for computation of characteristic sets.

Algorithm Ritt Input: $\Phi = \{\phi_1, \ldots, \phi_n\}$

Output: $\{\mathbf{A}_1, \ldots, \mathbf{A}_{n_A}\} = Ritt(\Phi)$, a set of autoreduced sets.

1. Compute $\mathbf{A} = R(\Phi)$.

2. If $Irred(\mathbf{A}) =$ "Yes", proceed to step 4.

3. Compute F, G according to Remark 4.5.7. Return the output

$$\{Ritt(\{\Phi, F\}), Ritt(\{\Phi, G\})\}$$

4. Return the output

$$\{\mathbf{A}, Ritt(\{\Phi, S_{i_1}\}), \ldots, Ritt(\{\Phi, S_{i_p}\}),$$

$$Ritt(\{\Phi, I_{j_1}\}), \ldots, Ritt(\{\Phi, I_{j_q}\})$$

where S_i and I_j are the separants and initials of $\mathbf{A}$ that do not belong to $\mathcal{F}$.

4.6.2 As defined above, Ritt's algorithm is recursive. It's finiteness depends on two facts. The first one is that Step 4 gives the result $\mathbf{A}$ when $H_A \in \mathcal{F}$. The second one is that a sequence of autoreduced sets with decreasing rank are produced.

4.6.3 It leads to the

Proposition Ritt's algorithm stops after a finite number of calls to itself.

Proof Suppose the algorithm calls itself at step 3. Then, since F, G are reduced with respect to $\mathbf{A}$, $R(\{\Phi, F\})$ and $R(\{\Phi, G\})$ are lower than $\mathbf{A}$. Similarly each of $R(\{\Phi, S_{i_1}\})$ and $R(\{\Phi, I_{i_1}\})$ are lower than $\mathbf{A}$. The algorithm can thus call itself an infinite number of times only if it produces a descending infinite sequence of autoreduced sets, which is impossible according to Proposition 4.2.5.

4.6.4 Another consequence is
Proposition When Ritt's algorithm stops, the returned autoreduced set-s are characteristic sets for prime differential ideals whose intersection is $\sqrt{[\Phi]}$, Φ being the input to the algorithm.
Proof Follows from a combination of Propositions 4.5.6, 4.5.9 and 4.5.10.

4.6.5 A potentially difficult step in Ritt's algorithm is number 2, the irreducibility test. Factorization of polynomials is a procedure with high complexity. In this case the complexity is further increased by the requirement to factorize over the extended fields $\mathcal{F}(\tau, \eta_1, \ldots, \eta_{j-1})$. How this can be done is described e.g. in [28]. The practical importance of the factorization difficulty is diminished by the following points.

- In many applications it is sufficient to know that (55) holds. This is true in some of the identifiability tests considered in [27].

- In some calculations, like those for the input output description discussed in [23], the result tends to be such that all polynomials in the characteristic set, except the first one, are of degree one in their leader. (This corresponds to observability.) In this case there is only an ordinary factorization test for the first polynomial.

- If the polynomials are too complex for a complete factorization test to be made, it is possible to run several "cheap" simpler tests, that pick up some types of factors. Practical experience suggests that in most cases the reducibility is actually due to obvious factorizations.

- If some simple factorization test is made, as suggested above or as in the modified Ritt algorithm suggested in [26], the full irreducibility test has the character of something that is done at the last step to make the result certain (and almost always has the answer "Yes").

5   Examples

5.1   A circuit example

Consider the circuit described in [16], see figure 5.1. The resistors are nonlinear elements described by

$$i = g(v) = \frac{v^3}{3} - v$$

where v is the voltage over the resistor and i is the current. Suppose for simplicity that $C = 1$. Then the circuit is described by

$$
\begin{aligned}
I &= g(v_1) + g(v_1 - v_2) \\
\dot{v}_2 &= g(v_1 - v_2)
\end{aligned}
\tag{56}
$$

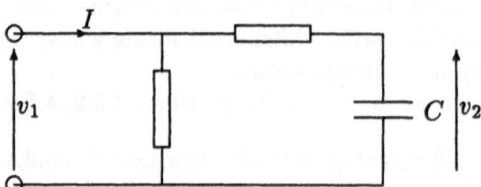

Figure 5.1: Electric circuit with nonlinear resistors

This is not a standard state space description. It can however easily be altered to a characteristic set for the ranking

$$I < \dot{I} < \ddot{I} < \cdots < v_2 < v_1 < \dot{v}_2 < \dot{v}_1 < \cdots$$

The leaders of the differential polynomials corresponding to (56) will then be v_1 and $\dot{v}_2$ respectively. We will not have an autoreduced set, since v_1 has the same power (3) in both polynomials. By using the first equation to eliminate v_1^3 from the second, we can get the description

$$A_1 = 0, \quad A_2 = 0 \tag{57}$$

with the differnetial polynomials

$$A_1 = g(v_1) + g(v_1 - v_2) - I$$

$$A_2 = \dot{v}_2 - \left(-v_2^3 + 3v_1 v_2^2 - 3v_1^2 v_2 + 3v_2 + 3I\right)/6$$

Now A_1, A_2 form a characteristic set (which is the same one we would get using Ritt's algorithm). We see that the differential transcendence degree is one (number of differential polynomials minus number of variables), with I as a possible input. We also see that the order of the system is one (the sum of the orders of all leaders).

We could get a different description by using Ritt's algorithm to eliminate v_1 from the second equation in (56). If we consider the case $I = 1/2$, we get, using the ranking $v_2^{(\cdot)} < v_1^{(\cdot)}$, the following characteristic set.

$$A_1 = 1728\dot{v}_2^3 - 1080v_2^3\dot{v}_2^2 + 2592v_2\dot{v}_2^2 - 1296\dot{v}_2^2 +$$

$$+144v_2^6\dot{v}_2 - 432v_2^4\dot{v}_2 + 540v_2^3\dot{v}_2 - 1296v_2\dot{v}_2 + 324\dot{v}_2 +$$

$$+8v_2^9 - 144v_2^7 - 36v_2^6 + 648v_2^5 + 108v_2^4 - 810v_2^3 + 162v_2 - 27$$

and

$$A_2 = -12v_1 v_2 - 12v_1 \dot{v}_2 + 3v_1 + 4v_1 v_2^3 - 2v_2^4 + 6v_2^2 - 6v_2 + 6v_2\dot{v}_2$$

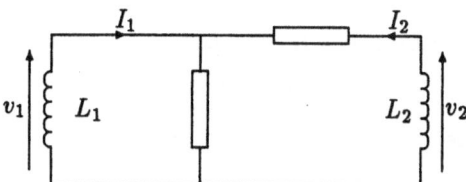

Figure 5.2: Electric circuit with nonlinear resistors and inductors

Note that we have been able to eliminate v_1 to get a differential equation for v_2 ($A_1 = 0$). This is however still an implicit differential equation. Note also that $A_2 = 0$ gives v_1 as a rational expression in v_2 (and its derivative). Thus v_2 is a differentially primitive element.

5.2 A different circuit example

A variation of the previous example which is also described in [16] is shown in figure 5.2. Here the system is described by the equations

$$
\begin{aligned}
g(L_1 \dot{I}_1) + I_1 + I_2 &= 0 \\
g(L_1 \dot{I}_1 - L_2 \dot{I}_2) - I_2 &= 0
\end{aligned}
\tag{58}
$$

This is not a standard state space description, since the derivatives are implicitly defined. Using the ranking

$$
I_1 < I_2 < \dot{I}_1 < \dot{I}_2 < \cdots
$$

we see that the corresponding differential polynomials have the leaders $\dot{I}_1$ and $\dot{I}_2$ respectively. We do not get an autoreduced set directly, since $\dot{I}_1$ is present to the same power in both equations. Using the first equation to eliminate the highest power of $\dot{I}_1$ from the second, we get

$$
A_1 = 0, \quad A_2 = 0
$$

where the differential polynomials A_1 and A_2 are defined by

$$
A_1 = L_1 \dot{I}_1 - \frac{L_1^3 \dot{I}_1^3}{3} - I_1 - I_2
$$

$$
A_2 = 3 L_1^2 \dot{I}_1^2 L_2 \dot{I}_2 - 3 L_1 \dot{I}_1 L_2^2 \dot{I}_2^2 + L_2^3 \dot{I}_2^3 - 3 L_2 \dot{I}_2 + 6 I_2 + 3 I_1
$$

We see that the differential transcendence degree is zero (same number of differential polynomials and variables) corresponding to no inputs. The order of the system is two (sum of the orders of the leaders). The separants are

$$S_1 = L_1(1 - L_1\dot{I}_1)(1 + L_1\dot{I}_1)$$

$$S_2 = 3L_2(L_1\dot{I}_1 - L_2\dot{I}_2 + 1)(L_1\dot{I}_1 - L_2\dot{I}_2 - 1)$$

or in terms of the voltages

$$S_1 = L_1(1 + v_1)(1 - v_1)$$

$$S_2 = 3L_2(-v_1 + v_2 + 1)(-v_1 + v_2 - 1)$$

For values of the voltages where the separants are nonzero, it is possible to solve for the derivatives $\dot{I}_1$ and $\dot{I}_2$ locally (by the implicit function theorem) and there is thus a standard state space description locally. Voltages where one of the separants is zero might correspond to so called *impasse points*, see [16].

5.3   The crane example

Consider the crane of example 1.1.3. Using Ritt's algorithm on the equations (6) gives the differential polynomials

$$A_1 = x^2 - 2xD + D^2 + z^2 - R^2$$

$$A_2 = -z^2R^2\ddot{z} + R^4\ddot{z} + R^2z\dot{z}^2 - 2z^2\dot{z}\,R\dot{R}$$

$$+z^3R\ddot{R} - R^3z\ddot{R} - z^3D\ddot{D} + xz^3\ddot{D} + R^2zD\ddot{D} - z^4 - R^4 - xR^2z\ddot{D} + z^3\dot{R}^2 + 2z^2R^2$$

They form a characteristic set under the ranking

$$D < \dot{D} < \ddot{D} < \cdots < R < \dot{R} < \ddot{R} < \cdots < z < x < \dot{z} < \dot{x} < \cdots$$

We see that the differential transcendence degree is two (number of variables minus number of differential polynomials) with D and R as possible inputs. The order of the system is two (sum of orders of the leaders).

5.4   An example of planar motion

Consider an object moving with constant velocity in the $x - y$-plane. If the velocities are denoted v_x and v_y respectively, one has the system description

$$\dot{x} - v_x = 0, \quad \dot{y} - v_y = 0, \quad \dot{v}_x = 0, \quad \dot{v}_y = 0$$

Suppose the distance to the object is measured with a device that gives only distance but no information of the direction. The output equation is then

$$r - x^2 + y^2 = 0$$

This is a description in state space form. As mentioned before it can also be regarded as a characteristic set under the ranking

$$x < y < v_x < v_y < \dot{x} < \dot{y} < \dot{v}_x < \dot{v}_y < \cdots < r < \dot{r} < \ddot{r} < \cdots$$

In this situation one can however also extract useful information by computing a characteristic set for the ranking

$$r < \dot{r} < \cdots < v_x < \dot{v}_x < \cdots < v_y < \dot{v}_y < \cdots < x < \dot{x} < y < \dot{y} < \cdots$$

The result is the following set of differential polynomials

$$A_1 = r^{(3)}, \quad A_2 = \dot{v}_x, \quad A_3 = 2\,v_y{}^2 - \ddot{r} + 2\,v_x{}^2$$

$$A_4 = 4\,\dot{r}\,v_x x - \dot{r}^2 + 2\,\ddot{r}\,r - 2\,\ddot{r}\,x^2 - 4\,v_x{}^2 r$$

$$A_5 = 2\,yv_y - \dot{r} + 2\,v_x x$$

We see that v_x is not observable, i. e. not algebraic over the differential field generated by r. If it were, then there would be a polynomial, nondifferential in v_x, with coefficients depending on r and its derivatives. This polynomial would then appear in the characteristic set instead of A_2. We also get information about the observability with v_x known. Then v_y would be observable apart from the sign, x would be observable as one of the solutions of a quadratic equation and y would be observable once one has decided upon the choice of solutions for v_y and x. Also note that r satisfies a differential equation of lower order (3 instead of 4) than the original state space description. This is also a consequence of the unobservability, see [24].

6   Conclusions

We have given an outline of a unified appraoach to linear and nonlinear systems, based on algebraic concepts. We have also shown that many of the algebraic concepts have their counterpart in constructive algorithms that can be implemented in computer algebra languages. Finally we have given some examples of the use of these concepts.

References

[1] J. Ackermann, *Sampled-Data Control Systems*, Springer-Verlag, Berlin, 1985.

[2] P. M. Cohn, *Free Rings and their Relations*, Academic Press, London, 1985.

[3] J.M. Coron, Linearized control systems and application to smooth stabilization, *SIAM J. Control Optimiz.*, 1993.

[4] E. Delaleau, Lowering orders of input derivatives in generalized state representations. *Proc. IFAC-Symposium NOLCOS'92, Bordeaux* (1992) 209–213.

[5] E. Delaleau and W. Respondek, Removing input derivatives and lowering their orders in generalized state-space representations. In *Proc. 31st IEEE Control Decision Conf., Tucson.* (1991)

[6] E. Delaleau and M. Fliess, Algorithme de structure, filtrations et découplage, *C.R. Acad. Sci. Paris*, I-315, pp. 101–106, 1992.

[7] S. Diop, Elimination in Control Theory, *Math. Control Signals Systems*, 4, 17–32, 1991.

[8] S. Diop, Differential-algebraic decision methods and some applications to system theory, *Theoretical Computer Science*, 98, 137–161, 1992.

[9] S. Diop and M. Fliess, On nonlinear observability, in C. Commault et al eds., *Proc. 1st European Control Conf.*, Hermës, Paris, 152–157, 1991.

[10] S. Diop and M. Fliess, Nonlinear observability, identifiability and persistent trajectories, in *Proc. 30th IEEE Conf. on Decision and Control*, IEEE Press, New York, 714–719, 1991.

[11] M. Fliess, Automatique et corps différentiels, *Forum Math.*, 227–238, 1989.

[12] M. Fliess, Generalized Controller Canonical Forms for Linear and Nonlinear Dynamics, *IEEE Trans. Automatic Control*, AC-35, 994–1001, 1990.

[13] M. Fliess, Some structural properties of generalized linear systems, *Systems Control Lett*, 15, 1990, pp. 391–396.

[14] M. Fliess, A simple definition of hidden modes, poles and zeros, *Kybernetika*, 27, pp. 186–189, 1991.

[15] M. Fliess, A remark on Willems' trajectory characterization of linear controllability, *Systems Control Lett.*, 19, pp. 43-45, 1992.

[16] M. Fliess and M. Hasler, Questioning the classical state space description via circuit examples. In M. A. Kashoek, J. H. van Schuppen and A. C. M. Ran, editors, *Realization and Modelling in System Theory, MTNS' 89*, Birkhuser, 1990 volume 1, 1–12.

[17] M. Fliess, J. Lévine and P. Rouchon, A simplified approach of crane control via a generalized state-space model. *Proceedings 30th IEEE Control Decision Conf., Brighton*, 1991, 736–741.

[18] M. Fliess, J. Lévine and P. Rouchon, Sur les systèmes non linéaires différentiellement plats, *C. R. Acad. Sci. Paris*, 315-I, 619–624, 1992.

[19] M. Fliess, J. Lévine and P. Rouchon, Index of an implicit time-varying linear differential equation: A noncommutative linear algebraic approach. *Linear Algebra Applications*, to appear.

[20] M. Fliess and F Messager, Vers une stabilisation non linéaire discontinue, in A. Bensoussan and J. L. Lions eds. *Analysis and Optimization of Systems*, Lect. Notes Control Inform. Sci., 144, Springer, Berlin, 778–787, 1990.

[21] M. J. Freedmann and J. C. Willems, Smooth representation of systems with differentiated inputs, *IEEE Trans. Automat. Control* 23 (1978) 16–21.

[22] K. Forsman, *Constructive Commutative Algebra in Nonlinear Control Theory*, Linkping Studies in Science and Technology, 261, 1991.

[23] S. T. Glad, Nonlinear state-space and input-output descriptions using differential polynomials. In J. Descusse, M. Fliess, A. Isidori and D. Leborgne, eds., *New Trends in Nonlinear Control Theory*, vol. 112 Lecture Notes in Control and Information Sciences, Springer verlag, (1989), 182–189.

[24] S. T. Glad, Differential algebraic modelling of nonlinear systems, In M. A. Kashoek, J. H. van Schuppen and A. C. M. Ran, editors, *Realization and Modelling in System Theory, MTNS' 89*, Birkhuser, 1990 volume 1, 97–105.

[25] S. T. Glad, Nonlinear regulators and Ritt's remainder algorithm, in *Colloque International sur l'Analyse des Systèmes Dynamiques Controlés, Lyon, July 3-6, 1990*

[26] S. T. Glad, Implementing Ritt's algorithm of differential algebra, *I-FAC Symposium on Control Systems Design, NOLCOS'92*, Bordeaux, France, 610–614, 1992.

[27] S. T. Glad and L. Ljung, Model Structure Identifiability and Persistence of Excitation, *Proc. 29th CDC, Honululu, Hawaii* , 1990, 3236-3240.

[28] H. M. Edwards, *Galois Theory*, Springer-Verlag, Berlin, 1984

[29] M. Hasler and J. Neirynck, *Circuits non linéaires*, Presses Polytechniques Romandes, Lausanne, 1985; in English: *Nonlinear Circuits*, Artech House, Boston, 1986.

[30] R. Hermann and A. J. Krener, Nonlinear Controllability and Observability, *IEEE Trans. Automat. Control*, 22 (1977) 728–740.

[31] D. Hilbert, Ueber den Begriff der Klasse von Differentialgleichungen,*Math. Ann.*, 73, 1912, pp. 95-108.

[32] A. Isidori, *Nonlinear Control Systems*, 2nd ed., Springer-Verlag, New York, 1989.

[33] J. Johnson, Kähler differentials and differential algebra, *Ann. of Math*, 89, 92–98, 1969.

[34] T. Kailath, *Linear systems*, Prentice-Hall, Englewood Cliffs, N. J., 1980.

[35] R. E. Kalman. Mathematical description of linear systems, *SIAM J. Control*, vol. 1, 152–192, 1963.

[36] R. E. Kalman. *Lectures on Controllability and Observability*, CIME Summer Course (Cremonese, Rome, 1968).

[37] R. E. Kalman, P. L. Falb and M. A. Arbib, *Topics in Mathematical System Theory*, McGraw-Hill, New York, 1969.

[38] E. R. Kolchin, Extensions of differential fields, III, *Bull. Amer. Math. Soc.*, 53, 1947, 397–401.

[39] E. R. Kolchin, *Differential Algebra and Algebraic Groups*, Academic Press, New York, 1973.

[40] E. Kunz, *Kähler differentials*, Vieweg, Braunschweig/Wiesbaden, 1986.

[41] J. W. Nieuwenhuis and J. C. Willems, Deterministic ARMA models, in A. Bensoussan and J. L. Lions eds. *Analysis and Optimization of Systems*, Lecture Notes Control Inform. Sci., 83, 429–439,Springer-verlag, Berlin, 1986.

[42] H. Nijmeijer and A. van der Schaft, *Nonlinear Dynamical Control Systems*, Springer-Verlag, New York, 1990.

[43] F. Ollivier, *Le problème de l'identifiabilité structurelle globale: approche théorique, méthodes effectives et bornes de complexité*, Thèse de Doctorat en Science, École Polytechnique, 1990.

[44] F. Ollivier, Generalized standard bases with applications to control, Proc. European Control Conference, ECC'91, 170–176, 1991.

[45] J. F. Pommaret, *Lie Pseudogroups and Mechanics*, Gordon and Breach, New York, 1988.

[46] J. F. Ritt, *Differential algebra*, American Mathematical Society, Providence, RI, 1950.

[47] C. E. Schrader and M. K. Sain, Research on system zeros: A Survey, *Internat. J. Control*, 50, 1407–1433, 1989.

[48] I.R. Shafarevitch, Basic Notions of Algebra, in A.I. Kostrikin and I.R. Shafarevitch eds, *Algebra I*, Encycl. Math. Sci., Springer-Verlag, Berlin, 1990.

[49] E.D. Sontag, Finite dimensional open loop control generator for nonlinear control systems, *Internat. J. Control*, 47, pp. 537–556. 1988.

[50] E.D. Sontag, Universal nonsingular controls, *Systems Control Lett.*, 19, pp. 221–224. 1992.

[51] H. J. Sussmann and V. Jurdjevic, Controllability of Nonlinear Systems, *J. Diff. Equations*, 12, 95–116, 1972.

[52] B. L. van der Waerden, *Algebra*, Springer-Verlag, Berlin, 1966.

[53] H. Weyl, *The classical groups*, Princeton University Press, Princeton, New Jersey, 1939.

[54] J. C. Willems, From time series to linear systems – Part I: Finite-dimensional linear time invariant systems. *Automatica* 22 (1986) 561–580.

[55] J. C. Willems, From time series to linear systems – Part II: Exact modelling. *Automatica* 22 (1986) 675–694.

[56] J. C. Willems, From time series to linear systems – Part III: Approximate modelling. *Automatica* 23 (1987) 87–115.

[57] J. C. Willems, Models for dynamics, *Dynamics Reported 2*, (1989) 171–269.

[58] J. C. Willems, Paradigms and puzzles in the theory of dynamical systems. *IEEE Trans. Automat. Control* 36 (1991) 259–294

9. Robust Multivariable Control Using H^∞ Methods: Analysis, Design and Industrial Applications

I. Postlethwaite[*]   S. Skogestad [†]

Abstract

The purpose of this paper is to introduce the reader to multivariable frequency domain methods including $\mathbf{H^\infty}$-design. These methods provide a direct generalization of the classical loop-shaping methods used for SISO systems. We also aim to provide a basic understanding of how robustness problems arise, and what analysis and design tools are available to identify and to avoid them.

As an introduction to the robustness problems in multivariable systems we discuss the control of a distillation column. Because of strong interactions in the plant, a decoupling control strategy is extremely sensitive to input gain uncertainty (caused by actuator uncertainty). These interactions are analyzed using singular value decomposition (SVD) and relative gain array (RGA) methods.

We then discuss possible sources of model uncertainty, and look at the traditional methods for obtaining robust designs, such as gain margin, phase margin and maximum peak criterions (M-circles). However, these measures are difficult to generalize to multivariable systems. In such cases a more detailed modelling of the uncertainty in terms of norm-bounded perturbations (Δ's) is used. The frequency-domain is particularly well suited for representing non-parametric (unstructured) uncertainty. To test for robust stability and performance in the presence of model uncertainty, the structured singular value, μ, provides a powerful tool.

The latter part of the paper is concerned with design and in particular multivariable loop shaping of singular values of appropriately specified transfer functions. One and two degrees of freedom controllers are considered, and the paper ends with a case study on advanced control of high performance helicopters.

---

[*]Engineering Department & Chemical Engineering, University of Leicester, Leicester LE1 7RH, UK. E-mail: ixp@le.ac.uk. Tel: +44-533-522546. Fax: +44-533-522619

[†]Dept. of Chem. Engineering, University of Trondheim, NTH, 7034 Trondheim, Norway. E-mail: skoge@kjemi.unit.no. Tel: +44-533-522546. Fax: +44-533-522619.

Contents

1   Introduction

The main goal of this introduction is to answer the following question: Why use the frequency domain ($\mathbf{H}^{\infty}$-norm) for defining performance and describing uncertainty? We will also discuss the two main approaches to $\mathbf{H}^{\infty}$-design, namely the loop-shaping and the signal-oriented approaches.

We use $\|M\|_{\infty}$ to denote the $\mathbf{H}^{\infty}$-norm of a linear transfer function $M(s)$. For the scalar case, $\|M\|_{\infty}$ is simply equal to the peak magnitude $\sup_{\omega}|M(j\omega)|$, where *sup* denotes *the least upper bound*, which for all practical purposes is equal to the maximum value. For the multivariable case we, "sum up" the channels using the singular value and we have

$$\|M\|_{\infty} \overset{\text{def}}{=} \sup_{\omega} \bar{\sigma}(M(j\omega)) \tag{1}$$

1.1   The loop shaping approach

This approach to control system design could also be called the classical approach, the engineering approach, the frequency domain approach or the transfer function approach.

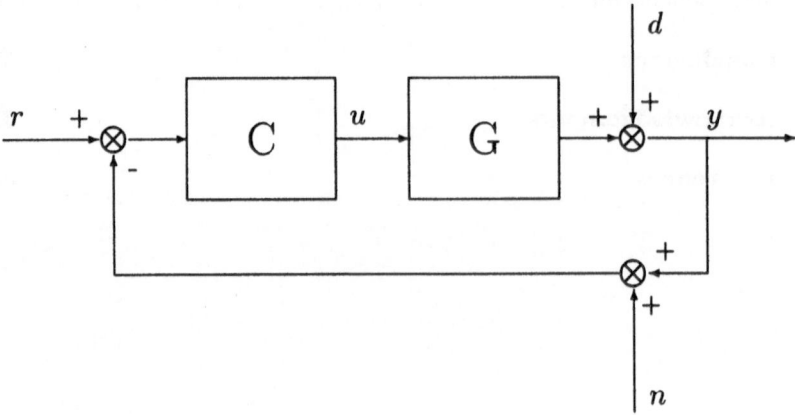

Figure 1.1: Conventional feedback control system.

Consider the conventional feedback system in Fig. 1.1 where $G(s)$ is the plant and $C(s)$ the feedback controller.

By "loop shaping" one traditionally means a design approach where one specifies directly the shape of the magnitude of the open-loop transfer function $L = GC$. However, we shall use the term in a wider sense and also allow the specification of closed-loop transfer functions such as $S = (I + GC)^{-1}$. We use the following definition: *Loop-shaping is any design method that involves directly specifying the magnitude of one or more nominal transfer functions*. To distinguish between various approaches we will talk about "shaping L" or "shaping S", and so on.

Shaping L. For single-input single-output (SISO) systems the specifications in terms of the *open-loop* transfer function $L = GC$ typically include:

i) Crossover frequency, ω_c (defined as $|L(j\omega_c)| = 1$).

ii) System type (defined as number of integrators in $L(s)$).

iii) The shape of $L(j\omega)$, e.g., in terms of the slope of $|L(j\omega)|$ in certain frequency ranges.

iv) Phase margin, PM (given by the phase of $L(j\omega_c)$).

v) Gain margin, GM (given by the gain of L at the frequency where its phase is $-180°$).

The first three specifications have to do with performance in terms of speed of response and allowable tracking error. The last two specifications are included to avoid some of the potential difficulties with feedback: 1) the closed-loop system may become unstable, 2) noise and disturbances in a certain frequency range close to the bandwidth frequency may get amplified.

Specifications directly on $L = GC$ make the design procedure simple as it is clear how changes in the controller affect $L(s)$, and this approach is well suited for non-formalized design procedure. Indeed, towards the end of this paper we shall discuss the MacFarlane-Glover loop-shaping procedure where the initial step in the controller design is to select a reasonable $L(s)$.

Shaping closed-loop transfer functions. In may cases one prefers to define specifications in terms of a *closed-loop* transfer function for the following reasons: 1) The final performance we want to evaluate is that of the closed-loop system. 2) The robustness specifications in terms of GM and PM are difficult to generalize to MIMO systems. 3) Synthesis is difficult if the specifications are in terms of L (one may have to resort to numerical procedures such as the Method of Inequalities (MOI) described towards the end of the paper).

Shaping S. The closed-loop sensitivity function, $S = (I + GC)^{-1}$, is a very good indicator of performance. Typical specifications in terms of S include:

i) Minimum bandwidth frequency ω_B (defined as the smallest frequency at which $\bar{\sigma}(S(j\omega)) = 0.707$)

ii) Allowable tracking error at selected frequencies.

iii) System type, or if system contains no integrators, the allowed static tracking error, A.

iv) The shape of S over selected frequency ranges.

v) Maximum allowed peak magnitude for S, $\|S(j\omega)\|_\infty = M_s$.

The peak specification prevents amplification of noise at high frequencies, and also introduces a margin of robustness; typically we select $M_s = 2$. Mathematically, these specifictions may be captured simply by an upper bound, $1/|w_P|^1$ on the magnitude of S, namely

$$\bar{\sigma}(S(j\omega)) \leq 1/|w_p(j\omega)|, \forall \omega \quad \Leftrightarrow \quad \|w_P S\|_\infty \leq 1 \qquad (2)$$

A typical upper bound is shown in Fig. 1.2. The weight illustrated on that plot may be represented as

$$w_P(s) = \frac{s/M_s + \omega_B}{s + \omega_B A} \qquad (3)$$

and we see that $|w_P(j\omega)|^{-1}$ is equal to $A \leq 1$ at low frequencies, is equal to $M_s \geq 1$ at high frequencies, and the asymptote crosses 1 at the bandwidth frequency, ω_B.

The loop shape $L = \omega_B/s$ yields an S which exactly matches the bound (3) at frequencies just below the bandwidth and easily satisfies the bound at other frequences. This L has a slope ("roll-off") in the frequency range below the bandwidth of about -1 on a log-log plot (-20 dB/decade). In many cases, in order to improve performance, we may want a steeper slope for L (and S) in some frequency range below the bandwidth, and a higher-order weight may be selected.

Stacked requirements. The specification $\|w_P S\|_\infty \leq 1$ does not allow us to specify an *upper* bound on the bandwidth or the "roll-off" of $L(s)$ above the bandwidth. To specify this one needs to make a specification on another transfer function, for example, on the complementary sensitivity $T = I - S = GCS$. Also, one may want to bound other transfer functions, to achieve robustness or to avoid too large input signals.

As an example, one may define an upper bound, $1/|w_T|$ on the magnitude of T to make sure the system behaves "nicely" at high frequencies,

---

[1] Subscript P stands for performance since S is mainly used as a performance indicator

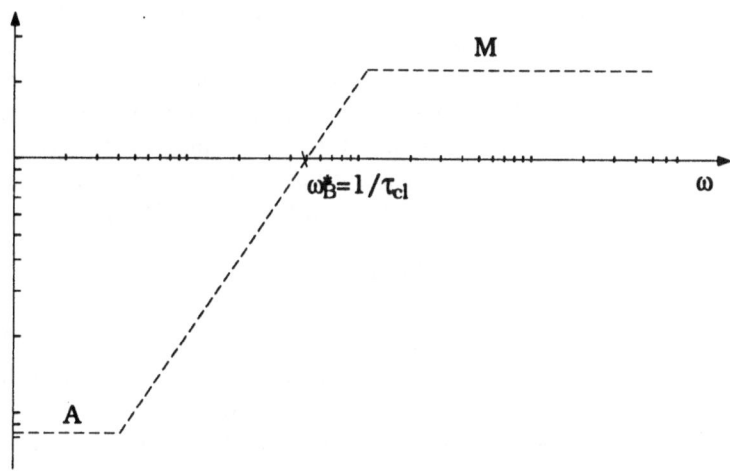

Figure 1.2: Upper bound on $\bar{\sigma}(S)$ given by weight $1/|w_P(j\omega)|$ in (3).

and an upper bound, $1/|w_u|$, on the magnitude of CS to avoid large input signals. To combine these specifications, the "stacking approach" is usually used, resulting in the following specification:

$$\|M\|_\infty = \sup_\omega \bar{\sigma}(M(j\omega)) \leq 1; \quad M = \begin{pmatrix} w_P S \\ w_T GCS \\ w_u CS \end{pmatrix} \tag{4}$$

The "mixed-sensitivity" specification with $M = \begin{pmatrix} w_P S \\ w_T GCS \end{pmatrix}$ is used, for example, by Chiang and Safonov (1988, 1992) in the Matlab Robust Control Toolbox Manual.

The "stacking procedure" is selected for mathematical convenience as it does not allow us to exactly specify the bounds on the individual transfer functions as was described above. For example, assume that $\phi_1(C)$ and $\phi_2(C)$ are two real scalars (here we could have $\phi_1(C) = \|w_P S\|_\infty$ and $\phi_2(C) = \|w_T GCS\|_\infty$) and that we want to achieve

$$\phi_1 \leq 1 \quad \text{and} \quad \phi_2 \leq 1 \tag{5}$$

This is not quite the same as the "stacked" requirement

$$\bar{\sigma}\begin{pmatrix} \phi_1 \\ \phi_2 \end{pmatrix} = \sqrt{\phi_1^2 + \phi_2^2} \leq 1 \tag{6}$$

The two requirements are quite similar when either ϕ_1 or ϕ_2 is small, but in the worst case when ϕ_1 and ϕ_2 are equal, we get from (6) that $\phi_1 \leq 0.707$

and $\phi_2 \leq 0.707$, that is, there is a possible "error" equal to a factor $\sqrt{2} \approx 3$ dB. In general, with n stacked requirements the resulting error is at most $\sqrt{n}$. This inaccuracy in the specifications is something we are probably willing to sacrifice in the interests of mathematical convenience. In any case, the specifications are in general rather rough, and are effectively knobs for the engineer to select and adjust until a satisfactory design is reached.

The $\mathbf{H}^\infty$-optimal controller is obtained by solving the problem

$$\min_C \|M(C)\|_\infty \tag{7}$$

Provided M can be written as a linear fractional transformation (LFT) of C, $M(C) = N_{11} + N_{21}C(I - N_{22}C)^{-1}N_{12}$, the solution is easily obtained with standard software (e.g. the Robust Control or Mu Toolboxes in Matlab). In practice, to be able to write M as an LFT of C, one must be able to represent M by a block diagram with the input (or output) at only one location. For example, the M in (4) may be represented in a block diagram with a single input entering at the output of the plant as shown in Fig.1.3.

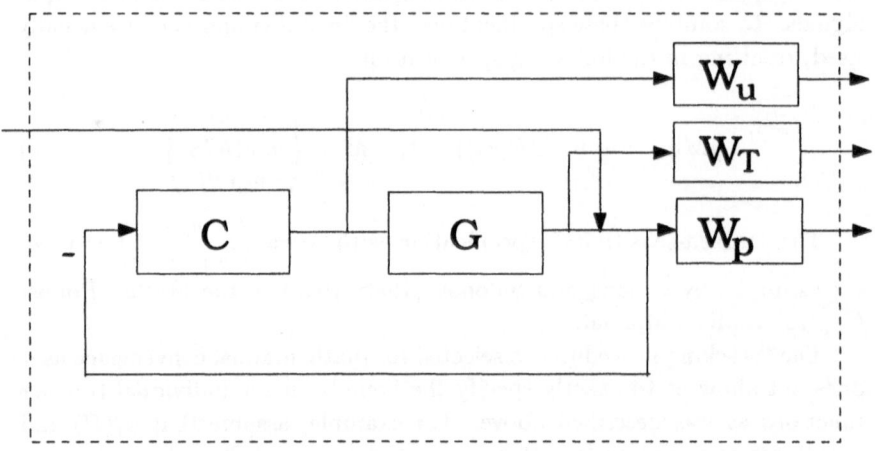

Figure 1.3: Block digram corresponding stacked requirement in (4).

Let $\gamma_0 = \min_C \|M(C)\|_\infty$ denote the optimal $\mathbf{H}^\infty$-norm for the problem in (7). An important property of $\mathbf{H}^\infty$-optimal controllers is that it will yield a flat frequency response, that is, it will yield $\bar{\sigma}(M(j\omega)) = \gamma_0$ at all frequencies. *The practical implication is that, except for at most a factor $\sqrt{n}$, the transfer functions resulting from solving (7) will be very close to γ_0 times the bounds selected by the designer.* This means, that the designer may almost exactly specify the final shape of, for example, $\bar{\sigma}(S)$, $\bar{\sigma}(GCS)$ or $\bar{\sigma}(CS)$.

Remark. For cases where M cannot be written as an LFT of C, which is a special case of the Hadamard-weighted $\mathbf{H}^{\infty}$-problem studied by van Diggelen and Glover (1991, 1992), the solution to the $\mathbf{H}^{\infty}$-problem in (7) remains intractable. Van Diggelen and Glover (1991,1992) do, however, present a solution for a similar problem where the Frobenius norm is used instead of the singular value to "sum up the channels" in the $\mathbf{H}^{\infty}$-norm.

Summary. The classical loop-shaping approach to controller design involves direct specifications of how the final solution should be in terms of the magnitude frequency response. It requires the engineer to be able to formulate bounds that lead to acceptable robustness and closed-loop performance. This approach is often preferred because it has few adjustable design parameters (knobs) and directly involves the engineer in the design. We shall return with a more detailed discussion on the physical significance of some transfer functions which the engineer may want to bound in Section 5.1.

1.2 The signal-oriented approach

The signal-oriented approach is very general, and may be more appropriate for multivariable problems in which a number of objectives must be taken into account simultaneously. Here we define the plant, including possibly the model uncertainty, define the class of external signals affecting the system and define the norm of the "error signals" we want to keep small. Direct bounds on selected transfer functions, such as the closed-loop bandwidth, cannot be specified in this case. On the other hand, one may argue that the concept of bandwidth is difficult to use for complex systems.

The "modern" state space methods from the 60's, such as LQG control, are based on the signal-oriented approach. Here the input signals are assumed to be stochastic (or alternatively impulses in a deterministic setting) and the output signals are measured in terms of the 2-norm. These methods may be generalized to include frequency dependent weights on the signals leading to what is called the Wiener-Hopf or H_2-norm design method.

Sinusoidal signals and the $\mathbf{H}^{\infty}$-norm. We may also consider the system response to persistent sinusoidal signals of varying frequency. This leads to the signal- oriented $\mathbf{H}^{\infty}$-norm approach used, for example, by Doyle et al. (1987) in their space-shuttle application. A signal-oriented problem specification with disturbances, commands and noise, and with bounds on both the input and outputs is shown in Fig. 1.4. The overall performance objective is that $\|E\|_{\infty} \leq 1$. For more details the reader is referred to Lundström et al. (1991).

The $\mathbf{H}^\infty$-norm may be interpreted in other ways, such as the induced 2-norm from inputs to outputs. In any event, as far as signals are concerned

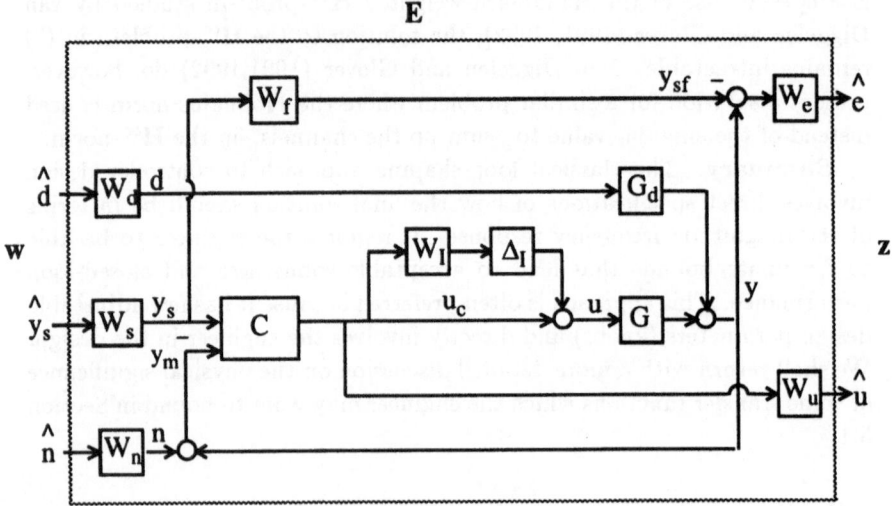

Figure 1.4: Typical block diagram for signal-oriented approach to $\mathbf{H}^\infty$-performance.

there does not seem to be an overwhelming case for using the $\mathbf{H}^\infty$-norm rather than the more traditional H_2-norm (LQG). When we begin to consider issues of model uncertainty, however, the frequency domain approaches such as $\mathbf{H}^\infty$ are preferred.

Model uncertainty. The traditional method of dealing with robustness and model uncertainty within the framework of "optimal control" (LQG) has been to introduce uncertain signals (noise and disturbances). One particular approach is loop transfer recovery (LTR) where unrealistic noise is added specifically to obtain a robust design. Of course, one may say that model uncertainty generates some sort of disturbance, but this disturbance depends on the other signals in the system and thus introduces an element of feedback. Therefore there is a fundamental difference between these two sources of "uncertainty" (at least for linear systems): model uncertainty may introduce instability, whereas signal uncertainty can not.

A more direct way to handle robustness issues within the signal approach, is to model the uncertainty explicitly. It appears that the frequency domain is very well suited for describing model uncertainty, and in particular for describing non-parametric uncertainty, resulting, for example, by using a simplified low-order model of a high-order plant. Indeed, Owen and Zames (1991) make the following observation in a recent paper: *"The design of feedback controllers in the presence of nonparametric and unstructured uncertainty ... is the raison d'être for* $\mathbf{H}^\infty$ *feedback optimization, for if disturbances and plant models are clearly parameterized* $\mathbf{H}^\infty$*-methods seem to offer no clear advantages over more conventional state-space and parametric methods."*

If the $\mathbf{H}^\infty$-norm is selected for the uncertainty, then, again for mathematical convenience, one may also want to select the $\mathbf{H}^\infty$-norm for performance. This leads to a robust performance (RP) problem, for which there exists a very efficient analysis tool, namely the structured singular value, μ. But, for controller synthesis there are difficulties: the μ-synthesis problem, in its full generality, is not yet solved mathematically; where solutions exist the controllers tend to be high order; the available algorithms may not always converge and design problems are sometimes difficult to specify directly.

1.3   Combined approaches

The loop shaping approach and the signal approach above may, of course, be combined. One such approach is "loop-shaping with uncertainty" as used, for example, by Skogestad et al (1988). Here performance is specified in terms of an upper bound on the sensitivity S, leading to the specification $\|w_P S\|_\infty \leq 1$. One then takes a "worst-case" approach and requires that this bound is satisfied for all plants as defined by the uncertainty description. For the case of input uncertainty of relative magnitude w_I this leads to the following robust performance analysis condition (see Eq.48)

$$RP \quad \text{iff} \quad \mu(N(j\omega)) < 1, \ \forall \omega; \quad N = \begin{pmatrix} w_I CSG & w_I CS \\ w_P SG & w_P S \end{pmatrix} \tag{8}$$

where μ is the structured singular value computed with respect to a special block-diagonal structure. This is similar to the $\mathbf{H}^\infty$-condition in (4), but in (8) the bounds on the transfer functions are specified indirectly, and it is not clear what nominal transfer functions CSG, CS, SG and S are allowed. Specifically, recall that the $\mathbf{H}^\infty$-optimal controller would essentially result in a controller which matches all the bounds at all frequencies (except for at most a factor $\sqrt{n}$). On the other hand, the μ-optimal design, may result in a design where, for example, $\bar{\sigma}(w_P S)$ is very small at some frequency and large at another frequency. It is therefore not clear from the specifications what the final (nominal) design will be.

However, the μ-approach does have definite advantages since we do know that the worst-case sensitivity function S_p (p stands for perturbed) will exactly satisfy our requirements, i.e., for all possible perturbed plants

$\|w_P S_p\|_\infty \leq 1$. Whereas for the $\mathbf{H}^\infty$-problem we only specify nominal transfer functions and must make sure by specifying these carefully that robustness is achieved. When applied to design, the approach (8) has the usual problems associated with μ-synthesis: computational difficulties, high-order controllers, and the indirect specification of individual transfer functions . From the above discussion then, it follows that μ-analysis may be very useful for checking the robustness of designs obtained, for example, by an $\mathbf{H}^\infty$ design procedure.

Another "combined approach" is the the "Glover-McFarlane loop-shaping procedure". In this, one first specifies the desired *open-loop* shape $L(s)$ for performance using simple pre- and post- compensators. One then "robustifies" this design by considering a particular robust stability condition, which involves solving an $\mathbf{H}^\infty$-problem. This procedure is further described in Section 5 and used in the design example in Section 6.

1.4 Summary

We have considered two alternative approaches to controller design: the loop- shaping approach and the signal approach. In both cases we find the frequency- domain to be the natural setting. The loop-shaping approach, with direct specifications on bandwidth etc, is directly based in the frequency domain so here there is no alternative. For the signal-oriented approach there are a variety of ways to define the signals. The reason why the frequency-domain ($\mathbf{H}^\infty$-norm) is again preferred is that it is very well suited to handling unstructured model uncertainty.

In a practical design situation, the above two approaches may be combined. For example, one may design the controller (Step B below) by some loop-shaping approach (involving $\mathbf{H}^\infty$-synthesis) and then analyze the solution (Step C below) using a signal-oriented approach with model uncertainty explicitly included (involving μ-analysis).

This paper is concerned with analysis and design of control systems for industrial plants. In this case the designer must usually go through the following steps:

Step A. Controllability analysis: This is where the plant is analysed and we discover what closed-loop performance can be expected, what the limitations are, how good the control might be.

Step B. Controller design: This is where the design problem is formulated and the controller synthesized.

Step C. Control system analysis: This is where the feedback control system is assessed by analysis and simulation to judge how well it might behave in practice.

With the above steps in mind, the following topics are covered in the remainder of this paper.

Section 2: Analysis of the plant - controllability

Section 3: Robustness problems - Introductory distillation example

Section 4: Tools for robustness analysis

Section 5: Robust control system design

Section 6: A case study.

1.5   Notation

G - nominal plant model
M - matrix used to test for robust stability (section 1-4) or coprime factor of G (section 5)
RGA - matrix of relative gains, $= G \times (G^{-1})^T$ where $\times$ represents element-by-element multiplication.
s - Laplace variable ($s = j\omega$ yields the frequency response)
$S = (I + GC)^{-1}$ - sensitivity function
$T = GC(I + GC)^{-1}$ - closed-loop transfer function
$T_I = CG(I + CG)^{-1}$ - closed-loop transfer function from the plant input
w and W_i - frequency-dependent weighting functions

Greek letters
Δ - overall perturbation block used to represent uncertainty
Δ_I - perturbation block for input uncertainty
$\gamma(A) = \bar{\sigma}(A)/\underline{\sigma}(A)$ - condition number of matrix A
$\mu(A)$ - structured singular value of matrix A
ω - frequency [rad/s or rad/min]
$\bar{\sigma}(A)$ - maximum singular value of matrix A
$\underline{\sigma}(A)$ - minimum singular value of matrix A

Subscripts
p - perturbed (with model uncertainty)
P - performance

2   Analysis of the plant - Controllability

Before attempting to start any controller design one should have some idea of how easy the plant actually is to control. Is it a difficult control problem? Indeed, does there even exist a controller which meets the required performance objectives? It appears that the frequency-domain is very well suited for answering such problems in a general setting. One reason for this is the very useful idea of "bandwidth" which is a purely frequency-domain concept. The concept of right half plane (RHP) zeros is also of fundamental importance in answering questions of the kind.

In this paper the term "controllability" (of a plant) has the meaning of "inherent control characteristics of the plant" or maybe better "achievable performance" (irrespective of the controller). This usage is in agreement with most persons intuitive feeling about the term, and was also how the term was used historically in the control literature. For example, Ziegler and Nichols (1943) define controllability as *"the ability of the process to achieve and maintain the desired equilibrium value"*.

Unfortunately, in the 60's Kalman defined the term "controllability" in the very narrow meaning of "state controllability". This concept is of interest for realizations and numerical calculations, but as long as we know that all the unstable modes are both controllable and observable, it has almost no practical significance.

It would be desirable to have a more precise definition of controllability, but on the other hand this is difficult and probably not useful. An exact definition would require selection of a certain norm to measure the control error, and would also require a detailed specification of all external signals such as noise, reference signals and disturbances. Indeed, Ziegler and Nichols (1943) note in their paper that although they "took the area under a recovery curve as one measure of controllability ... this is only one of many possible bases for comparison of control results". They also stress that it is difficult to narrow controllability down to one single attribute of the plant. They say: "Unfortunately, the authors are not able to give a formula for controllability. It appears that when such a factor is devised it will consist of several factors. One might be called the "recovery factor", the ability of the process to recover from the maximum change in demand or load. Another, a "load factor" must take into account the point in the process at which the disturbance occurs". Later in the paper they state that the total integrated control error, $\int |e(t)|dt$, is equal to: (Load Factor) · (Recovery Factor).

Essentially, the "recovery factor" depends on the process model, $g(s)$, and recovery is poor (and thus the recovery factor is large) if it contains large time delays or if the plant gain is small. The "load factor" expresses the effect of the disturbances and thus depends on the disturbance model, $g_d(s)$. These concepts are very similar to the ideas summarized below in terms of upper and lower bandwidth limitations.

2.1 Summary of controllability results for SISO plants

Consider the control system in Fig. 1.1 for the case when all blocks are scalar. The control error $e = y - r$ may be written

$$e(s) = g(s)u(s) + g_d(s)d(s) - r(s) \tag{9}$$

We assume that the signals are persistent sinusoids, and assume that g and g_d are scaled, such that at each frequency the allowed input $|u(j\omega)| < 1$, the expected disturbance $|d(j\omega)| < 1$, and the outputs are scaled such that the expected reference signal $|r(j\omega)| < 1$.

Below we have given some "controllability" requirements which apply to the closed-loop bandwidth, ω_B. The requirements are fundamental, although the expressions given in terms of bounds on ω_B are not exact. However, in practice they must be fulfilled.

i) *Disturbances. Must require $\omega_B > \omega_d$. Here ω_d is the frequency at which $|g_d(j\omega_d)|$ crosses 1 from above. Below this frequency the error will be unacceptable ($|e| > 1$) for a disturbance $d = 1$ unless control is used. More specifically, we must for feedback control require at frequencies lower than ω_d: $|gc(j\omega)| > |g_d(j\omega)|$.*

ii) *Commands (setpoints). Specify directly minimum required ω_B. This requirement comes in addition to the bandwidth requirement imposed by the disturbances, and is usually relatively easy to specify.*

iii) *Open-loop unstable pole at $s = p$. Must require $\omega_B > 1/|p|$. We need fast control to stabilize the system, and the bandwidth must approximately be greater than $1/|p|$ where $|p|$ is the distance of the RHP-pole from the origin.*

iv) *Input constraints, must require $|g(j\omega)| > 1, \forall \omega < \omega_B$. This is needed to avoid input constraints ($|u(j\omega)| < 1$) for perfect tracking of $r(j\omega) = 1$.*

v) *Input constraints, must require $|g(j\omega)| > |g_d(j\omega)|$, $\forall \omega < \omega_d$. This is needed to avoid input constraints for perfect rejection of disturbance $d(j\omega) = 1$.*

The above two conditions are requirements that the plant must satisfy in order to be able to apply tight control in a certain frequency reange.

They are independent of the controller, and can therefore be affected only by changing the plant $g(s)$.

In the frequency range up to the bandwidth ω_B there should not be any time delays, RHP-zeros and high-order plant dynamics that need to be counteracted. We get

vi) *Time delay θ. Must require $\omega_B < 1/\theta$.*

vii) *RHP-zero at $s = z$. Must require $\omega_B < |z|$.*

Note that RHP-zeros close to the origin are the worst. LHP-zeros pose no fundamental limitation, but a LHP-zero close to the origin yields an "overshoot" in the open-loop response which may be difficult to counteract. Therefore, to simplify controller design and avoid robustness problems, it is often best to have the LHP-zeros as far away from the origin as possible.

The above two constraints for time delays and RHP-zeros are fundamental, but the above relationships are rather approximate. Also, if there are combinations of both RHP-zeros and time delays then they must be considered combined, because they all make feedback control difficult (simply consider the overall phase lag).

viii) *In most practical cases: $\omega_B < \omega_{180}$.*

Here ω_{180} is the frequency at which the phase of $g(j\omega)$ is -180°. This condition is not a fundamental limitation, but more of a practical limitation. In particular it applies if the phase drops rather quickly around the frequency ω_{180}. The condition follows since in most cases the plant is not known sufficiently accurately to place zeros to counteract the poles at high frequency.

2.2 Controllability analysis for multivariable plants

We do not have space to go into detail about the controllability analysis of multivariable plants, but most of the ideas presented above may be generalized, e.g., see Wolff et al. (1992). Instead we will summarize some of the main tools which are used. All of them are based on the plant model $G(s)$ and the disturbance model $G_d(s)$.

i) Compute the multivariable RHP-poles and RHP-zeros and their associated directions. Test for functional controllability (the rank of G should equal the number of outputs).

ii) Perform a frequency-dependent SVD-analysis to understand the multivariable directions.

iii) Perform a frequency-dependent RGA-analysis to check for fundamental limitations due to inherently coupled outputs. Compute the plant condition number.

iv) Evaluate disturbance sensitivity. For decentralized control the use of the CLDG-matrix, $G_{diag}G^{-1}G_d$, directly generalizes the SISO results. Here G_{diag} is a diagonal matrix consisting of the diagonal elements of G. For the general case it is more complicated, but an SVD-analysis of G_d and $G^{-1}G_d$ yields useful information about which disturbances are difficult, and the bandwidth requirement in certain directions.

The above tools for controllability analysis are simple indicators which are easy to compute, and help the engineer to obtain insight into what the control problems are for the plant in question. In some cases a more detailed analysis which includes finding the optimal controller may be desirable. A suitable tool is the structured singular value μ (which must be minimized over all controllers to find the achievable performance for the problem). However, the use of such methods requires a careful definition of the performance specifications and model uncertainty which is often not available or which requires a significant effort to obtain.

Although, there has been good progress during the last few years, the area of controllability analysis is still a very interesting area for future research.

3 Robustness - Introductory distillation column example

An idealized distillation column example will be used to introduce the reader to the adverse effects of model uncertainty, in particular for multivariable plants. The example is taken mainly from Skogestad et al. (1988)

Before considering the example a short introduction to robustness and uncertainty seems in order.

3.1 Robustness and model uncertainty

A control system is robust if it is insensitive to differences between the actual system and the model of the system which was used to design the controller. Robustness problems are usually attributed to differences between the plant model and the actual plant (usually called model/plant mismatch or simply model uncertainty). Uncertainty in the plant model may have several origins:

i) There are always parameters in the linear model which are only known approximately or are simply in error.

ii) Measurement devices have imperfections. This may even give rise to uncertainty on the manipulated inputs, since the actual input is often measured and adjusted in a cascade manner. For example, this is often the case with valves where we measure the flow. In other cases limited valve resolution may cause input uncertainty.

iii) At high frequencies even the structure and the model order is unknown, and the uncertainty will exceed 100% at some frequency.

iv) The parameters in the linear model may vary due to nonlinearities or changes in the operating conditions.

v) In addition, the controller implemented may differ from the one obtained by solving the synthesis problem, and one may include uncertainty to allow for controller order reduction and implementation inaccuracies.

Other considerations for robustness include measurement and actuator failures, constraints, changes in control objectives, opening or closing other loops, etc. Furthermore, if a control design is based on an optimization then robustness problems may also be caused by the mathematical objective function, that is, how well this function describes the real control problem.

In the somewhat narrow use of the term used in this paper, we shall consider robustness with respect to model uncertainty, and assume that a fixed (linear) controller is used. Intuitively, to be able to cope with large changes in the process, this controller has to be detuned away from the best response we might have achieved if the process model was exact.

To consider the effect of model uncertainty, the uncertainty needs first to be quantified in some way. There are several ways of doing this. One powerful method is the frequency domain (so-called H-infinity uncertainty description) in terms of norm-bounded perturbations (Δ's). With this approach one can also take into account unknown or neglected high-frequency dynamics.

The following terms are useful:

- Nominal stability (NS). The system is stable with no model uncertainty.

- Nominal Performance (NP). The system satisfies the performance specifications with no model uncertainty.

- Robust stability (RS). The system is stable for all perturbed plants about the nominal model up to the worst-case model uncertainty.

- Robust performance (RP). The system satisfies the performance specifications for all perturbed plants about the nominal model up to the worst-case model uncertainty.

3.2   The distillation column model

We consider two-point (dual) composition control of a distillation column. The overhead composition of a distillation column is to be controlled at $y_D = 0.99$ (output 1) and the bottom composition at $x_B = 0.01$ (output 2), with reflux L (input 1) and boilup V (input 2) as manipulated inputs for composition control, i.e.,

$$y = \begin{pmatrix} \Delta y_D \\ \Delta x_B \end{pmatrix}, \quad u = \begin{pmatrix} \Delta L \\ \Delta V \end{pmatrix}$$

By linearizing the steady-state model and assuming that the dynamics may be approximated by a first order response with time constant $\tau = 75$ min, we derive the following linear model in terms of deviation variables

$$\begin{pmatrix} y_1 \\ y_2 \end{pmatrix} = G \begin{pmatrix} u_1 \\ u_2 \end{pmatrix}, \quad G(s) = \frac{1}{\tau s + 1} \begin{pmatrix} 87.8 & -86.4 \\ 108.2 & -109.6 \end{pmatrix} \tag{10}$$

Here we have scaled the inputs and outputs to be less than 1 in magnitude (this corresponds to the outputs in 0.01 mole fraction units, and the inputs scaled relative to the feed rate). The gains are much larger than 1 indicating no problems with input constraints, but this is somewhat deceiving as the gain in the the low-gain direction (corresponding to the smallest singular value) is actually just above 1.

This is admittedly a very crude model of a distillation column. Specifically, a) the parameters may vary drastically with operating point, b) there should be a high-order lag in the transfer function from u_1 to y_2 to represent the liquid flow down to the column, and c) higher-order composition dynamics should also be included. However, the model is simple and displays important features of the distillation column behavior. The RGA-matrix for this model is at all frequencies

$$RGA(G) = \begin{pmatrix} 35.1 & -34.1 \\ -34.1 & 35.1 \end{pmatrix} \tag{11}$$

The large elements in this matrix indicate that this process is fundamentally difficult to control (see section 4.2).

3.2.1 Interactions and ill-conditioned plants

¿From (10) we get

$$y_1(s) = \frac{87.8}{75s + 1} u_1(s)$$

Thus an increase in u_1 by only 0.01 (with u_2 constant) yields a steady- state change in y_1 of 0.878, that is, the outputs are very sensitive to changes in u_1. Similarly, an increase in u_2 by only 0.01 (with u_1 constant) yields $y_1 = -0.864$. Again, this is a very large change, but in the opposite direction of that for the increase in u_1.

We therefore see that changes in u_1 and u_2 counteract each other, and if we increase u_1 and u_2 simultaneously by 0.01, then the overall steady-state change in y_1 is only $0.878 - 0.864 = 0.014$. Physically, the reason for this small change is that the compositions in the column are only weakly dependent on changes in the *internal flows* (i.e., simultaneous changes in the internal flows L and V).

Summary: Since both u_1 and u_2 affect both outputs, y_1 and y_2, we say that the process is *interactive*. This is quantified by relatively large off-diagonal elements in $G(s)$. Furthermore, the process is *ill-conditioned*, that is, some combinations of u_1 and u_2 have a strong effect on the outputs, whereas other combinations of u_1 and u_2 (corresponding to $u_1 \approx u_2$) have a weak effect on the outputs. This is quantified by the condition number; the ratio between the gains in the strong and weak directions; which is large for this process (as seen below it is 141.7).

3.2.2 Singular Value Analysis of the Model

The above discussion shows that this distillation column is an ill- conditioned plant, where the effect (the gain) of the inputs on the outputs depends strongly on the *direction* of the inputs. To see this better, consider the SVD of the steady-state gain matrix

$$G = U\Sigma V^T \tag{12}$$

or equivalently since $V^T = V^{-1}$

$$G\bar{v} = \bar{\sigma}(G)\bar{u}, \quad G\underline{v} = \underline{\sigma}(G)\underline{u}$$

where

$$\Sigma = diag\{\bar{\sigma}, \underline{\sigma}\} = diag\{197.2, 1.39\}$$

$$V = (\bar{v}\ \underline{v}) = \begin{pmatrix} 0.707 & 0.708 \\ -0.708 & 0.707 \end{pmatrix}$$

$$U = (\bar{u} \ \underline{u}) = \begin{pmatrix} 0.625 & 0.781 \\ 0.781 & -0.625 \end{pmatrix}$$

The large plant gain, $\bar{\sigma}(G) = 197.2$, is obtained when the inputs are in the direction $\begin{pmatrix} u_1 \\ u_2 \end{pmatrix} = \bar{v} = \begin{pmatrix} 0.707 \\ -0.708 \end{pmatrix}$. From the direction of the output vector $\bar{u} = \begin{pmatrix} 0.625 \\ 0.781 \end{pmatrix}$, we see that these inputs cause the outputs to move in the same direction, that is, they mainly affect the average output $\frac{y_1 + y_2}{2}$. The low plant gain, $\underline{\sigma}(G) = 1.39$, is obtained for inputs in the direction $\begin{pmatrix} u_1 \\ u_2 \end{pmatrix} = \underline{v} = \begin{pmatrix} 0.708 \\ 0.707 \end{pmatrix}$. From the output vector $\underline{u} = \begin{pmatrix} 0.781 \\ -0.625 \end{pmatrix}$ we see that the effect then is to move the outputs in different directions, that is, to change $y_1 - y_2$. Thus, it takes a large control action to move the compositions in different directions, that is, to make both products purer simultaneously. Indeed, we see that in this direction it may be possible that one could be limited by input constraints (corresponding to $|u| > 1$). The condition number of the plant, which is the ratio of the high and low plant gain, is

$$\gamma(G) = \bar{\sigma}(G)/\underline{\sigma}(G) = 141.7 \tag{13}$$

The RGA is another indicator of ill-conditionedness, which is generally better than the condition number, because it is scaling independent. The sum of the absolute value of the elements in the RGA (denoted $\|RGA\|_{sum} = \Sigma|RGA_{ij}|$) is approximately equal to the minimized (with respect to input and output scaling) condition number, $\gamma^*(G) = \min_{D_1, D_2} \gamma(D_1 G D_2)$ where D_1 and D_2 are real diagonal "sacling" matrices. In our case we have $\|RGA\|_{sum} = 138.275$ and $\gamma^*(G) = 138.268$. (We note that the minimized condition number is quite similar to the condition number in this case, but this does not hold in general.)

3.3   Control of the column

3.3.1   Decoupling control

For "tight control" of ill-conditioned plants the controller should compensate for the strong directions by applying large input signals in the directions where the plant gain is low, that is, a "decoupling" controller similar to G^{-1} in directionality is desired. However, because of uncertainty, the direction of the large inputs may not correspond exactly to the low plant-gain direction, and the amplification of these large input signals may be much larger than expected. As shown in the simulations below, this will result in large values of the controlled variables y, leading to poor performance or even instability. Consider the following decoupling controller (or equivalently a steady-state decoupler combined with a PI controller):

$$C_1(s) = \frac{k_1}{s}G^{-1}(s) = \frac{k_1(1+75s)}{s}\begin{pmatrix} 0.39942 & -0.31487 \\ 0.39432 & -0.31997 \end{pmatrix}, \quad k_1 = 0.7\text{min}^{-1}$$

(14)

We have $GC = 0.7/sI$. In theory, this controller should counteract all the directions of the plant and give rise to two decoupled first-order responses with time constant $1/0.7 = 1.43$ min. This is indeed confirmed by the solid line in Fig.3.1 which shows the simulated response to a setpoint change in y_1. *We thus conclude that the decoupling controller satisfies the nominal performance (NP) requirement.*

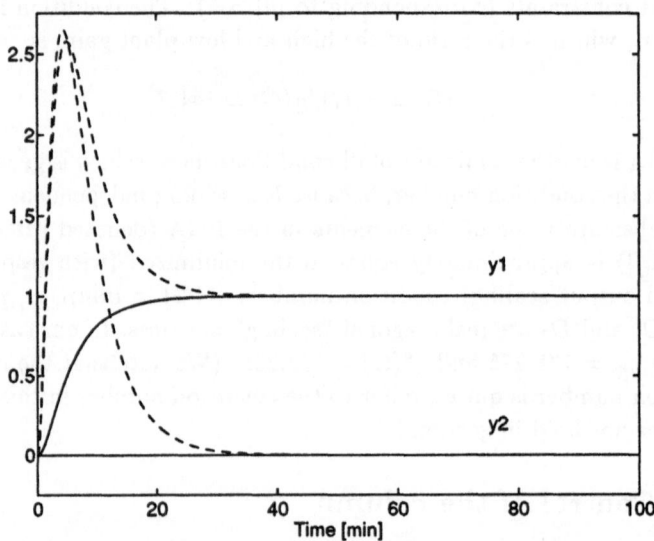

Figure 3.1: Response for decoupling controller to a unit setpoint change in y_1 with time constant 5 min, i.e, $r_1 = 1/(5s+1)$. Solid line: Nominal response with no uncertainty. Dotted line: 20% gain uncertainty as defined by Equation 15.

3.3.2 Robustness of decoupling control

We also note that this simple design yields an infinite gain margin (GM) and a phase margin (PM) of 90° in both channels. For multivariable systems such margins are however misleading as we shall see in the following.

To be specific consider the case with 20% error (uncertainty) in the gain in each input channel ("diagonal input uncertainty"):

$$u_1 = 1.2u_{1c}, \quad u_2 = 0.8u_{2c} \tag{15}$$

Note that this expression is in terms of deviation variables. Here u_1 and u_2 are the actual changes in the manipulated flow rates, while u_{1c} and u_{2c} are the desired changes (what we believe the inputs are) as specified by the controller. It is important to stress that this diagonal input uncertainty, which stems from our inability to know the exact values of the manipulated inputs, is *always* present. Note that the uncertainty is on the *change* in the inputs (flow rates), and not on their absolute values. A 20% error is reasonable for process control applications (some reduction may be possible, for example, by use of cascade control using flow measurements, but there will still be uncertainty because of errors in measurement sensitivity). Anyway, the main objective of this paper is to demonstrate the effect of uncertainty, and its exact magnitude is of less importance.

It is straightforward to see that the uncertainty in (15) does not by itself yield instability, thus *we have robust stability (RS) for the decoupling controller*. However, whereas for SISO systems we generally have that NP and RS imply robust performance (RP) this is often not the case for MIMO systems.

This is clearly shown from the dotted lines in Fig.3.1 which shows the response with the uncertainty in (15). It differs drastically from the nominal response represented by the solid line, and even though it is stable the response is clearly not acceptable; it is no longer decoupled, and $y_1(t)$ and $y_2(t)$ reach a value of about 2.5 before settling at their desired values of 1 and 0. *Thus RP is not satisfied for the decoupling controller.*

There is a simple reason for the observed poor response to the setpoint change in y_1. To accomplish this change, which occurs mostly in the "bad" direction corresponding to the low plant gains, the inverse-based controller generates large changes in u_1 and u_2, while trying to keep the $u_1 - u_2$ very small. However, uncertainty with respect to the actual values of u_1 and u_2 makes it impossible to make them both large while at the same time keeping their difference small – the result is an undesired large change in the actual value of $u_1 - u_2$, which subsequently results in large changes in y_1 and y_2 because of the large plant gain in this direction.

Remark. The system satisfied RS because the uncertainty only occurs at the input to the plant. In practice, with for example a small time delay added to one of the outputs, this controller would give an unstable response.

3.3.3 A robust controller: Single-loop PID

Unless special care is taken, most multivariable design methods (MPC, DMC, QDMC, LQG, LQG/LTR, DNA/INA, IMC, etc.) yield similar inverse-based controllers, and do not generally yield acceptable designs for ill-conditioned plants. This follows since they do not explicity take uncertainty into account, and the optimal solution is then to use a controller which tries to remove the interactions by inverting the plant model.

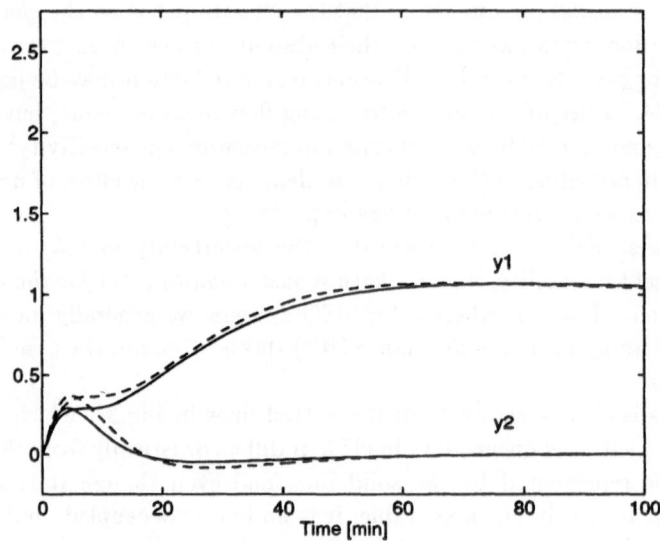

Figure 3.2: Response for PID controller.

The simplest way to make the closed-loop system insensitive to input uncertainty is to use a *simple* controller, for example two single-loop PID controllers, which does not try to make use of the details of the directions in the plant model. The problem with such a controller is that little or no correction is made for the strong interactions in the plant, and then even the nominal response (with no uncertainty) is relatively poor. This is shown in Fig.3.2 where we have used the following PID controllers (Lundström et al., 1991)

$$y_1 - u_1 : \quad K_c = 1.62; \tau_I = 41 \text{ min}; \tau_D = 0.38 \text{ min} \qquad (16)$$

$$y_2 - u_2 : \quad K_c = -0.39; \tau_I = 0.83 \text{ min}; \tau_D = 0.29 \text{ min} \qquad (17)$$

The controller tunings yield a relatively fast response for y_2, and a slower response for y_1. As seen from the dotted line in Fig.3.2 the response is not very much changed by introducing the model error in Eq.15.

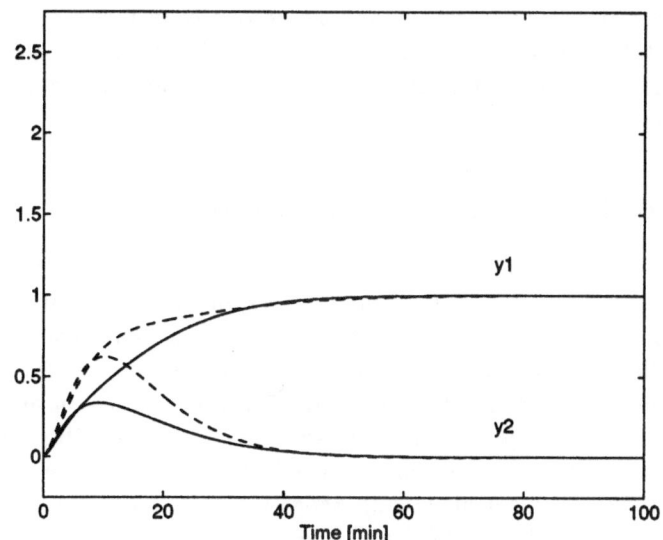

Figure 3.3: Response for μ-optimal controller.

In Fig.3.3 we show the response with the μ-optimal controller (see Lundström et al., 1991) which is designed to optimize the worst-case response (robust performance) as discussed towards the end of this paper. Although this is a multivariable controller, we note that the response is not too different from that with the simple PID controllers, although the response settles faster to the new steady-state.

3.3.4 Limitations with the example: Real columns

It should be stressed again that the column model used above is not representative of a real column. In a real column the liquid lag, θ_L, from the top to the bottom, makes the initial response less interactive and the column is easier to control than found above. It turns out that the important parameter to consider for controllability is *not* the RGA at steady-state (with exception of the sign), but rather the RGA at frequencies corresponding to the closed-loop bandwidth. For a model of a real distillation column the RGA is large at low frequencies (steady-state), but it drops at high frequencies and the RGA-matrix becomes close to the identity matrix at frequencies greater than $1/\theta_L$.

Thus, since the interactions are much less at high frequencies, control is simple, even with single-loop PI or PID controllers, if we are able to achieve very tight control of the column. However, if there are significant measurement delays (these are typically 5 min or larger), then we are forced to operate at a low bandwidth, and the responses in Figs.3.1-3.3 are more representative. Furthermore, it holds in general that one should *not* use a steady-state decoupler if the steady-state RGA-elements are large (typically larger than 5).

4   Tools for robustness analysis

In this section we will first introduce some simple tools, such as the frequency-dependent RGA, to understand the poor responses observed in the distillation example in the last section. Then we consider more general methods, which allow for a detailed description of the model uncertainty. This leads into a discussion of the structured singular value, μ, as an analysis tool for evaluating whether a system satisfies robust stability (RS) and robust performance (RP). Readers who want to learn more about μ are referred to Doyle (1982), Doyle at al. (1982), Skogestad et al. (1987), or to the texts by Morari and Zafiriou (1989) and Maciejowski (1989).

4.1   Simple tools for robustness analysis

4.1.1   SISO systems

For single-input-single-output (SISO) systems one has traditionally used gain margin (GM) and phase margin (PM) to avoid problems with model uncertainty. Consider a system with open-loop transfer function $g(s)c(s)$, and let $gc(j\omega)$ denote the frequency response. The GM tells by what factor the loop gain $|gc(j\omega)|$ may be increased before the system becomes unstable. The GM is thus a direct safeguard against steady-state gain uncertainty (error). Typically we require GM > 1.5.

The phase margin tells how much negative phase we can add to $gc(s)$ before the system becomes unstable. The PM is a direct safeguard against time delay uncertainty: If the system has a crossover frequency equal to ω_c, (defined as $|gc(j\omega_c)| = 1$), then the system becomes unstable if we add a time delay of $\theta = PM/\omega_c$. For example, if PM = 30° and $\omega_c = 1$ rad/min, then the allowed time delay error is $\theta = (30/57.3)[\text{rad}]/1[\text{rad/min}] = 0.52$ min.

Maximum peak criterions. In practice, we do not have pure gain and phase errors. For example, in a distillation column the time constant will usually increase when the steady-state gain increases. A more general way to specify stability margins is to require the Nyquist locus of $gc(j\omega)$, to stay outside some region of the -1 point (the "critical point") in the complex plane. Usually this is done by considering the maximum peak, M_t of the closed-loop transfer function T or the peak M_s of the sensitivity function. The reader may be familiar with M-circles drawn in the Nyquist plot or in the Nichols chart. Typically, we require that M_s and M_t are less than 2 (6 dB). $1/M_s$ is simply the minimum distance between $gc(j\omega)$ and the -1 point. In most cases the values of M_t and M_s are closely related, especially when the peak is large, There is a close relationship between M_t/M_s and PM and GM. Specifically, for a given M_s we are guaranteed

$$GM \geq \frac{M_s}{M_s - 1}; \qquad PM \geq 2\arcsin(\frac{1}{2M_s}) \geq \frac{1}{M_s}[\text{rad}] \qquad (18)$$

For example, with $M_s = 2$ we have GM ≥ 2 and PM $\geq 29.0° > 1/M_s$ [rad] $= 28.6°$. Similarly, for a given value of M_t we are guaranteed GM $\geq 1 + \frac{1}{M_t}$ and PM $\geq 2\arcsin(\frac{1}{2M_t}) > \frac{1}{M_t}$.

4.1.2 MIMO systems

It is difficult to generalize GM and PM to MIMO systems. On the other hand, the maximum peak criterions may be generalized easily. The most common generalization is to replace the absolute value by the maximum singular value, for example, by considering

$$M_t = \max_{\omega} \bar{\sigma}(T(j\omega)); \quad T = GC(I + GC)^{-1} \qquad (19)$$

Even though we may easily generalize the maximum peak criterion to mul-tivariable systems, it is often not useful for the following three reasons:

1) In contrast to the SISO case, it may be not sufficient to look at only the transfer function T. Specifically, for SISO systems $GC = CG$, but this does not hold for MIMO systems. This means that although the peak of T (in terms of $\bar{\sigma}(T(j\omega))))$ is low, the peak of $T_I = CG(I + CG)^{-1}$ may be large.

2) The singular value may be a poor generalization of the absolute value. There may be cases where the maximum peak criterion , eg. in terms of $\bar{\sigma}(T)$, is *not* satisfied, but in reality the system may be robustly stable. The reason is that the uncertainty generally has "structure", whereas the use of the singular value assumes unstructured uncertainty. As shown below one should rather use the structured singular value, i.e. $\mu(T)$.

3) In contrast to the SISO case, the response with model error may be poor (RP not satisfied), even though the stability margins are good (RS is satisfied) and the response without model error is good (NP satisfied).

For example, recall the distillation example above where for the decoupling controller $GC(s) = CG(s) = 0.7/sI$, and the values of M_t and M_s are both 1. Yet, the response with only 20% gain error in each input channel is extremely poor. To handle such effects in general one has to define the model uncertainty and compute the structured singular value for RP.

The conclusion of this section is that most of the tools developed for SISO systems, and also their direct generalizations such as the peak criterions, are not sufficient for MIMO systems.

4.2   The RGA as a simple tool to detect robustness problems

4.2.1   RGA and input uncertainty

We have seen that a decoupler performed very poorly for the distillation model. To understand this better consider the loop gain GC. The loop gain is an important quantity because it determines the feedback properties of the system. For example, the transfer function from setpoints, r, to control error, $e = y - r$, is given by $e = -Sr = -(I + GC)^{-1}r$. We therefore see that large changes in GC due to model uncertainty will lead to large changes in the feedback response. Consider the case with diagonal input uncertainty, Δ_I. Let Δ_1 and Δ_2 represent the relative uncertainty on the gain in each input channel. Then the actual ("perturbed") plant is

$$G_p(s) = G(s)(I + \Delta_I); \quad \Delta_I = \begin{pmatrix} \Delta_1 & 0 \\ 0 & \Delta_2 \end{pmatrix} \tag{20}$$

Note that Δ_i is *not* normalized to be less than 1 in this case. The perturbed loop gain with model uncertainty becomes

$$G_pC = G(I + \Delta_I)C = GC + G\Delta_I C \tag{21}$$

If a diagonal controller $C(s)$ (eg., two PI's) is used then we simply get (since Δ_I is also diagonal) $G_pC = GC(I + \Delta)$ and there is no particular sensitivity to this uncertainty. On the other hand, with a perfect decoupler (inverse- based controller) we have

$$C(s) = k(s)G^{-1}(s) \tag{22}$$

where $k(s)$ is a scalar transfer function, for example, $k(s) = 0.7/s$, and we have $GC = k(s)I$ where I is the identity matrix, and the perturbed loop gain becomes

$$G_pC = G(I + \Delta_I)C = k(s)(I + G\Delta_I G^{-1}) \tag{23}$$

For the distillation model (10) studied above the error term becomes

$$GΔ_I(G)^{-1} = \begin{pmatrix} 35.1Δ_1 - 34.1Δ_2 & -27.7Δ_1 + 27.7Δ_2 \\ 43.2Δ_1 - 43.2Δ_2 & -34.1Δ_1 + 35.1Δ_2 \end{pmatrix} \quad (24)$$

This error term is worse (largest) when $Δ_1$ and $Δ_2$ have opposite signs. With $Δ_1 = 0.2$ and $Δ_2 = -0.2$ as used in the simulations (Eq.15) we find

$$GΔ_I G^{-1} = \begin{pmatrix} 13.8 & -11.1 \\ 17.2 & -13.8 \end{pmatrix} \quad (25)$$

The elements in this matrix are much larger than one, and the observed poor response with uncertainty is not surprising.

The observant reader may have noted that the RGA-elements appear on the diagonal in the matrix $GΔ_I G^{-1}$ in (24). This turns out to be true in general as diagonal elements of the error term prove to be a direct function of the RGA (Skogestad and Morari, 1987)

$$(GΔG^{-1})_{ii} = Σ_{j=1}^{n} λ_{ij}(G)Δ_j \quad (26)$$

Thus, if the plant has large RGA elements and an inverse-based controller is used, the overall system will be extremely sensitive to input uncertainty.

Control implications. Consider a plant with large RGA-elements in the frequency-range corresponding to the closed-loop time constant. A diagonal controller (eg., single-loop PI's) is robust (insensitive) with respect to input uncertainty, but will be unable to compensate for the strong couplings (as expressed by the large RGA- elements) and will yield poor performance (even nominally). On the other hand, an inverse-based controller which corrects for the interactions may yield excellent nominal performance, but will be very sensitive to input uncertainty and will not yield robust performance. In summary, plants with large RGA-elements around the crossover-frequency are fundamentally difficult to control, and decouplers or other inverse-based controllers should never be used for such plants (The rule is never to use a *decoupling controller* for a plant with large RGA-elements). However, one-way decouplers may work satisfactorily.

4.2.2   RGA and element uncertainty/identification

Above we introduced the RGA as a sensitivity measure with respect to input gain uncertainty. In fact, the RGA is an even better sensitivity measure with respect to element-by-element uncertainty in the matrix.

Consider any complex matrix G and let $λ_{ij}$ denote the ij'th element in it's RGA-matrix. The following result holds (Hovd and Skogestad, 1992):

The (complex) matrix G becomes singular if we make a relative change $-1/\lambda_{ij}$ in its ij-th element, that is, if a single element in G is perturbed from g_{ij} to $g_{pij} = g_{ij}(1 - \frac{1}{\lambda_{ij}})$.

Thus, the RGA-matrix is a direct measure of sensitivity to element-by-element uncertainty and matrices with large RGA-values become singular for small relative errors in the elements.

Example. The matrix G in (10) is non-singular. The 1,2-element of the RGA is $\lambda_{12}(G) = -34.1$. Thus the matrix G becomes singular if $g_{12} = -86.4$ is perturbed to $g_{p12} = -86.4(1 - 1/(-34.1)) = -88.9$.

The result above is primarily an important algebraic property of the RGA, but it also has some important control implications:

1) Consider a plant with transfer matrix $G(s)$. If the relative uncertainty in an element at a given frequency is larger than $|1/\lambda_{ij}(j\omega)|$ then the plant may be singular at this frequency. This is of course detrimental for control performance. However, the assumption of element-by-element uncertainty is often poor from a physical point of view because the elements are usually always *coupled* in some way. In particular, this is the case for distillation columns: We know that the elements are coupled such that the model will not become singular due to small individual changes in the elements. The importance of the result above as a "proof" of why large RGA-elements imply control problems is therefore not as obvious as it may first seem.

2) However, for process identification the result is definitely useful: Models of multivariable plants, $G(s)$, are often obtained by identifying one element at the time, for example, by using step or impulse responses. From the result above it is clear this method will most likely give meaningless results (eg., the wrong sign of the steady-state RGA) if there are large RGA- elements within the bandwidth where the model is intended to be used. Consequently, identification must be combined with first principles modelling if a good multivariable model is desired in such cases.

Example. Assume the true plant model is

$$G = \begin{pmatrix} 87.8 & -86.4 \\ 108.2 & -109.6 \end{pmatrix}$$

By extremely careful identification we obtain the following model:

$$G_p = \begin{pmatrix} 87 & -88 \\ 109 & -108 \end{pmatrix}$$

This model seems to be very good, but is actually useless for control purposes since the RGA-elements have the wrong sign (the 1,1-element in the RGA is -47.9 instead of $+35.1$). A controller with integral action based on G_p would yield an unstable system.

To learn more about the RGA the reader is referred to Hovd and Skogestad (1992) where additional references can be found.

4.3 Advanced tools for robustness analysis: μ

So far in this paper we have pointed out the special robustness problems encountered for MIMO plants, and we have used the RGA as our main tool to detect these robustness problems. We found that plants with *large* RGA-elements are 1) fundamentally difficult to control because of sensitivity to input gain uncertainty, and decouplers should not be used, and 2) very difficult to identify because of sensitivity to element-by-element uncertainty.

We have not yet addressed the problem of analyzing the robustness of a given system with plant $G(s)$ and controller $C(s)$. In the beginning of this section we mentioned that the peak criterions in terms of M were useful for robustness analysis for SISO systems both in terms of stability (RS) and performance (RP). However, for MIMO systems things are not as simple. We shall first consider uncertainty descriptions and robust stability and then move on to performance. The calculations and plots in the remainder of this paper refer to the simple distillation model (10), using as a controller a steady-state decoupler plus PI-control.

4.3.1 Uncertainty modelling

Before considering how to analyze uncertain systems, we will consider the $\mathbf{H}^\infty$-approach to modelling plant uncertainty.

Linear Fractional Transformations (LFT) provide a general framework for modelling uncertainty (Doyle, 1984). A LFT may be written in the following form (see Fig. 4.1)

$$z = F_u(P, \Delta)w = (P_{22} + P_{21}\Delta(I - P_{11}\Delta)^{-1}P_{12})w \qquad (27)$$

Here P_{22} is the nominal mapping from w to z and Δ is a $\mathbf{H}^\infty$-norm bounded perturbation,

$$\|\Delta\|_\infty = \sup_\omega \bar{\sigma}(\Delta(j\omega)) \leq 1 \qquad (28)$$

Several sources of uncertainty may be combined and then $\Delta = diag\{\Delta_i\}$ is a block-diagonal matrix with perturbation blocks Δ_1, Δ_2 etc. These blocks may represent parametric uncertainty, in which case they are scalars Δ_i , possibly repeated, or they may represent unstructured uncertainty in which case they may be matrix-valued.

Each of these perturbations is bounded in terms of its $\mathbf{H}^\infty$-norm. For parametric uncertainty this is actually not very convenient as it would

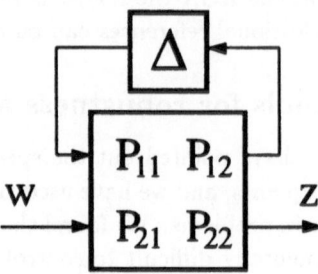

Figure 4.1: Uncertainty represented as linear fractional transformation (LFT).

allow for complex variations in the parameter, $|\Delta_i| \leq 1$. Therefore for parametric uncertainty we generally restrict Δ_i to be real. Thus, it clear that the frequency domain does not offer any advantage for parametric uncertainty. On the other hand, the frequency bounds come in nicely when handling non-parametric uncertainty such as neglected dynamics. Also, it is very convenient for lumping several sources of uncertainty, although this must be done with some care to avoid being too conservative (when the uncertainty description allows unrealistic plants).

For unstructured uncertainty we have to make a choice of where to place the perturbation representing the uncertainty in question. Some alternatives are shown in Fig.4.2. These may all be represented by the LFT in Eq. (27).

There is no definite rule on which unstructured uncertainty to use, but the following may be useful: 1) Use the multiplicative (relative) uncertainty to represent neglected and uncertain dynamics occuring *between* the plant and the controller (e.g., neglected or uncertain actuator and measurement dynamics). 2) Use the "feedforward" (additive) forms when the zero uncertainty is large (in particular if a zero may go from the LHP to RHP) 3) Use the "feedback" forms when the pole uncertainty is large (in particular if a pole may cross the $j\omega$-axis). One particular combination of the feedforward and feedback forms, which appears to be useful, is the coprime uncertainty used in the Glover-McFarlane loop shaping procedure described in the next section.

However, care must be taken when representing uncertainty in an unstructured form. For example, for our distillation column example, it may be tempting to add some unstructured additive uncertainty to the plant. It turns out that this uncertainty description would be extremely conservative for this plant as the sign of the plant (represented by the sign of det $G(s)$

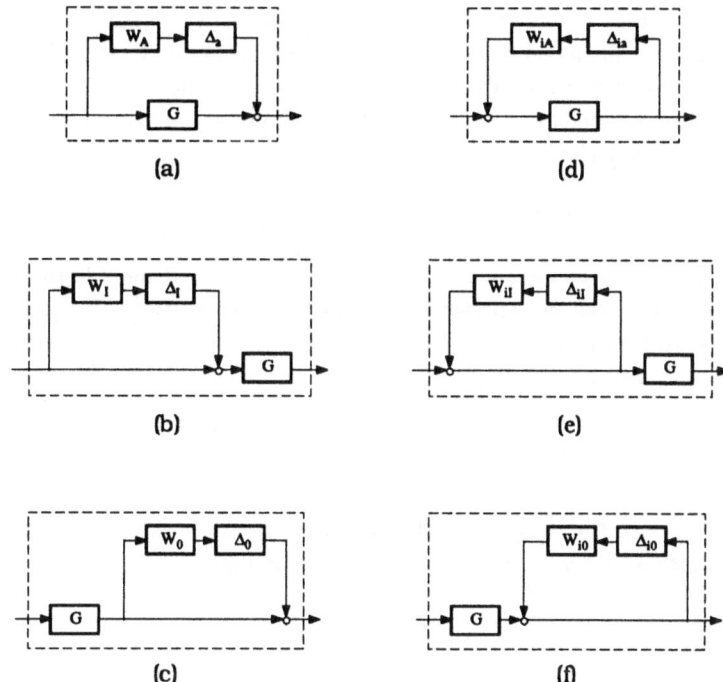

Figure 4.2: Alternative ways of representing unstructured uncertainty. (a) Additive uncertainty, (b) Multiplicative input uncertainty, (c) Multiplicative output uncertainty, (d) Inverse additive uncertainty, (e) Inverse multiplicative input uncertainty, (f) Inverse multiplicative output uncertainty.

or by the signs in the RGA-matrix) is extremely sensitive to such changes. In practice, as noted earlier, this kind of uncertainty does not occur for distillation columns as there are strong couplings between the elements in $G(s)$.

Two examples illustrate the usefulness of the general uncertainty description given above.

Neglected dynamics. Assume that the real set of plants is something like

$$\text{Real plant}: \quad g' = k'e^{-\theta s}, \quad k' \in [0.8k, 1.2k] \tag{29}$$

where k is the nominal ("average") gain, and we allow for gain variations of $\pm 20\%$. To simplify the controller design we want to use a simple nominal model with no delay, i.e.,

$$\text{Nominal model}: \quad g = k \tag{30}$$

The uncertainty in the gain may be handled directly as parametric uncertainty, but the neglected delay must clearly be represented in a non-parametric manner. In order to simplify the uncertainty description we choose to lump together gain variations and the neglected delay as unstructured multiplicative (relative) uncertainty:

Set of possible models : $g_p(s) = k(1 + w_I \Delta_I); \quad |\Delta_I(j\omega)| \leq 1, \quad \forall \omega$ (31)

Here Δ_I is a *complex* scalar. The modelled set g_p must include the real set of plants $g'(s)$. Let r_k represent the relative uncertainty in the gain. Then the following approximation for the weight is derived using a first-order Pade-approximation

$$(1 + r_k)e^{-\theta s} - 1 \approx (1 + r_k)\frac{1 - \frac{\theta}{2}s}{1 + \frac{\theta}{2}s} - 1 \qquad (32)$$

Since it is only the magnitude that matters we make this expression minimum phase and derive the following simple weight

$$w_I(s) = r_k \frac{1 + (\frac{1}{r_k} + \frac{1}{2})\theta s}{1 + \frac{1}{2}\theta s} \qquad (33)$$

The weight is somewhat optimistic (too small) at intermediate frequencies. In our case with $r_k = 0.2$ the magnitude of the weight is $r_k = 0.2$ at low frequencies, crosses 1 at about frequency $1/\theta$ and approaches $2(1 + r_k/2) = 2.2$ at high frequencies.

Note that even though the uncertainty weight only has 1 state it will allow for an infinite number of plants of arbitrary high order. On the other hand, (31) is *not* an exact representation of the original set of plants $g'(s)$ and may be conservative for that reason. For a scalar case it is probably not very conservative as the delay is generally the "worst case". However, in the multivariable case this may not always be true.

Pole variations represented as parametric uncertainty. Consider the set of plants $g' = 1/(s + a')$ where $-1 \leq a' \leq 3$. This may be exactly represented as

$$g_p = \frac{1}{s + a + 2\Delta}; \quad , a = 1, |\Delta| \leq 1 \qquad (34)$$

where Δ is a *real* scalar perturbation. This is in fact an inverse additive uncertainty (see Fig.4.2) with nominal model $g(s) = 1/(s + a)$ and $w_{ia} = 2$. Note also that poles crossing from the left to the right half plane may be modelled tightly with this uncertainty.

4.3.2   Conditions for Robust stability

By Robust Stability (RS) we mean that the system is stable for all possible plants as defined by the uncertainty set (using the Δ's as discussed above). This is a "worst case" approach, and for this reason one must be careful about not including unrealistic or impossible parameter variations. With this caution in mind, it turns out that the $\mathbf{H}^{\infty}$-norm (for completely unstructured uncertainty) and the structured singular value (for diagonally structured uncertainty) provide an exact way of analyzing robust stability,

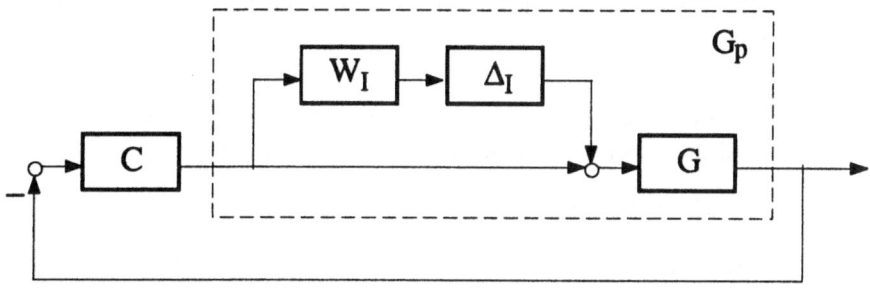

Figure 4.3: Multiplicative input uncertainty.

As an example, consider the the case with multiplicative input uncertainty shown in Fig. 4.3. We assume that the system without uncertainty ($\Delta = 0$) is stable (we have NS). Instability may then only be caused by the "new" feedback paths caused by the Δ-block. Therefore, to test for robust stability (RS) we rearrange the feedback system with uncertainty into the standard form in Fig.4.4 with the two blocks Δ and M. Here the *interconnection matrix* M is the transfer function from the output, u_Δ, to the input, y_Δ, of the Δ-block. For the case of multiplicative input uncertainty we have $\Delta = \Delta_I$ and obtain $M = wC(I+GC)^{-1}G = w_I T_I = w_I CSG$ (the negative sign has been dropped as it does not matter). To test for stability we make use of the "small gain theorem". Since the Δ-block is normalized to be less than 1 at all frequencies, this theorem says that the system is stable if the M-block is less than 1 at all frequencies. Robust stability is then satisfied if

$$\bar{\sigma}(M) = \bar{\sigma}(w_I T_I(j\omega)) < 1, \quad \forall \omega \qquad (35)$$

Unstructured uncertainty. One crucial point is that this condition is also necessary (it is clearly sufficient) for RS provided we allow for *all* Δ's satisfying $\bar{\sigma}(\Delta) \leq 1, \forall \omega$. That is, we have for the general block diagram in Fig.4.4:

$$RS \; \forall \; \|\Delta\|_\infty \leq 1 \quad \text{iff} \quad \|M\|_\infty < 1 \qquad (36)$$

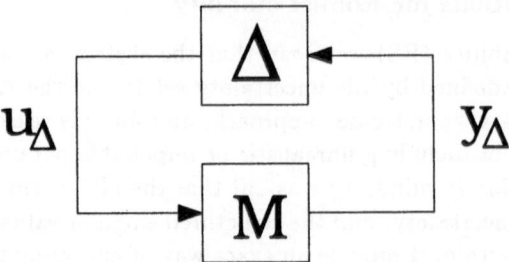

Figure 4.4: General block diagram for studying robust stability and robust performance.

The same robust stability condition applies for each of the six forms of unstructured uncertainty shown in Fig. 4.2 when we use

$$M_a = W_A CS, \quad M_b = W_I CSG, \quad M_c = W_O GCS \quad (37)$$

$$M_d = W_{iA} SG, \quad M_e = W_{iI}(I + CG)^{-1}, \quad M_f = W_{io}S \quad (38)$$

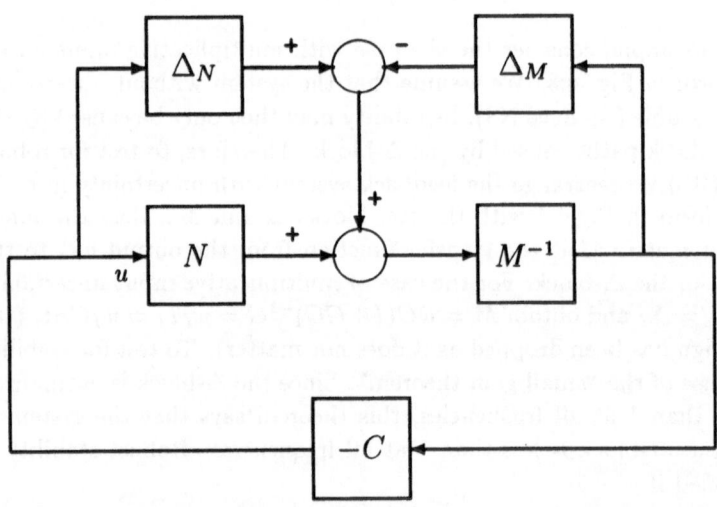

Figure 4.5: Coprime uncertainty description

However, even though (36) is mathematically correct it will generally be conservative for the following two reasons: 1) It allows for Δ to be complex, 2) It allows for Δ to be a full matrix.

It is actually the second point which is the main problem in most cases. However, before discussing it we shall introduce the coprime uncertainty description which will be used in the next section.

Coprime uncertainty description. Consider the uncertainty description in Fig.4.5 (note that the M in that figure denotes one coprime factor of the plant and not the interconnection matrix). This uncertainty description is rather general, as it allows for both zeros and poles crossing into the right half plane, and has proved to be very useful in applications. To test for RS we rearrange the block diagram to match Fig.4.4 with

$$\Delta = \begin{pmatrix} \Delta_N \\ \Delta_M \end{pmatrix}; \quad M_{RS} = \begin{pmatrix} C \\ I \end{pmatrix} (I + GC)^{-1} M^{-1} \tag{39}$$

where M_{RS} is the interconnection matrix. We get

$$RS \ \forall \ \left\| \begin{matrix} \Delta_N \\ \Delta_M \end{matrix} \right\|_\infty \leq 1 \quad \text{iff} \quad \|M_{RS}\|_\infty < 1 \tag{40}$$

The reason why we get a tight condition in terms of the $\mathbf{H}^\infty$-norm even though we have two uncertainty blocks is that the blocks enter into the same point in the block diagram.

Structured uncertainty. We will now consider the general case where (36) does not provide a tight bound because we have several Δ-blocks caused by individual sources of uncertainty.

For example, if the input uncertainty represents neglected dynamics in the the individual channels then the set of possible plants is given by

$$G_p(s) = G(I + w_I \Delta_I); \quad \Delta_I = \begin{pmatrix} \Delta_1 & 0 \\ 0 & \Delta_2 \end{pmatrix} \tag{41}$$

where Δ_i represents the independent uncertainty in each input channel such that the overall Δ_I is a diagonal matrix (it has "structure"). ((41) is identical to Eq.(20), except that w_I yields the magnitude, since Δ_i is now normalized to be less than 1.)

Also, for multivariable plants it makes a difference whether the uncertainty is at the input or the output of the plant. Thus, we may want to consider combined input and output uncertainty. This may be represented in the general form in Fig.4.4 with M as 2×2 block matrix and $\Delta = diag\{\Delta_I, \Delta_O\}$. Again, we note that Δ has a diagonal structure and (36) is conservative.

To improve the tightness of condition (36) we first note that the issue of stability should be independent of scaling. We then have the improved condition

$$RS \quad \text{if} \quad \min_{D(\omega)} \bar{\sigma}(DMD^{-1}) < 1, \forall \omega \tag{42}$$

where D is a real block-diagonal scaling matrix with structure corresponding to that of Δ, such that $\Delta D = D\Delta$. A further refinement of this
idea led to the introduction of the structured singular value, $\mu(M)$ (Doyle,
1982). We have (essentially this is the definition of μ)

$$RS \; \forall \; \text{structured} \; \Delta \quad \text{iff} \quad \mu_{\Delta}(M) < 1 \tag{43}$$

This is a tight condition provided the uncertainty description is tight. Note
that for computing μ we have to specify the block-*structure* of Δ and also if
Δ is real or complex. Today there exists very good software for computing
μ when Δ is complex. The most common method is to approximate μ by
a "scaled" singular value as introduced in (42):

$$\mu_{\Delta}(M) \leq \min_{D} \bar{\sigma}(DMD^{-1}) \tag{44}$$

This upper bound is exact when Δ has three or fewer "blocks", and the
largest deviation found so far for more blocks is 10-15% (Doyle, 1982).

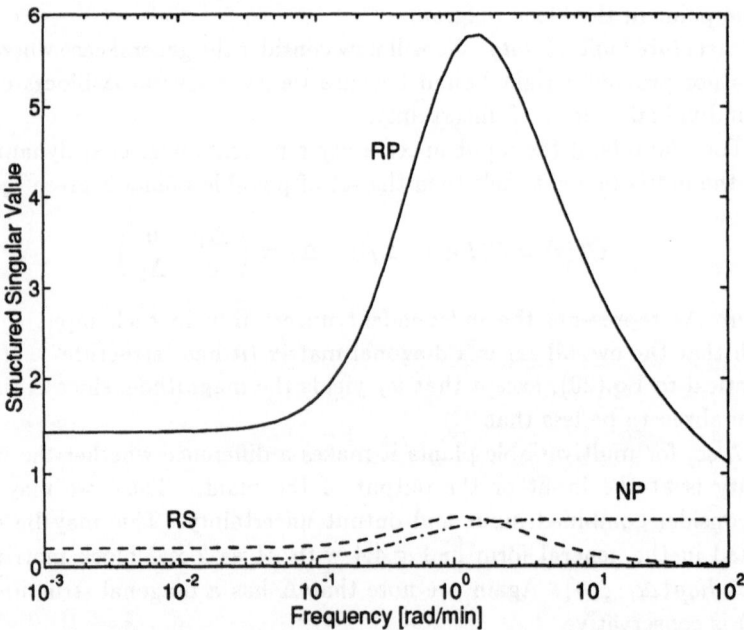

Figure 4.6: μ-plots for distillation example with decoupling controller.

Distillation example revisited. Consider the distillation example
from the previous section and consider multiplicative input uncertainty in

each of the two input channels

$$w_I(s) = 0.2 + \frac{0.9s}{0.5s+1} = 0.2\frac{5s+1}{0.5s+1} \tag{45}$$

With reference to (33) we see that this corresponds to 20% gain error and a neglected time delay of about 0.9 min. The weight levels off at 2 (200% uncertainty) at high frequency. The dotted line in Fig.4.6 shows $\mu(M) = \mu(w_I T_I)$ for RS with this uncertainty using the decoupling controller. The μ-plot for RS shows the inverse of the margin we have with respect to our stability requirement. For example, the peak value of $\mu_{\Delta_I}(M)$ as a function of frequency is about 0.53. This means that we may increase the uncertainty by a factor $1/\mu = 1.89$ before the worst-case model yields instability. This means that we tolerate about 38% gain uncertainty and a time delay of about 1.7 min before we get instability.

Remark: For the decoupling controller we have $GC = \frac{0.7}{s}I$, and $T_I = T = \frac{1}{1.43s+1}I$. For this particular case it turns out that the structure of Δ does not matter, and we get $\mu_\Delta(M) = \bar\sigma(w_I T_I) = |0.2\frac{5s+1}{(0.5s+1)(1.43s+1)}|$. However, in other cases it may be critical to use the right structure, e.g., see Fig. 16 in Skogestad et al. (1988).

4.3.3 Conditions for Robust Performance

An additional bonus of using the $\mathbf{H}^\infty$-norm both for uncertainty and performance is that the robust performance (RP) problem may be recast as a special case of the RS-problem (Doyle et al, 1982) with the performance specification represented as a fake uncertainty block, Δ_P. To test for RP one then considers the interconnection matrix M_{RP} from the the outputs to the inputs of *all* the Δ-blocks, including the Δ_P-block for performance. Note that Δ_P is a "full" matrix (no diagonal structure). This follows since performance is defined using the singular value and we have $\bar\sigma(A) = \mu_\Delta(A)$ when Δ is a full matrix. M_{RP} depends on the plant G, the controller C and the weights used to define uncertainty and performance. The condition for robust performance within the $\mathbf{H}^\infty$-framework then becomes

$$RP \quad \text{iff} \quad \mu_{\tilde\Delta}(M_{RP}) < 1; \quad \forall\omega, \quad \tilde\Delta = \begin{pmatrix} \Delta & 0 \\ 0 & \Delta_P \end{pmatrix} \tag{46}$$

Distillation example revisited. Let us now check if RP is satisfied for the distillation example. To do this performance must first be defined.

Nominal performance. NP is defined such that at each frequency the value of the weighted sensitivity, $\bar\sigma(w_P S)$, should be less than 1. We select the weight

$$w_P(s) = \frac{s/2 + 0.05}{s} \tag{47}$$

With reference to Eq.(3) we see that this requires integral action, a closed-loop bandwidth of about 0.05 [rad/min] (which of course is relatively slow when the allowed time delay is only about 0.9 min) and a maximum peak for $\bar{\sigma}(S)$ of $M_s = 2$.

As discussed above we may define μ for NP as $\mu_{\Delta_P}(w_P S) = \bar{\sigma}(w_P S)$ where Δ_P is a full matrix. As expected, we see from the dashed line in fig.4.6 that the NP-condition is easily satisfied with the decoupling controller: $\bar{\sigma}(w_P S)$ is small at low frequencies and approaches $1/M_s = 0.5$ at high frequency because of the maximum peak requirement on $\bar{\sigma}(S)$.

Robust Performance. RP means that the performance specification is satisfied for the worst-case uncertainty. The most efficient way to test for RP is to compute μ for RP. If this μ-value is less than 1 at all frequencies then the performance objective is satisfied for the *worst case*. Although our system has good robustness margins and excellent nominal performance we know from the simulations in Fig.3.1 that the performance with uncertainty (RP) may be extremely poor. This is indeed confirmed by the μ-curve for RP in Fig.4.6 which has a peak value of about 6. This means that even with 6 times less uncertainty, the performance will be about 6 times poorer than what we require. μ for robust performance was computed as $\mu_{\tilde{\Delta}}(M_{RP})$ where the matrix $\tilde{\Delta}$ in this case has a block-diagonal structure with Δ_I (the true uncertainty) and Δ_P (the fake uncertainty stemming from the performance specification) along the main diagonal, and

$$M_{RP} = \begin{pmatrix} w_I CSG & w_I CS \\ w_P SG & w_P S \end{pmatrix} \tag{48}$$

The derivation of M_{RP} follows by representing the performance as an inverse multiplicative perturbation similar to that in Fig.4.2d, and rearranging the block to match Fig.4.4 (see Skogestad et al, 1988).

The μ-optimal controller is the controller which minimizes μ for RP. The present approach to designing the μ-optimal controller ('D-K iteration') is a rather tedious procedure which involves solving a number of scaled H^∞-problems. The iterations are not guaranteed to converge and generally result in high-order controllers.

For our example Lundström et al. (1991) obtained a μ-optimal controller with 22 states which yields an essentially flat μ-curve with a "peak" of μ of 0.978. The simulation in Fig.3.3 shows that the response even with this controller is relatively poor (taken into account that the only obvious limitation is a delay of about 1 min). The reason is that the combined effect of large interactions (as seen from the large RGA-values) and input uncertainty makes this plant fundamentally difficult to control.

Comment: In the time domain our RP-problem specification may be formulated *approximately* as follows: Let the plant be

$$G_p(s) = G(s) \begin{pmatrix} k_1 e^{-\theta_1 s} & 0 \\ 0 & k_2 e^{-\theta_2 s} \end{pmatrix} \qquad (49)$$

where $G(s)$ is given in (10). Let $0.8 \leq k_1 \leq 1.2$, $0.8 \leq k_2 \leq 1.2$, $0 \leq \theta_1 \leq 0.9$ [min], and $0 \leq \theta_2 \leq 0.9$ [min]. The response to a step change in setpoint should have a closed-loop time constant less than about 20 minutes. Specifically, the error of each output to a unit setpoint change should be less than 0.37 after 20 minutes, less than 0.13 after 40 minutes, and less than 0.02 after 80 minutes, and with no large overshoot or oscillations in the response.

Conclusion. The structured singular value, μ, provides an excellent tool for analyzing the robustness of control systems. Within the $\mathbf{H}^\infty$-framework it is possible to consider most sources of model uncertainty, including parametric and unstructured uncertainty, and with help of μ one can essentially directly pick out the worst-case plant and see if it satisfies the specifications for RS or RP. However, for a number of reasons μ seems to be best suited for analysis, i.e, to answer "what if" questions. It may also be suited for evaluating the upper bound on achievable performance, i.e., as a kind of ultimate controllability tool. However, for actual controller design it seems like simpler methods, as the ones described in the next section, are more appropriate.

5 Robust Control System Design

In this section, we will focus on a loop shaping methodology for the design of robust multivariable control systems.

The classical loop shaping approach to control system design has been applied to industrial systems over several decades. For single-input single-output systems and loosely coupled systems, the approach has worked well. But for truly multivariable systems it has only been in the last decade that a reliable generalization of the approach has emerged. Multivariable loop shaping is based on the idea that a satisfactory definition of gain (range of gain) for a matrix transfer function is given by the singular values of the transfer function. By multivariable loop shaping, therefore, we mean the shaping of singular values of appropriately specified transfer functions.

5.1 Trade-offs in multivariable feedback design

In February 1981, the IEEE Transactions on Automatic Control published a Special Issue on Linear Multivariable Control Systems, the first six papers of which were on the use of singular values in the analysis and design of multivariable feedback systems. The paper by Doyle and Stein (1981) was particularly influential: it was primarily concerned with the fundamental question of how to achieve the benefits of feedback in the presence of unstructured uncertainty, and through the use of singular values it showed how the classical loop shaping ideas of feedback design could be generalized to multivariable systems. To see how this was done, consider the one degree of freedom configuration shown in figure 5.1.

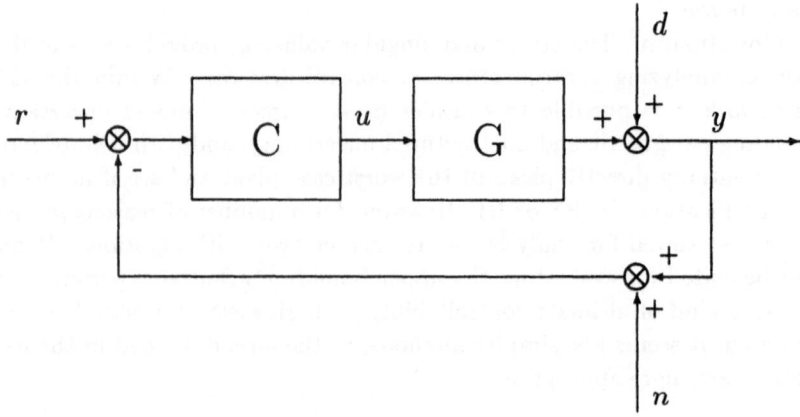

Figure 5.1: One degree of freedom feedback configuration

The plant G and controller C interconnection is driven by reference commands r, output disturbances d, and measurement noise n. y are the outputs to be controlled, and u are the control signals. In terms of the sensitivity function $S = (I + GC)^{-1}$ and the closed-loop transfer function $T = GC(I+GC)^{-1} = I - S$, we have the following important relationships:

$$y(s) = T(s)r(s) + S(s)d(s) - T(s)n(s) \qquad (50)$$

$$u(s) = C(s)S(s)[r(s) - n(s) - d(s)] \qquad (51)$$

These relationships determine several closed-loop objectives, in addition to the requirement that C stabilizes G; namely:

1. For *disturbance rejection* make $\bar{\sigma}(S)$ small.

2. For *noise attenuation* make $\bar{\sigma}(T)$ small.

3. For *reference tracking* make $\bar{\sigma}(T) \cong \underline{\sigma}(T) \cong 1$.

4. For *control energy reduction* make $\bar{\sigma}(CS)$ small.

If the unstructured uncertainty in the plane model G is represented by an additive perturbation i.e. $G_p = G + \Delta$, then a further closed-loop objective is (recall (37)):

5. For *robust stability* make $\bar{\sigma}(CS)$ small.

Alternatively, if the uncertainty is modelled by a multiplicative output perturbation such that $G_p = (I + \Delta)G$, then we have:

6. For *robust stability* make $\bar{\sigma}(T)$ small.

The closed-loop requirements 1 to 6 cannot all be satisfied simultaneously. Feedback design is therefore a trade-off over frequency of conflicting objectives. This is not always as difficult as it sounds because the frequency ranges over which the objectives are important can be quite different. For example, disturbance rejection is typically a low frequency requirement while noise mitigation is often only relevant at higher frequencies.

In classical loop-shaping, it is the magnitude of the open-loop transfer function GC which is shaped, whereas the above design requirements are all in terms of closed-loop transfer functions. However, it is relatively easy to convert the closed-loop requirements into the following open-loop objectives:

1. For *disturbance rejection* make $\underline{\sigma}(GC)$ large.

2. For *noise attenuation* make $\bar{\sigma}(GC)$ small.

3. For *reference tracking* make $\underline{\sigma}(GC)$ large.

4. For *control energy reduction* make $\bar{\sigma}(C)$ small.

5 & 6. For *robust stability* make $\bar{\sigma}(C)$ small.

Typically, requirements 1 and 3 are important at low frequencies, while 2, 4, 5 and 6 are high frequency conditions as illustrated in Figure 5.2.

To shape the gains (singular values) of GC by selecting C is a relatively easy task but to do this in a way which also guarantees closed-loop stability is in general non-trivial. Doyle and Stein (1981) suggested that an LQG controller could be used in which the regulator is designed using a "sensitivity recovery" procedure of Kwakernaak (1969) to give desirable

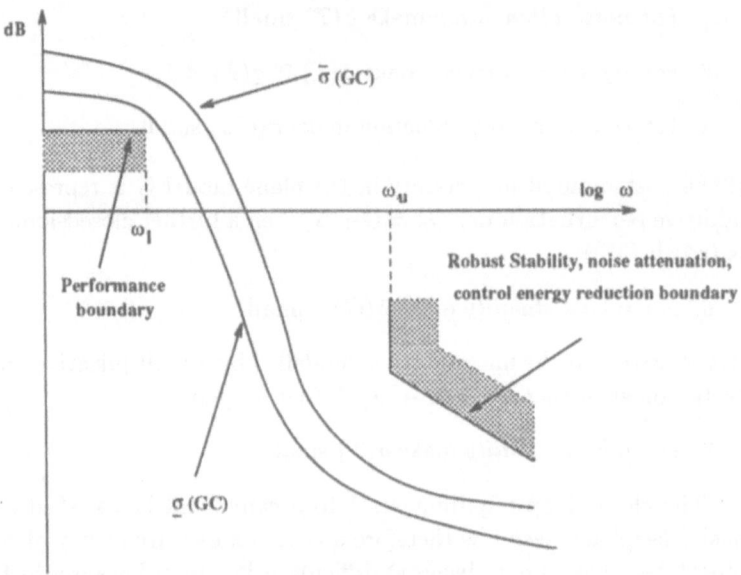

Figure 5.2: Design tradeoffs for the multivariable loop transfer function GC

properties (gain and phase margins) in GC. They also gave a dual "robust-ness recovery" procedure for designing the filter in an LQG controller to give desirable properties in CG. Recall that CG is not in general equal to GC which implies that stability margins vary from one break point to another in a multivariable system. Both these loop transfer recovery procedures had problems:

- they were unsuitable for directly achieving specified loop shapes

- the gauranteed stability margins were only gauranteed as limiting properties in the design

- in the limit the controllers effectively inverted the plant and so the procedure broke down for nonminimum phase systems.

It was not until 1990, that a satisfactory loop shaping design procedure was developed by McFarlane and Glover (1990). This will be described in section 5.3, but first it will be necessary to consider a related robust stablization problem.

5.2 Robust Stabilization

As previously discussed in this paper, gain and phase margins are unreliable indicators of robust stability for multivariable systems because they do not take account of the coupling between loops. In section 4, several uncertainty descriptions were presented in which the uncertainty was captured by a norm bounded perturbation. Robustness levels could then be quantified in terms of the maximum singular values of various closed-loop transfer functions.

For example, in the feedback configuration of figure 5.1, if G is replaced by $G_p = G + \Delta$, where $\bar{\sigma}[\Delta(jw)] < \varepsilon(w)$, then the closed-loop remains stable if $\bar{\sigma}[C(jw)S(jw)] < \varepsilon^{-1}(w)$ for all w. A design objective, for robust stabilization, might therefore be to find a C which stabilizes G and minimizes $\|CS\|_{\infty}$. A more general uncertainty description, which allows for both poles and zeros crossing into the RHP, is the coprime uncertainty description used by Glover and McFarlane (1989). This leads to an attractive robust stabilization problem formulated in an H^{∞} framework. The main results are summarized below.

5.2.1 Normalized coprime factorization

The plant model

$$G = M^{-1}N, \tag{52}$$

is a normalized left coprime factorization (LCF) of G if $M, N \in RH_{\infty}$ (the set of stable real rational transfer function matrices) and $MM^* + NN^* = I$ where for a real rational function of s, X^* denotes $X^T(-s)$.

With the notation

$$G(s) = D + C(sI - A)^{-1}B \overset{s}{=} \left[\begin{array}{c|c} A & B \\ \hline C & D \end{array}\right] \tag{53}$$

a state-space realization of a normalized coprime factorization of G is given (Vidyasagar, 1988) by

$$[N \quad M] \overset{s}{=} \left[\begin{array}{c|cc} A + HC & B + HD & H \\ \hline R^{-1/2}C & R^{-1/2}D & R^{-1/2} \end{array}\right] \tag{54}$$

where

$$H = -(BD^T + ZC^T)R^{-1} \tag{55}$$

$$R = I + DD^T \tag{56}$$

and the matrix $Z \geq 0$ is the unique stabilizing solution to the algebraic Riccati equation (ARE)

$$(A - BS^{-1}D^TC)Z + Z(A - BS^{-1}D^TC)^T - ZC^TR^{-1}CZ + BS^{-1}B^T = 0 \tag{57}$$

where

$$S = I + D^TD. \tag{58}$$

5.2.2   Perturbed plant model

A perturbed model G_p can be defined as

$$G_P = (M + \Delta_M)^{-1}(N + \Delta_N) \qquad (59)$$

where $\Delta_M, \Delta_N \in RH_\infty$ and $\left\| \begin{array}{c} \Delta_N \\ \Delta_M \end{array} \right\|_\infty < 1$, as illustrated in figure 5.3.

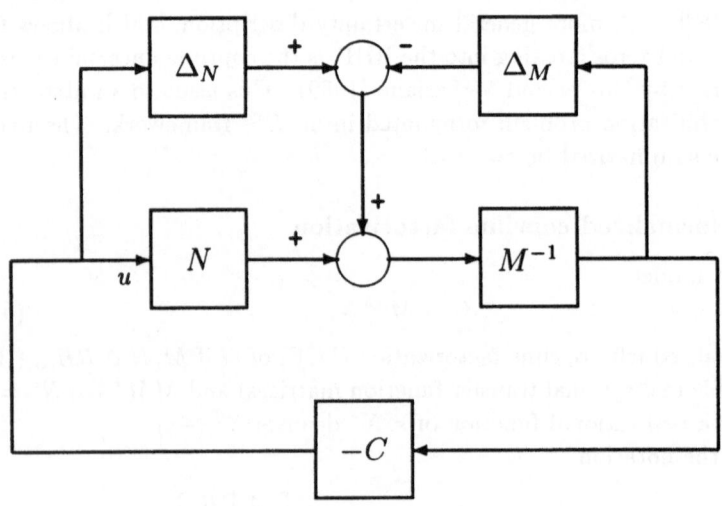

Figure 5.3: Perturbed plant model and controller

5.2.3   Robust stabilization

The robust stability condition for the class of perturbed models defined by (59) was derived previously in (40). For $\left\| \begin{array}{c} \Delta_N \\ \Delta_M \end{array} \right\|_\infty < 1$ we have

$$RS \quad \text{iff} \quad \gamma \overset{\text{def}}{=} \left\| \begin{bmatrix} C \\ I \end{bmatrix} (I + GC)^{-1} M^{-1} \right\|_\infty \leq 1 \qquad (60)$$

A reasonable objective is therefore to find the stabilizing controller that minimizes γ and thus allows for the largest perturbations. This is the problem of robust stabilization of normalised coprime factor plant descriptions as introduced by Glover and McFarlane (1989). The minimum value of γ

for all stabilizing controllers C is

$$\gamma_0 = \inf_{C \text{ stabilising}} \left\| \begin{bmatrix} C \\ I \end{bmatrix} (I + GC)^{-1} M^{-1} \right\|_{\infty} \tag{61}$$

and is given in Glover and McFarlane (1989) by

$$\gamma_0 = \left(1 - \| [N, M] \|_H^2 \right)^{-1/2}, \tag{62}$$

where $\| \cdot \|_H$ denotes the Hankel norm. From (Glover and McFarlane, 1989)

$$\| [N, M] \|_H^2 = \lambda_{\max} \left(ZX(I + ZX)^{-1} \right), \tag{63}$$

where $\lambda_{\max}(.)$ represents the maximum eigenvalue, and $X \geq 0$ is the unique stabilizing solution of the ARE

$$(A - BS^{-1} D^T C)^T X + X(A - BS^{-1} D^T C) - XBS^{-1} B^T X + C^T R^{-1} C = 0. \tag{64}$$

Hence, it can be shown that

$$\gamma_0 = (1 + \lambda_{\max}(ZX))^{1/2}. \tag{65}$$

A controller which achieves γ_0 is given in (McFarlane and Glover, 1990) by

$$C \overset{s}{=} \left[\begin{array}{c|c} A + BF + \gamma_0^2 (Q^T)^{-1} ZC^T (C + DF) & \gamma_0^2 (Q^T)^{-1} ZC^T \\ \hline B^T X & -D^T \end{array} \right], \tag{66}$$

where

$$F = -S^{-1}(D^T C + B^T X), \tag{67}$$

and

$$Q = (1 - \gamma_0^2)I + XZ. \tag{68}$$

The above results on robust stabilization are particularly attractive because the optimal γ and the corresponding optimal controller can be found without an iterative search on γ which is normally required to solve H^{∞} problems.

In the next section, it is shown how the robust stabilization problem can be used in conjunction with the ideas of Doyle and Stein on singular value loop shaping to arrive at a reliable multivariable loop shaping design procedure.

5.3   Loop shaping design

Robust stabilization alone is not much used in practice because the designer is not able to specify the desired performance requirements. To do this McFarlane and Glover (1990) proposed pre- and post-compensating the plant to shape the open-loop singular values prior to robust stabilization of the "shaped" plant.

If W_1 and W_2 are the pre- and post-compensators respectively, then the shaped plant G_S is given by

$$G_S = W_2 G W_1 \qquad (69)$$

as shown in figure 5.4. The controller C is synthesised by solving the robust stabilization problem of section 5.2 for the shaped plant G_S with a normalized left coprime factorization $G_s = M_s^{-1} N_s$. The feedback controller for the plant G is then $C = -W_1 C_s W_2$.

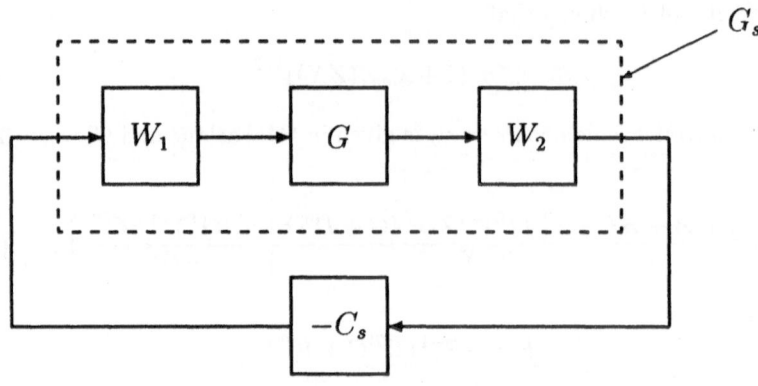

Figure 5.4: The shaped plant and controller

The above procedure contains all the essential ingredients of classical loop shaping, and can be easily implemented using reliable algorithms in, for example, Matlab. Skill is required in the selection of weights, but experience on real applications has shown that robust controllers can be designed with relatively little effort by following a few simple guidelines. Hyde (1991) offers a step by step procedure for weights selection developed during his Ph.D work with Glover on the robust control of VSTOL aircraft. These guidelines are summarised below in subsection 5.3.1. Successful application of the procedure has also been reported by Postlethwaite and Walker (1992) in their work on advanced control of high performance helicopters some of which will be described in section 6.

5.3.1 A loop shaping design procedure

The following procedure is a summary of that found in (Hyde, 1991):

1. Scale the outputs so that the same amount of cross coupling into each of the outputs is equally undesirable.

2. Scale the inputs to reflect the relative actuator capabilities or expected usage. This may involve a few iterations based on the control signals which result from successive designs.

3. The inputs and outputs should be ordered so that the plant is as diagonal as possible. The relative gain array can be useful here.

4. Select the elements of diagonal pre- and post-compensator weights W_1 and W_2 so that the roll off rates of the singular values are approximately 20 dB/decade at the desired bandwidths. Some trial and error is involved here.

   W_2 is often chosen as a constant reflecting the relative importance of the outputs to be controlled while W_1 contains the dynamic shaping.

   Integral action (for steady-state accuracy) and high frequency roll off (for noise attenuation and robustness) should be placed in W_1 if desired.

5. Sometimes it is found useful to "align "the singular values at the desired bandwidth using a further constant weight W_A cascaded with W_1. This is effectively a decoupler and should not be used if the plant is ill-conditioned.

6. Robust stabilization of the shaped plant is carried out as described in section 5.2. If the optimal gamma, γ_0, is less than about 4, then the design is usually successful. That is, the shape of the open-loop singular values will not have changed much after robust stabilization. A large value of γ indicates that the specified singular value shapes are incompatible with robust stability requirements.

7. Analysis of the design may prompt further modifications of the weights if all the specifications are not met.

8. When implementing the controller, the configuration shown in figure 5.5 has been found useful when compared with the conventional set

up in figure 5.1. This is because the references do not directly excite the dynamics of C_s which can result in large amounts of overshoot (classical derivative kick). The prefilter ensures a steady state gain of 1 between r and y.

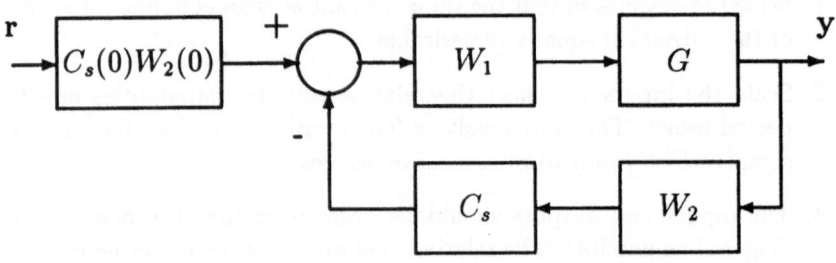

Figure 5.5: A practical implementation of the loop shaping controller

5.3.2   Loop shaping design and the method of inequalities

In Whidborne et al (1992), it has recently been shown how the method of inequalities (Zakian and Naib, 1973) can be used in the loop shaping design procedure to select the weights W_1 and W_2 to satisfy a given set of performance inequalities. Although computationally demanding the technique has proved useful when stringent performance specifications are required to be met.

   The method of inequalities (MOI) introduced by Zakian (Zakian and Naib, 1973) is a computer-aided multi-objective design approach, where desired performance is represented by a set of algebraic inequalities, and the aim of the design is to simultaneously satisfy these inequalities. The design problem is expressed as

$$\phi_i(p) \leq \varepsilon_i \quad \text{for} \quad i = 1 \ldots n \tag{70}$$

where ε_i are real numbers, $p \in \mathcal{P}$ is a real vector $(p_1, p_2, \ldots, p_q)$ chosen from a given set $\mathcal{P}$ and ϕ_i are real functions of p. The functions ϕ_i are performance indices, the components of p represent the design parameters and ε_i are chosen by the designer and represent tolerable values of ϕ_i. The aim is the satisfaction of the set of inequalities in order that an acceptable design is reached.

The functions $\phi_i(p)$ may be functionals of the system step response, for example the rise time, overshoot or the integral absolute error, or functionals of the frequency response, such as the bandwidth. They can also represent measures of the system stability. The actual solution to the set of inequalities (70) may be obtained by means of numerical search algorithms, such as the moving boundaries process (Zakian and Naib, 1973).

In some previous applications of the MOI, the design parameter has parameterized a controller with a particular structure. For example, $p = (p_1, p_2)$ could parameterize a PI controller $p_1 + p_2/s$. This has meant that the designer has had to choose the structure of the control scheme and the order of the controllers. In general, the smaller the size of the design vector p, the easier it is for the numerical search algorithm to find a solution, if one exists. While this does give the designer some flexibility and leads to simple controllers, and is of particular value when the structure of the controller is constrained in some way, it does mean that better solutions may exist with more complicated and higher order controllers. A further limitation of using the MOI in this way is that a stability point must be located as a pre-requisite to searching the parameter space to improve the index set ϕ, that is a point such that $\phi_i < \infty$ for $i = 1, 2, \ldots, n$ must be found initially.

Two aspects of design using the loop shaping design procedure (LSDP) make it amenable to combine this approach with the MOI. Firstly, unlike most H_∞-optimization problems, the H_∞-optimal controller for the weighted plant can be synthesised from the solution of just two ARE's and does not require time-consuming γ-iteration. Secondly, in the LSDP, the weighting functions are chosen by considering the open-loop response of the weighted plant, so effectively the weights W_1 and W_2 are the design parameters. This means that the design problem can be formulated as in the method of inequalities, with the weighting parameters used as the design parameters p to satisfy some set of *closed-loop* performance inequalities.

Such an approach to the MOI overcomes the limitations to the MOI described earlier. The designer does not have to choose the order or structure of the controller, but instead chooses the structure and order of the weighting functions. With low-order weighting functions, high order controllers are synthesised which often leads to significantly better performance or robustness than if simple low order controllers were used. Additionally, the problem of finding a stability point is simply a case of choosing sensible initial weighting functions; an easy matter if the open-loop singular value plots are studied.

For more details of loop shaping design and the method of inequalities see Whidborne et al (1992).

5.4   Two degrees of freedom controllers

Most control design problems naturally possess two degrees of freedom (DOF). In general this arises from the existence of, on the one hand, measurement or feedback signals and on the other, commands or references. Quite often, one degree of freedom is forsaken in the design, and the controller is driven by, for example, an error signal (i.e. the difference between command and output). Other *ad hoc* means may also be used to arrive at a 1-DOF implementation. A general 2-DOF feedback control scheme is depicted in figure 5.6. The commands and feedbacks enter the controller separately where they are independently processed.

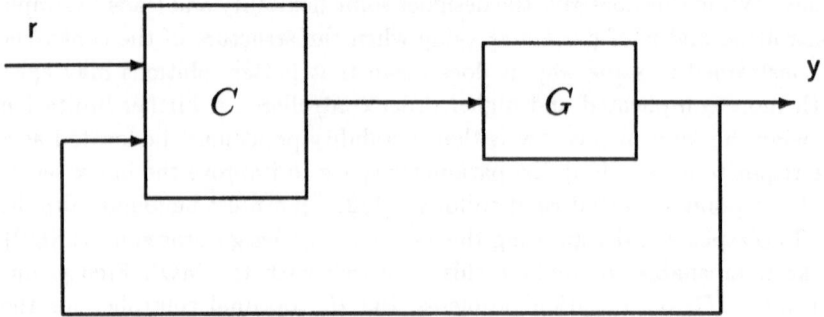

Figure 5.6: General 2-DOF feedback control scheme

5.4.1   An extended loop shaping design procedure

Limebeer et al (1993) have recently proposed an extension of McFarlane and Glover's loop shaping design procedure (LSDP) which uses a 2-DOF scheme to enhance the model matching properties of the closed-loop system. The feedback part of the controller is designed to meet robust stability and disturbance rejection requirements in a manner identical to the 1-DOF LSDP. That is, weights are first selected to produce a shaped plant with desirable singular values. An additional prefilter part of the controller is then introduced to force the response of the closed loop system to follow that of a specified model T_0.

The design problem to be solved is illustrated in figure 5.7. The scalar parameter ρ is used to adjust the emphasis that is placed on model-matching in the optimization. For $\rho = 0$, the problem reverts to the standard LSDP. As ρ is increased, more emphasis is placed on model following.

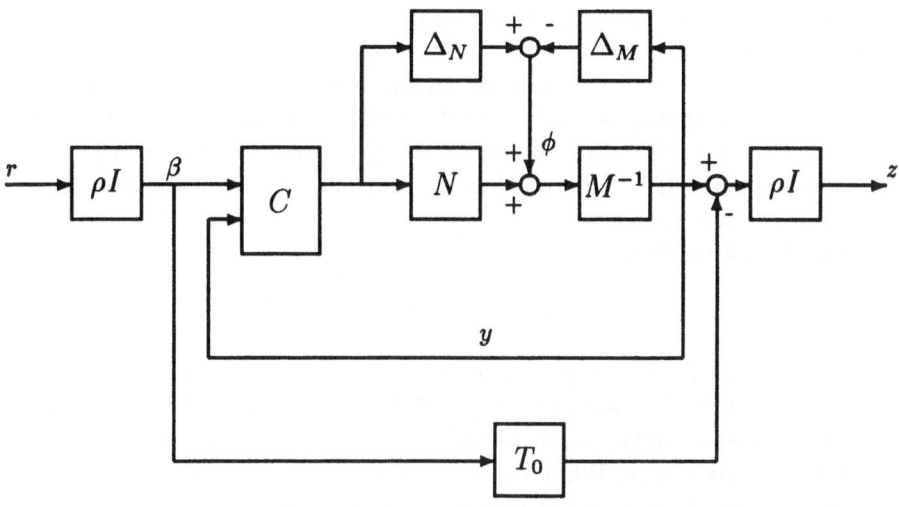

Figure 5.7: Two degrees-of-freedom design problem

The H^∞ optimization problem to be solved is that of finding a controller C which stabilizes G and which minimizes the H^∞ norm of the transfer function between the signals $(r^T \phi^T)^T$ and $(u^T y^T z^T)^T$ as defined in figure 5.7. Note that the robust stabilization problem alone involves minimizing the H^∞ norm of the transfer function between ϕ and $(u^T y^T)^T$. The two degrees of freedom design problem is easily solved using standard routines, in for example Matlab, but as a standard H^∞ optimization problem an iterative approach is required. In practice, sub-optimal controllers are often used which satisfy a given bound γ on the transfer function being minimized.

The two degrees of freedom approach will usually also involve a few iterations on ρ to achieve the desired balance between robust stabilization and model matching.

5.4.2 A further extension using the method of inequalities

In section 5.3, we saw how the one degree of freedom loop shaping design procedure could be enhanced by using the method of inequalities to select the weights W_1 and W_2. It is tempting therefore to think that the same could be done for the 2-DOF approach. However, the latter requires γ-iteration for its solution which makes it too slow computationally to be effectively combined with the MOI. Whidborne et al (1993), therefore, proposed an alternative strategyy based on fixing the structure of the prefilter part of a 2-DOF controller.

The proposed approach involves adding a prefilter C_p to a feedback controller C designed, using the 1-DOF LSDP, as illustrated in figure 5.8. C_p is parameterised with a subset of design parameters while C is the solution to the LSDP with weights W_1 and W_2 parameterised with the remaining design parameters.

Functional constraints

$$\phi_i(W_1, W_2, C_p) \leq \varepsilon_i \quad \text{for} \quad i = 1 \ldots n \quad (71)$$

can then be defined to represent performance requirements and a MOI approach used to find the parameters of W_1, W_2 and C_p. Given W_1 and W_2, C_p follows straight from a 1-DOF LSDP with no iterations.

For more details see Whidborne et al (1993), where the MOI approach has been successfully used to design a 2-DOF controller for the distillation column benchmark example.The results of this case study will be presented at the minicourse on Robust Multivariable Control using H^∞ Methods (2nd European Control Conference, Groningen, 1993) for which this paper has been prepared. In the next section, a second case study on helicopter control will be considered. It is a straightforward application of the loop shaping 2 DOF design methodology without MOI.

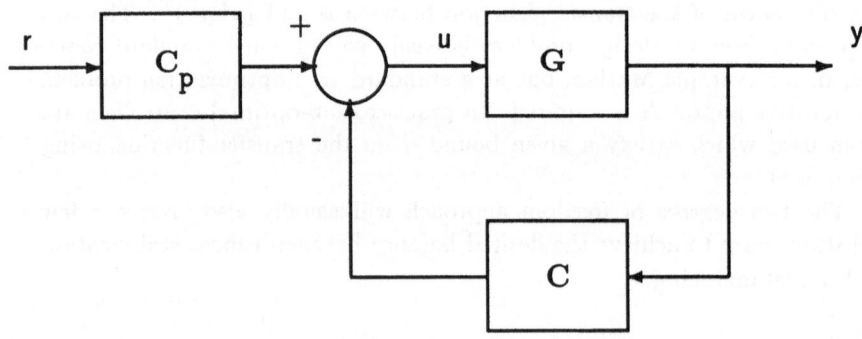

Figure 5.8: A 2 degrees of freedom controller configuration

6   Advanced control of high performance helicopters: a case study

To fly a high performance helicopter in low-level or nap-of-the-earth flight currently demands a high pilot workload; so high in fact as to limit the potential of the aircraft to a level below what is theoretically possible. Thus in order to enable the next generation of helicopters to fulfil the challenging specifications that are emerging, an automatic flight control system will be an essential ingredient. In this section, we will report on the findings of an on-going study into the role of advanced multivariable control theory in the design of full-authority full-flight-envelope control laws.

6.1   Background

The Control Systems Research Group at Leicester University has several years experience in the design of advanced control laws for future generation helicopters. For the past three years a major research project has been undertaken, funded by the UK Defence Research Agency (DRA) Bedford, formerly the Royal Aerospace Establishment, to investigate the role of advanced multivariable control theory in the design of full-authority full-flight-envelope control laws.

This section outlines some of the salient features and results arising out of this DRA-funded research. The work has enabled an in-depth study using computer simulation to help assess the impact that advanced control systems might play in improving the handling qualities of future military helicopters. The main achievements have been to extend and to improve upon the results of earlier work (Yue and Postlethwaite, 1990) and to demonstrate that multivariable design methods using H^∞-optimization, and in particular the loop shaping methodology of section 5, provide a valuable way forward in the design of robust full-flight-envelope control systems.

In May 1992, an important goal of the research project was achieved, with the successful piloted simulation, using the Large Motion System simulator at DRA Bedford, of a multivariable control system designed for wide envelope use. The system was tested over a period of three days by two experienced helicopter test pilots, one from the Royal Navy, the other from the Army. The testing consisted of two phases: the first, a familiarisation phase, during which the pilots could accustom themselves to the response types available from the control system and generally gain a feel for how to fly the aircraft via the controller; the second, the test phase, during which the pilots were asked to perform a set of specified tasks, each designed to highlight certain characteristics of the aircraft's response. Each pilot completed an in-cockpit assessment of the system's response using the Cooper-Harper pilot rating scale (Cooper and Harper, 1969). With this the pilot can classify the desirable and unsatisfactory handling aspects on a points system, scaled from 1 to 10, where 1 represents the most satisfactory qualities. A rating of 10 represents major and unacceptable system deficiences, where control may be lost during part of the flight envelope. Cooper-Harper ratings of 1 to 3 are said to conform to a level 1 handling qualities rating and are a goal of any helicopter flight control system.

The control system tested in May 1992 received level 1 ratings in a large majority of the runs made. The controller was also tested over a wide range of speeds, from hover to well in excess of 100 knots. During the three days of tests, it became clear that in simulation at least, the multivariable controller was able to provide robust stability and decoupled performance. Both pilots agreed that, in spite of certain deficiencies in the primary yaw response and collective to yaw coupling, the control law provided excellent stability and control.

Following the May trials a redesign was undertaken to increase the yaw axis bandwidth and to slightly reduce the heave axis bandwidth. This was a very simple matter with the 2-DOF loop shaping design technique being used. The new design was fully tested in piloted simulation in December 1992. At these trials, the previously identified deficiencies were no longer present and all the mission task elements performed were given level 1 Cooper-Harper ratings.

6.2   The helicopter model

The aircraft model used in our work is representative of the Westland LYNX, a twin-engined multi-purpose military helicopter, approximately 9000 lbs gross weight, with a four-blade semi-rigid main rotor. The unaugmented aircraft is unstable, and exhibits many of the cross-couplings characteristic of a single- main-rotor helicopter. These characteristics have been captured by a computer model known as the Rationalized Helicopter Model (RHM) (Padfield, 1981) that was used in our study. This model has been developed at DRA Bedford over a number of years and is a mature and fairly accurate (though by no means definitive) nonlinear model of the Lynx. In addition to the basic rigid body, engine and actuator components, it also includes second order rotor flapping and coning modes for off-line use. The model has the advantage that essentially the same code can be used for the real-time piloted simulation as for the workstation-based off-line handling qualities assessment.

The equations governing the motion of the helicopter are complex and difficult to formulate with high levels of precision. For example, the rotor dynamics are particularly difficult to model. A robust design methodology is therefore essential for high performance helicopter control.

The starting point for our designs was a set of five eighth-order linear differential equations modelling the small-perturbation rigid body motion of the aircraft about five trimmed conditions of straight-and-level flight in the range 0 to 80 knots. The controller designs were first evaluated on the eighth-order models used in the design, then on twenty-one state linear models, and finally using the full nonlinear model. The robust design methodology used in the controller design did turn out to provide excellent robustness with respect to nonlinearities and time delays which, although simulated, were not explicitly included in the linear design process.

6.3   Design objectives

The main objectives were to design a full-authority control system that:

- Robustly stabilized the aircraft with respect to changes in flight condition, and model uncertainty and non-linearity.

- Provided high levels of decoupling between primary controlled variables.

- Achieved compliance with the Level 1 criteria given in the US Army Aeronautical Design Standards, (ADS-33C, 1989).

6.4   Design method

The two degrees of freedom loop shaping design procedure of section 5.4 was used to robustly stabilize the aircraft over a wide range of flight conditions, whilst simultaneously forcing the closed-loop system to approximate the behaviour of a specified transfer function model T_0. The overall control law was actually comprised of five controllers, designed at a range of flight conditions between 0 and 80 knots, each one having a Kalman filter-like structure. The latter is a property of the LSDP and is very useful when scheduling between controllers over the flight envelope. As the dynamics of the open-loop aircraft vary with speed, so too do the controllers obtained at each operating point. Therefore, these controllers can be scheduled with forward speed if required, to give wide-envelope performance.

The aim of the design was to synthesize a full-authority controller that robustly stabilized the aircraft and provided a decoupled Attitude-Command/ Attitude-Hold (ACAH) response type that closely approximated the behaviour of a simple transfer-function model.

The outputs to be directly controlled were:

- Heave velocity

- Pitch attitude

- Roll attitude

- Heading rate

With a full authority control law such as that proposed here, the controller has total control over the blade angles, and is interposed between the pilot and the actuation system. The pilot flies the aircraft by isuing appropriate demands to the controller. These demands, together with the sensor feedback signals, are fed to the flight control computer which generates appropriate blade angle demands. Other than that we make no assumptions about the implementation details.

The controller was designed to operate on six feedback measurements: the four controlled outputs listed above and the body-axis pitch and roll rate signals. The other inputs to the controller consisted of the 4 pilot inceptor inputs.

The control law output consisted of four blade angle demands:

- Main rotor collective

- Longitudinal cyclic

- Lateral cyclic

- Tail rotor collective

These demands were passed directly to the actuator model.

6.5 Weighting function selection, the design parameter ρ, and the desired transfer function T_0

The same loop shaping weights W_1 and W_2, and the same desired transfer function T_0, were used for all 5 operating point designs. It was found that the ADS-33C bandwidth requirements impact directly on the "crossover "frequency of the weight W_1 which was chosen to have the first order diagonal form

$$W_1 = diag\left\{\frac{s+a}{s}, \frac{s+b}{s}, \frac{s+c}{s}, \frac{s+d}{s}\right\}. \tag{72}$$

A static diagonal W_2 was chosen as

$$W_2 = diag\left\{1, 1, 1, 1, 0.1, 0.1\right\} \tag{73}$$

That is, the roll rate and pitch rate signals are weighted less than the other four outputs which are required to be controlled. The rates are included as extra measurements because it is well known that they will make the control problem easier.

The desired closed-loop transfer function T_0 is chosen to be diagonal with second order transfer functions on each of the four channels: heave velocity, pitch and roll attitudes and heading rate. The damping and natural frequencies of these transfer functions were selected to give what were considered to be adequate responses. The selection of ρ is a compromise

ρ	0	0.1	0.2	0.4	0.75	1.0	1.5	2.0	3.0
γ_o	2.89	2.90	2.92	2.99	3.23	3.46	3.98	4.59	6.35

Table 1: Relationship between ρ and γ_o for a hover design

between robust stability and model matching. Table 1, shows the relationship between ρ and γ_0 for a hover design, where γ_0 is the minimum H^∞-norm of the transfer function being minimised. The reciprocal of γ_0 is roughly proportional to the multivariable stability margin. For the hover design in question a value of $\rho = 1.5$ was used, together with a suboptimal value of $\gamma = 4.2$.

6.6 Controller scheduling

The controller was designed to run in either of two modes: (i) fixed gain, (ii) interpolated. In fixed gain mode, the closest controller for the given flight condition would be switched in to provide control. This controller would remain operative until the mode was de-selected. If the interpolated mode was engaged, the controllers would be interpolated smoothly as a function of air-speed to compensate for variation in dynamics. To implement for real would require an accurate measurement (or estimate) of forward air-speed.

6.7 Outer-loop modes

To enhance the handling qualities provided by the basic ACAH response of the inner loop H^∞ controller, three outer loop modes were also implemented.

- Turn coordination: this was provided by augmenting the heading rate demand as a function of bank angle at moderate/high speed. This enabled a coordinated turn to be effected as a single axis task.

- Automatic trimming: this was achieved using a trim-map to offset the linear inner loop controller with the appropriate trim attitude.

- Hover acquisition/hold: this mode enabled the pilot to acquire and hold hover automatically. Longitudinal and lateral velocty state estimates were needed to achieve this.

During the piloted trials, the first two modes were used continuously, but insufficient time was available to evaluate the hover acquisition utility.

6.8   Step response analysis

The response of the closed-loop system (comprising controller and full nonlinear model) to step input demands on pitch and roll channels are shown in figures 6.1 and 6.2. These show, respectively, an acceleration from hover and the commencement of a coordinated turn at 60 knots. In both cases there is seen to be minimal cross-coupling.

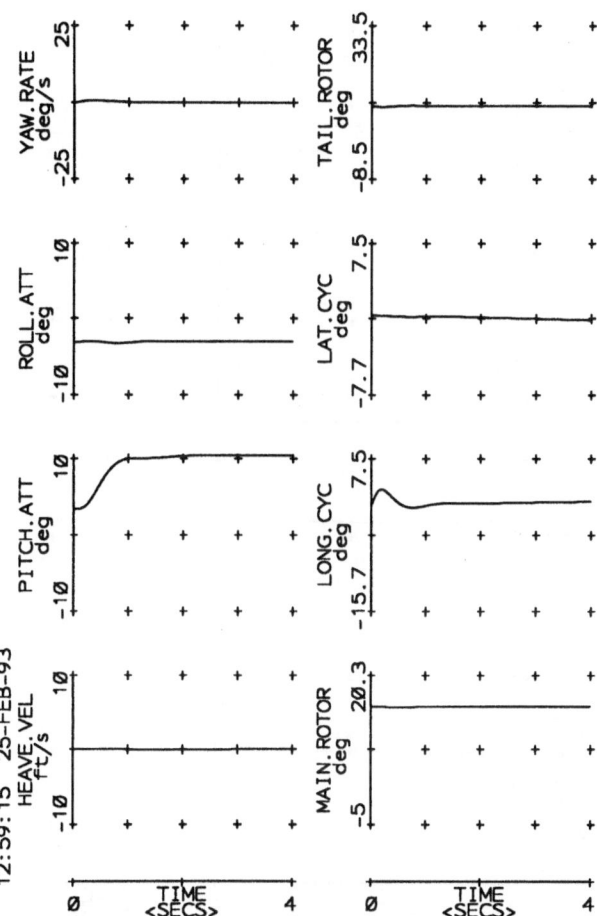

Figure 6.1: Pitch axis step response: outputs and actuators

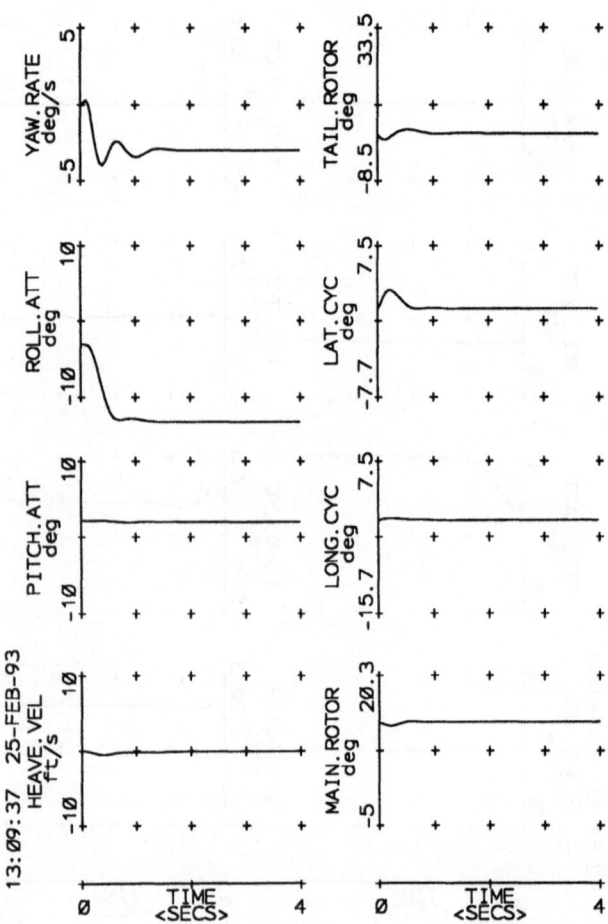

Figure 6.2: Roll axis step response: outputs and actuators

6.9   Handling qualities assessment: off-line analysis

ADS-33C (1989) details the latest requirements' specification for military helicopters which is intended to ensure that mission effectiveness will not be compromised by deficient handling qualities. The requirements are stated in terms of three limiting " levels "of acceptability of one or more given parameters. The levels indicate performance attributes that equate with pilot ratings on the Cooper-Harper scale. A Matlab Handling Qualities Toolbox (Howitt, 1991) was used as a supplement to existing computer aided control system design packages in order to integrate handling qualities assessment into the complete design and analysis cycle. The dynamics of the closed loop vehicle were assessed against the dynamic response requirements specified in sections 3.3 and 3.4 of ADS-33C using the off-line simulation model. A selection of the results are given in Walker et al (1993). In summary, the performance provided by the control law led to level 1 handling quality ratings for almost all of the mission tasks performed.

6.10   Handling qualities assessment: piloted simulation on the DRA Bedford large motion simulator

The simulation model was written in Fortran and run on an Encore Concept-32 computer with an integration step of 20 mS. A Lynx-like single seat cockpit was used, mounted on the large motion system which provides ±30 degrees of pitch, roll and yaw, ±4 metres of sway and ±5 metres of heave motion. Also, the pilot's seat was dynamically driven to give vibration and sustained normal acceleration cues. The visual display was generated by a Link-Miles Image IV CGI system and gave approximately 48 degrees field of view (FOV) in pitch and 120 degrees FOV in azimuth with full daylight texturing. A three axis side-stick was used to control pitch, roll and yaw together with a conventional collective for heave.

Handling qualities were assessed for three hover/low speed mission task elements (sidestep, quick-hop, bob-up) and three moderate/high speed tasks (lateral jinking, hurdles, yaw pointing) using CGI databases developed by DRA Bedford.

Two DRA test pilots took part in the trials (of May and December 1992), both with significant experience of Lynx and the simulator. For each task in turn, the pilot performed two or three familiarisation runs before performing a definitive evaluation run, at the end of which the simulation was paused so that comments and handling qualities ratings could be recorded. The six tasks are briefly described below.

Sidestep: With reference to figure 6.3a, the objective was to translate sideways through 150 ft from a hover at a height of 30 ft above ground level in front of one diamond and square sighting arrangement, to acquire and maintain a stable hover in front of the next sighting system.

Quick-hop: The quick-hop task (figure 6.3b) is the corresponding longitudinal task to the sidestep, requiring a re-position from hover over a distance of 500 ft. Again, similar levels of initial pitch attitude were used to determine the task aggression. The task was flown down a walled alley to give suitable height and lateral position cues.

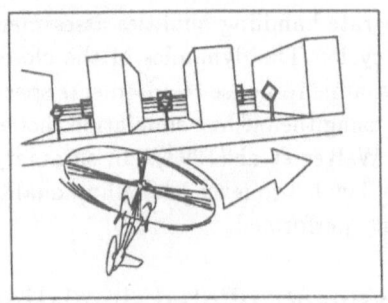
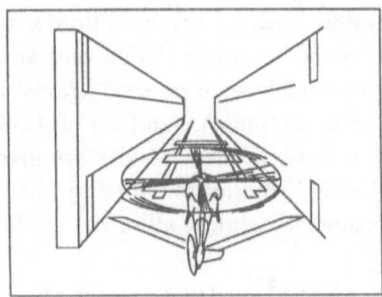

(a) Sidestep task   (b) Quick-hop task

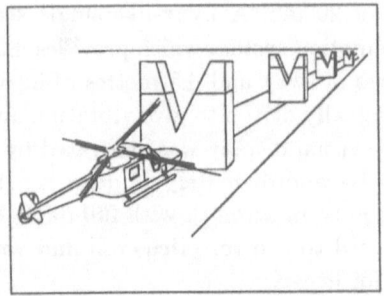
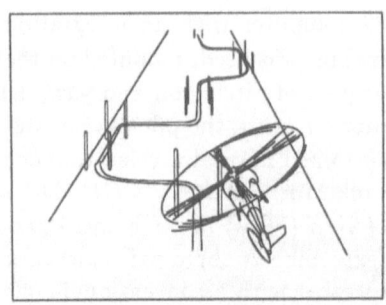

(c) Hurdles / Bob-up task   (d) Lateral jinking task

Figure 6.3

Bob-up: The bob-up task was performed in front of one of the V-notch hurdles (figure 6.3c). From a hover aligned with the bottom of the V-notch, the pilot had to acquire and maintain a new height denoted by a mark on the notch.

Lateral jinking: The lateral jinking task concerned a series of 'S' turns through slalom gates followed by a corresponding line tracking phase (figure 6.3d). The task had to be flown whilst maintaining a speed of 60 knots and a height of 25 ft.

Hurdles: Using the same V-notch hurdles as seen for the bob-up task, a collective-only flight path re-positioning task was flown at 60, 75 and 90 knots to represent increasing task aggression. From an initial height aligned with the bottom of the V-notch, the pilot had to pass through each hurdle at the height denoted by a mark on the notch and then regain the original speed and height as quickly as possible.

Yaw pointing: Whilst translating down the runway centre line at 60 knots, the pilot was required to yaw to acquire and track one of a number of offset posts.

Table 2 is a detailed compilation of one of the pilot's questionnaires based on the May 1992 trials. The primary response in heave, pitch and roll was excellent. But the primary response in yaw was sluggish and there was some undesirable collective-to-yaw coupling.

A redesign was undertaken to increase the yaw axis bandwidth and to slightly reduce the heave axis bandwidth. This was done very simply in the two degrees of freedom loop shaping design procedure by simply modifying the desired closed-loop transfer function T_0.

The new design was tested in December 1992 and achieved pilot ratings of level 1 for all six tasks.

6.11   Conclusion

The results of this case study have demonstrated that multivariable design techniques can play a significant role in the design of control systems for high performance helicopters.

7   Conclusions

The paper has provided an introduction to frequency domain methods for the analysis and design of multivariable control systems. Particular attention was given to H^∞ methods and to problems of robustness which arise when plant models are uncertain, which is always the case. The additional problems associated with the control of ill-conditioned plants were also considered.

The relative gain array, the singular value decomposition and the structured singular value were shown to be invaluable tools for analysis.

For multivariable design, emphasis was given to the shaping of the singular values of the loop transfer function. The technique of McFarlane and Glover and its extension to two degrees of freedom controllers were considered in detail. The power of the approach was demonstrated by its application to the design of a full-authority wide-envelope control system for a high performance helicopter.

8   Acknowledgements

The authors are grateful to Mr Petter Lundström (Trondheim University) and Mr Neale Foster (Leicester University) for their comments and assistance in preparing these notes.

The helicopter controller design was largely completed by Dr. Daniel Walker (Leicester University) and section 6 draws heavily on the papers by Postlethwaite and Walker (1992) and Walker et al (1993).

The helicopter case study was conducted with the support of the UK Ministry of Defence through Extramural Research Agreement No. 2206/32/RAE(B).

9   References

Anon., Aeronautical Design Standard ADS-33C "Handling Qualities Requirements for Military Rotorcraft", *US Army AVSCOM* , August, 1989.

Chiang And Safonov, *Robust Control toolbox for Matlab. User's guide.* The Math-Works, South Natick, MA, USA (1988, 1992).

C.E.Cooper and R.P.Harper, "The use of pilot rating scale in the evaluation of aircraft handling qualities", NASA TM-D-5133, 1969.

D.J.Walker, I.Postlethwaite, J.Howitt, N.P.Foster, "Rotorcraft Flying Qualities Improvement Using Advanced Control", *Proc. American Helicopter Society/NASA Conf.* , San Francisco, January, 1993.

J.C. Doyle, "Analysis of Feedback Systems with Structured Uncertainties", *IEE Proc*, **129** (D), 242-250 (1982).

J.C.Doyle and G.Stein, "Multivariable Feedback Design: Concepts for a Classical/Modern Synthesis", *IEEE Trans. AC* , **26, 1** , 4-16, 1981.

J.C. Doyle, J.E. Wall and G. Stein, "Performance and Robustness Analysis for Structured Uncertainty", *Proc. IEEE Conf. on Decision and Control*, Orlando, Florida, Dec. 1992.

J.C. Doyle, Lecture Notes *ONR/Honeywell workshop on Adcances in Multivariable Control*, Minneapolis, MN (1984).

J.C. Doyle, K.Lenz, and A.K. Packard, "Design examples using μ-synthesis: Space shuttle lateral axis FCS during reentry", in NATO ASI Series, **F34**,

Modelling, Robustness and Sensitivity Reduction in Control Systems, R.F. Curtain (Ed.), Springer-Verlag (1987).

K.Glover and D.C.McFarlane, "Robust stabilization of normalised coprime factor plant descriptions with H^∞-bounded uncertainty", *IEEE Trans. AC*, **34, 8** , 821-830, 1989.

M. Hovd and S. Skogestad, "Simple Frequency-Dependent Tools for Control System Analysis, Structure Selection and Design", *Automatica*, **28**, 5, 989-996 (1992).

J.Howitt, "Matlab toolbox for handling qualities assessment of flight control laws", *Proc. IEE Control '91*, Scotland, 1991.

R.A. Hyde, *The Application of Robust Control to VSTOL Aircraft*, Ph.D. Thesis, Cambridge University, 1991.

H.Kwakernaak, "Optimal Low-Sensitivity Linear Feedback Systems" *Automatica* , **5**, 279, 1969.

D.J.N.Limebeer, E.M.Kasenally, and J.D.Perkins, "On the Design of Robust Two Degree of Freedom Controllers", *Automatica* , **29**, **1**, 157-168, 1993.

P. Lundström, S. Skogestad and Z-Q. Wang, "Performance Weight Selection for H-infinity and mu-control metods", *Trans. Inst. of Measurement and Control*, **13**, 5, 241-252, 1991.

P. Lundström, S. Skogestad and Z.Q. Wang, "Uncertainty Weight Selection for H-infinity and Mu-Control Methods", Proc. IEEE Conf. on Decision and Control (CDC), 1537-1542, Brighton, UK, Dec. 1991b.

J.M. Maciejowski, *Multivariable Feedback Design*, Addison-Wesley (1989).

D.C.McFarlane and K.Glover, *Robust Controller Design Using Normalised Coprime Factor Plant Descriptions*, Springer-Verlag , Berlin, 1990.

M. Morari and E. Zafiriou, *Robust Process Control*, Prentice-Hall (1989).

Owen and Zames, "Unstructured uncertainty in $\mathbf{H}^\infty$", In: *Control of uncertain dynamic systems*, Bhattacharyya and Keel (Eds.), CRC Press, Boca Raton, FL , 3-20 (1991).

G.D.Padfield, "Theoretical model of helicopter flight mechanics for application to piloted simulation", *RAE* , TR 81048, 1981.

I.Postlethwaite and D.J.Walker, "Advanced Control of High Performance Rotorcraft", *Proc. IMA Conf. on Aerospace Vehicle Dynamics and Control*, Cranfield Inst. of Technology, September, 1992.

S. Skogestad and M. Morari, "Implication of Large RGA-Elements on Control Performance", *Ind. Eng. Chem. Res.*, **26**, 11, 2323-2330 (1987).

S. Skogestad, M. Morari and J.C. Doyle, "Robust Control of Ill- Conditioned Plants: High-Purity Distillation", *IEEE Trans. Autom. Control*, **33**, 12, 1092-1105 (1988). (Also see *correction* to μ-optimal controller in **34**, 6, 672 (1989)).

F. van Diggelsen and K. Glover, "Element-by-element weighted $\mathbf{H}^\infty$-Frobenius and H_2 norm problems", it Proc. 30 th IEEE Conf. on Decision and Control (CDC), Brighton, England, 923-924, 1991.

F. van Diggelsen and K. Glover, "A Hadamard weighted loop shaping design procedure", it Proc. 31 th IEEE Conf. on Decision and Control (CDC), Tuscon, Arizona, 2193-2198, 1992.

M.Vidyasager, "Normalised coprime factorizations for non strictly proper systems", *IEEE Trans. AC*, **33**,300-301, 1988.

J.F.Whidborne, I.Postlethwaite, D.W.Gu, "Robust controller design using H^∞ loop shaping and the method of inequalities", *Leicester University Engineering Department (LUED)*, Report No. 92-33, 1992.

E.A. Wolff, S. Skogestad, M. Hovd and K.W. Mathisen, "A procedure for controllability analysis", *Preprints IFAC workshop on Interactions between process design and process control, London, Sept. 1992*, Edited by J.D. Perkins, Pergamon Press, 1992, 127-132.

C.C. Yu and W.L. Luyben, "Robustness with Respect to Integral Controllability", *Ind. Eng. Chem. Res.*, **26**, 1043-1045 (1987).

A.Yue and I.Postlethwaite, "Improvement of helicopter handling qualities using H^∞ optimization". *IEE Proc.*, Part D, 137, 115-129, 1990.

V. Zakian and U.Al-Naib, "Design of dynamical and control systems by the method of inequalities", *Proc. IEE*, **120**, 11 , 1421-1427, 1973.

Ziegler and Nichols, "Process lags in automatic-control circuits", *Trans. of the A.S.M.E.*, **65**, 433-444 (1943).

Task	Level of Aggression	Comments	HQR	Level
Side Step	Low	Loads of spare capacity	2	1
	Mid	Task workload still minimal, response perfect.	2	1
	High	Increased level of aggression does not increase workload. Very easy	2(low)	1
Quick Hop	Low	Desired performance easily achieved. Slight right drift. 3-axis task. A lot of inertia in model. Control law good.	2	1
	Mid	Easier at higher aggression because less anticipation required. No problems.	2	1
Hurdles	Low	Desired performance achieved satisfactorily. Yaw coupling only problem, but spare capacity.	3	1
	High	At top of hurdle, control activity high and little spare capacity. > 10° coupling into heading.	5	2
Lateral jinking	Low	Stacks of spare capacity. Minimal control activity. Single axis task. No cross-coupling.	2	1
	Mid	As above	2	1
	High	As above	3(low)	1
Yaw Pointing	V.Low	Adequate performance achieved with difficulty. Control activity high. Not much spare capacity.Precision difficult.	5	2
	Low	PIO problems. Very high yaw inertia. Low sensitivity, possibly some lag. Maximum rate O.K. but needs to be tighter	7	3

Table 2: Pilot comment from the May 1992 trials

10. Neural Networks for Control

E.D. Sontag *

Abstract

This paper starts by placing neural net techniques in a general nonlinear control framework. After that, several basic theoretical results on networks are surveyed.

1 Introduction

The field commonly referred to as "neuro-" or "connectionist" control has been the focus of a tremendous amount of activity in the past few years. As emphasized in [20], work in this and related areas represents to a great extent a return to the ideas of Norbert Wiener, who in his 1940s work on "cybernetics" drew no boundaries between the fields now called artificial intelligence, information theory, and control theory.

In this presentation, requested by the conference organizers, the goal is definitely *not* to provide an exhaustive or even a representative survey. Most recent major control conferences have had introductory courses devoted to the topic, and, in addition, many good overviews are available in the general literature; see for instance the papers [5], [20], [7], [16], and other papers in the books [27] and [39]. Rather, the first objective is to explain the general context in which this work has taken place, namely the resurrection of some very old ideas dealing with numerical and "learning" techniques for control —a rebirth due more than anything else to the availability of raw computational power, in amounts which were unimaginable when those ideas were first suggested. Starting from a very general control paradigm, the framework is progressively specialized until the particular case of neural networks is arrived at. After this, the paper deals with an aspect which is perhaps not adequately covered by other expository efforts: the study of the question *what is special about neural nets?*, understood in

---
*Department of Mathematics, Rutgers University, New Brunswick NJ 08903, USA. Research described here partially supported by US Air Force Grant AFOSR-91-0343. This paper was written in part while visiting Siemens Corporate Research, Princeton.

the sense of the search for mathematical properties which are not shared by arbitrary parametric models.

In keeping with the program outlined above, there will not be any discussion of implementation details, simulation results, or any other "practical" issues; once more, the reader is referred to the above survey papers for pointers to the relevant literature. Moreover, the availability of these papers and their extensive bibliographies has the advantage of allowing indirect citations; a reference such as: [20:175] will mean "item [175] in the bibliography given in the paper [20]."

1.1   Learning and Adaptive Control

One of the main advantages claimed for neurocontrollers is their alleged ability to "learn" and "generalize" from partial data. What is meant by this, roughly, is that after being exposed to the "correct" control actions in several situations, the learning system should be able to react appropriately, by interpolation or extrapolation, to new situations. (Later the paper reviews one of the ways in which this idea can be framed, using the terminology of computational learning theory, but almost nothing has been done in applying such a formalization to control systems, so for now the use of terms such as learning is done in a completely informal manner.) The "learning" may be thought of as being performed "off-line," by a numerical algorithm during a training phase, or "on-line," during actual closed-loop operation. But the distinction between off-line and on-line is essentially one of modeling rather than a mathematical one; it mainly reflects the time span considered and the information available from the environment. Especially in the latter case, learning is closely related to adaptation in the usual sense of the word in control theory. (Actually, the term "adaptive control" is itself ill-defined; when studying nonlinear systems, adaptive control can often be seen simply as a form of control of partial states.)

Still, when studying very specific types of systems, and using particular controller structures, many authors differentiate between adaptation and on-line learning. In practice, adaptive control tends to refer to the control of slowly changing systems (after a modeling distinction is made between state variables and parameters), and sudden changes in parameters can lead to transient behavior during which adaptation occurs and performance is degraded. Learning control, on the other hand, is a term that tends to denote controllers that adapt to various regions of the state or the parameter space, and which store the control laws that have been found to be appropriate for that region, to be later retrieved when the same operating circumstances are encountered, with no need for readaptation (but requiring "pattern recognition" capabilities to classify such previously learned situations in order to allow later for fast recognition). Clustering algorithms, neural networks, and large amounts of memory are usually associated with learning controllers.

2 Plants, Controllers, and Models

In order to provide some unity to the description of neurocontrol, the paper starts with a general paradigm, which is progressively specialized. This discussion should be understood as an informal one. It is *not* the purpose to give rigorous mathematical definitions, but rather to establish a language in which to frame the various applied issues that are being considered. It is far too early, given the current poor state of knowledge and level of results obtained, to attempt to provide a general theoretical setting for neurocontrol. So this section should be read as a philosophical, informal discussion. In addition, the reader will note that the term "neural network" will be used several times but is left undefined. The discussion depends on *nothing more* than the fact that nets provide a certain particular class of dynamical systems (or, in static situations, maps). By not defining the term, this fact is emphasized.

2.1 The Initial Paradigm

Take as a starting point the following basic control paradigm. The object to be controlled, called the "plant," interacts with its environment by means of two types of input and output channels.

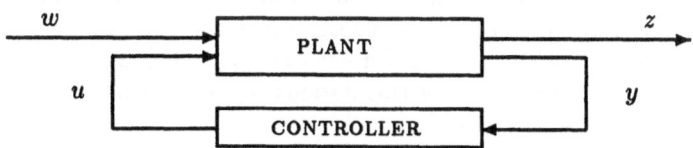

The signal w contains all inputs that cannot be directly affected by the controller. This includes disturbances, measured or unmeasured, as well as signals to be tracked and other reference information. If the controller is to have access to some of these inputs, a pass-through connection is assumed to be part of the box labeled "plant" so that the respective information appears again in the measured output variable y; this variable contains as well all other data which is immediately available to the controller, such as values of state variables. The signal u is the part of the input that can be manipulated, and the signal z encodes the control objective. The objective may be for instance to drive this variable (which might be the difference between a certain function of the states and a desired reference function) to zero. If needed, one may assume that the control values themselves appear in y, and this is useful to keep in mind when the controller will consist of separate subsystems, as done later.

This setup can be formalized in various ways, using abstract definitions of control systems as in [33], but the current discussion will be completely intuitive.

2.2   Learning Control Structure

The first refinement of the basic control paradigm that one may consider is the splitting of the controller into a fixed part and a "learning" or adaptable part; see the Figure (the signal flow is drawn from left to right, for ease of reference).

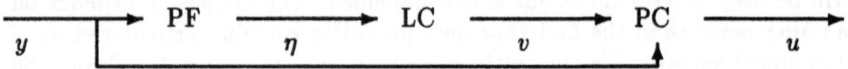

The box labeled LC represents a "learning controller" as discussed later. The other two parts —which may be missing— are used to indicate that part of the controller which does not change. The box labeled PF represents a "prefiltering" device; it may perform for instance one of the following (not necessarily distinct) functions:

Feature Extraction. This may be based on Fourier, wavelet, or other numeric transforms, or on symbolic procedures such as edge detection in images. One may also include in this category the computation of basic operations on measured signals y; for example, obtaining all pairwise products $y_i y_j$ of coordinates allows correlations to be used as inputs by the subsequent parts of the controller.

Sampling. Quantization, time-sampling, and general A/D conversion.

Template Matching. The output of PF may be a binary signal which indicates if y equals a specific value, or more generally, if this signal belongs to a predetermined set. More generally, its output may be a discrete signal that indicates membership in various, possibly overlapping, receptive fields, that is, a vector of the form $(\chi_{S_1}(y), \ldots, \chi_{S_k}(y))$, where each S_l is a subset of the space where instantaneous values of y lie. When these sets consist of just one point each, pure template matching results. More interestingly, one may have spheres centered at various points in y-space, which allows a certain degree of interpolation. Instead of characteristic functions, a more "fuzzy" weighting profile may be used, corresponding to degrees of membership. The possible overlap of regions of influence allows for distributed representations of the input; CMAC-type controllers (see e.g. [7:2] for many references) employ such representations together with hashing memory addressing systems for associative recall. Localized receptive fields play a role quite analogous, for approximation problems, to splines, and this connection is explored in [20:108].

The box labeled PC represents a "precomputed (part of the) controller" which also is invariant in structure during operation. Included here are, as examples:

Gain Scheduled Control Laws. A finite number of precomputed compensators, to be switched on according to the value of a discrete signal v, or to be blended according to weights indicated by a continuous v ("fuzzy" control applications).

A State Feedback Law. For instance, v may be a state estimate produced by LC (which then plays the role of an observer), and PC may be a state feedback law. Recall that the observation y may be assumed to include, by making the "plant" larger, information on the input signal u; or one may include a backward path from PC into LC, for use by the observer.

Sampling. As with the prefiltering box, one may include here smoothers, D/A conversion, hold devices, and so forth.

This structure is only given for expository purposes; in practice the boundaries between prefiltering, learning, and precomputed controller are often blurred. (For instance, one may use various clustering algorithms for the PF part, but these algorithms may in turn be "learned" during operation of the controller.)

2.3 Refining the LC Block

Refining further the general control paradigm in order to arrive to neurocontrol techniques, assume that the block LC splits into two parts, a "tuner" and a box representing a system (dynamical, or a static mapping) with parameters determined by the tuner.

The block labeled Σ_λ^k corresponds to a system or static mapping parameterized by a vector λ of real parameters and operating on the input signal η. There is as well a discrete parameter k, so that the size of the vector λ is a function of k. For instance, k may determine a choice of number of state variables in a controller, or number of neurons —or more generally an encoding of an "architecture"—for a neural network, and λ encompasses all free parameters in the corresponding model. The tuner (see arrow) selects the parameters of Σ_λ^k according to some "learning procedure" which uses all available information (included as part of η).

In adaptive control, one might include as models Σ_λ^k linear systems of order k, parameterized by λ. The tuner would then include an adaptation rule for this controller. Or, the tuner might be used to set the parameters in an identification model Σ_λ^k, and the result of identification could be a signal used to drive a precomputed controller. The term "precomputed" could be taken to mean, in this last case, "to be computed by a standard algorithm" such as LQG design.

This is the place where neural nets are most often used in applications. Typically the family Σ_λ^k will include a class of systems and/or controllers that is sufficiently rich to represent a wide variety of dynamical behaviors (if used as an identification model, or for representing controllers with memory), or static maps (for pattern recognition applications, or for implementing static controllers). This motivates the search for such rich classes of models, and the study of their representational and approximation properties. Many choices are possible. For dynamical systems, one may pick linear or bilinear systems, or more generally systems of the form

$$\dot{x} = f(x, \lambda, u),\, y = h(x, \lambda) \tag{1}$$

(or analogously in discrete time) where the functional form of f and h is predetermined, for each state space dimension n. This functional form may simply be a linear combination of a set of basic functions (polynomials, splines with fixed nodes, and so forth), with the λ variables providing the coefficients of the combination. Or, the parameters may appear nonlinearly, as in rational parameterizations, splines with free nodes, and neural nets (whatever this last term means). Instead of state-space models, one may use parameterized classes of input/output behaviors; in that case, the parameters might correspond to coefficients in transfer functions or in coprime factorizations, or kernels in Volterra or Chen-Fliess series expansions.

2.4   A (Mini-) Illustration

The following is an example of a parametric adaptation rule used in neuro-control applications. Although very simplified and for a very special case, it illustrates some of the typical ideas. (If desired, one could formulate in various manners this example in terms of the above general paradigms, but as explained earlier, those structures have no formal meaning and are only given to help understand the literature. In this particular case, the methodology will be clear.) Assume that the plant to be controlled is described, in discrete time, by the scalar equation

$$x^+ = f(x) + u$$

(use the superscript $+$ to indicate time shift). The mapping f is unknown but x can be observed. The objective is to track a known reference signal r; more precisely, to obtain

$$x(t) - r(t - 1) \rightarrow 0 \text{ as } t \rightarrow \infty$$

by means of an appropriate control. If the mapping f were known, one would use the control law $u(t) = -f(x(t)) + r(t)$ in order to achieve perfect tracking. (Note that this system is easy to control; most applications of neurocontrol have been indeed for systems that would be easy to control if perfectly known, most often for feedback-linearizable systems. Later, the paper considers some problems that arise when the system itself is truly nonlinear, and the constraints that this imposes on neural controller structure.) Since f is unknown, one proposes a parametric model such as

$$x^+ = F(x, \lambda) + u \tag{2}$$

where F is "rich enough" and attempts to find a $\hat{\lambda}$ so that $F(x, \hat{\lambda}) \approx f(x)$ for all x. Once such an approximation is found, the certainty equivalence controller $u(t) = -F(x(t), \hat{\lambda}) + r(t)$ is used, possibly updating $\hat{\lambda}$ during operation. For instance, F may be a generic polynomial in x of some large degree k, and λ is a vector listing all its coefficients.

How should the parameter(s) λ be estimated? One possibility is to assume that (2) is the true model but it is subject to driving and observation noise:

$$x^+ = F(x, \lambda) + u + \xi_1, \quad y = x + \xi_2$$

where ξ_1, ξ_2 are independent 0-mean Gaussian variables and y is being observed. This leads to an extended Kalman filtering formulation, if one assumes also a stochastic model for the parameters: $\lambda^+ = \lambda + \xi_3$ where ξ_3 is independent from the other noise components. One may then use the a posteriori EKF estimate $\hat{\lambda}[t|t]$ as $\hat{\lambda}$. The estimate $\hat{x}[t|t]$ can be used instead of x. (If only partial observations are available, an output mapping $y = h(x) + \xi_2$ may be assumed instead of the identity, and a similar procedure may be followed.) See for instance [25] and [23] for this type of parametric nonlinear adaptive control approach.

When parameterizations are linear in λ, this procedure, or one based on Lyapunov techniques for parameter identification —possibly in conjunction with sliding mode robust control, to handle the regions where the true f differs considerably from the best possible model of the form $F(x, \lambda)$— can be proved to converge in various senses; see especially [32], as well as for instance [30].

In general, however, there are absolutely no guarantees that such a procedure solves the tracking problem, and only simulations are offered in the literature to justify the approach. (An exception are certain local convergence results; see e.g. [12].) Moreover, the control problem itself was capable of being trivially solved by feedback linearization, once that the plant was identified. Note also that the nonlinear adaptive control approach is independent of the neural nets application; by and large there have been *no theoretical results* given in the literature, for adaptive control, that would *in any way use properties particular to neural nets.* This is typical of most neural networks applications at present, and it is the main reason for concentrating below instead on the basic representation, learnability, and feedback control questions involved.

2.5   Other Techniques for Control

When systems are not feedback linearizable, nonlinear control becomes a very hard problem, even leaving aside identification issues (as an illustration, see [33], Section 4.8, and the many references given there, for stabilization questions). There are several adaptation approaches which have been popular in neurocontrol and which correspond to various combinations of tuners and parametric models. Many of these approaches are very related to each other, and they are really independent of the use of neural networks; in fact, they represent methods of numerical control that have been around since at least the early 1960s. Briefly described next are some of the ideas that have appeared in this context, with the only objective of helping the reader navigate through the literature.

Identifying a State Feedback Law. Here one assumes a parametric form for a state feedback law $u = k(\lambda, x)$ and the parameter λ is chosen so as to minimize some cost criterion. Obviously, the class of feedback functions that can be represented in this manner should be rich enough to contain at least one solution to whatever the ultimate control objective is. (In practice, this is simply assumed to be the case.) A cost criterion that depends on the parameter and the state is picked to be minimized, such as for instance:

$$J(\lambda, x_0) = \int_0^\infty Q(x(t), u(t)) \, dt \, ,$$

where $x(\cdot), u(\cdot)$ are the state and input obtained when using $u = k(\lambda, x)$ as a controller for the given system $\dot{x} = f(x, u)$ and $x(0) = x_0$. Now one may attempt to minimize $\max_x J(\lambda, x)$ over λ. This is the type of approach taken, for example, in [4], in which the viability problem (make the state stay in a desired set) is attacked using neural net controllers. (The minimax character of the problem makes it suitable for nondifferentiable optimization techniques.)

Estimate a Lyapunov or Bellman Function. A basic problem in feedback control theory, which also arises in many other areas (it is known as the "credit assignment problem" in artificial intelligence) is that of deciding on proper control actions at each instant in view of an overall, long-term, objective. Minimization of an integrated cost, as above, leads implicitly to such good choices, and this is the root of the Lagrangian or variational approach. Another possibility, which underlies the dynamic programming paradigm, is to attempt to minimize at each instant a quantity that measures the overall "goodness" of a given state. In optimal control, this quantity is known as the *Bellman* (or the cost-to-go, or value, function); in stabilization, one talks about *Lyapunov* (or energy) functions; game-playing programs use *position evaluations* (and subgoals); in some work that falls under the rubric of learning control, one introduces a *critic*. Roughly speaking, all these are variants of the same basic principle of assigning a cost (or, dually, an expectation of success) to a given state (or maybe to a state and a proposed action) in such a way that this cost is a good predictor of eventual, long-term, outcome. Then, choosing the right action reduces to a simpler, nondynamic, pointwise-in-time, optimization problem: choose a control that leads to a next state with least cost.

Finding a suitable measure in the above sense is a highly nontrivial task, and neural network research has not contributed anything conceptually new in that regard. The usual numerical techniques from dynamic programming (value or policy iteration) can be used in order to compute the Bellman function for an optimization problem. Other work has been based on posing a parametric form for a Lyapunov ("critic", etc) function $V(\lambda, x)$. In this latter mode, one attempts to fit parameters λ for both the Lyapunov function and the proposed feedback law $K(\lambda, x)$ simultaneously. This is done by adjusting λ after a complete "training" event, by means of, typically, a gradient descent step. The "adaptive critic" work by Barto and others (see [7] and the many references there) is one example of this approach, which is especially attractive when the overall goal is ill-defined in quantitative terms. For instance, the technique is often illustrated through the problem of controlling an inverted pendulum-on-a-cart system, in which the parameters are unknown and the only training signal allowed is the knowledge that the car has achieved a certain maximum displacement or that the pendulum angle has reached a maximum deviation. This type of work is also closely related to relatively old literature in learning control by researchers such as Mendel, Fu, and Sklansky in the mid to late 1960s and Narendra in the 1970s. See for instance [7:(64,74)] and [16:54].

Local Adaptive Control. One may consider several selected operating conditions, and design an adaptive (or a robust) controller for the linearization around each of them. (Operating conditions close to one of the selected ones give rise to linear models that are thought of as parameter variations.) The "learning part" consists in building an association between the current state and the appropriate controller. This is conceptually just a variant on the idea of gain-scheduling and use of a pattern recognition device to choose the appropriate gain, and it appears in many neurocontrol applications. There is nothing special in the use of "neural nets" as pattern recognizers or associative memories; any other reasonable technique would in principle apply equally well. Again, no mathematical analysis is ever given of overall performance.

Expert-Systems (Mimicking) Approach. Often a good controller may be available, but various reasons may make it worth to simulate its action with a neural net. For instance, a human expert may be good at a given control task and one may want to duplicate her behavior. In this case, one may be able to fit a parameterized class of functions, on the basis of observations of typical controller operation; see e.g. [7:111].

Feedforward Control. A popular technique in neurocontrol is closely related to model reference adaptive control. Here a controller, in the form for instance of a recurrent network (see below), is trained in such a manner that when cascaded with the plant the composite system emulates a desired model. The training is often done through gradient descent minimization of a cost functional.

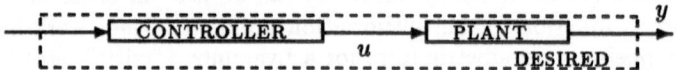

A particular case that has been much explored experimentally —see for instance [7:(3,4,56,58,67,79,90)]— is the one where the desired model is a pure multiple delay (in discrete time) or integrator (continuous time), that is, the case of system inversion. A training set is obtained by generating inputs $u(\cdot)$ at random, and observing the corresponding outputs $y(\cdot)$ produced by the plant. The inverse system is then trained by attempting to fit the reversed pairs (y, u). Once that an inverse is calculated, a desired output $y_d(\cdot)$ can be obtained for the overall system (possibly delayed or integrated) by feeding y_d to the controller. This approach is already subject to serious robustness problems even for linear systems. But in the nonlinear case there are major additional difficulties, starting with the fact that nice inverses can be assumed to exist at best locally, so training on incomplete data cannot be expected to result in a good interpolation or "generalization" capability, which after all is the main objective of learning control.

If only certain particular outputs $y_d(\cdot)$ are of interest, then generating training data in the random manner just sketched is very inefficient. It is more reasonable in that case to train the controller so that when given those *particular* targets y_d it generates inputs to the plant that produce them. Given such a more limited objective, the procedure described above amounts to exhaustive search, and steepest descent is preferable. In order to apply steepest descent, one sets up an appropriate cost to be minimized. Since the derivative of this cost involves a derivative of the plant, this may lead to serious errors if the plant is not perfectly known, but only an experimentally obtained approximation is available. A bit more formally, the problem is as follows. One wishes to find a value λ so that, for certain y_d,

$$P(C(y_d, \lambda)) = y_d$$

where the "plant" transformation P is approximately known and the "controller" C is a function of outputs and parameters. Equivalently, one attempts to minimize $F(\lambda) = \|P(C(y_d, \lambda)) - y_d\|^2$. The gradient flow in this case is $\dot{\lambda} = -\nabla_\lambda F(\lambda)$. Now, if instead of P one must work with an approximation P_1 of P, the gradient of F will be computed using derivatives of P_1 (understood in a suitable functional sense, for dynamical systems) rather than of P. Thus it is essential that the approximation of P by P_1 had been previously done in a topology adequate for the problem, in this case a C^1 topology, in other words, so that the derivative of P_1 is close to the derivative of P. This issue was pointed out in the neural nets literature, and used in deriving algorithms, in for instance [7:(46,104,105,106)].

2.6 Modeling via Recurrent Nets

An especially popular architecture for systems identification (models Σ_λ^k) has been that of recurrent neural nets. Described later are some theoretical results for these, but in practice the question that has attracted the most attention has been that of fitting parameters to observed input/output behavior. The work done has consisted mostly of a rediscovery of elementary facts about sensitivity analysis. Essentially, given a system as in Equation (1), an input $u(\cdot)$ on an interval $[0, T]$, and a desired final output $y_d(T)$, one wants to find parameters λ such that the output $y(T)$, say for a given initial state, differs as little as possible from $y_d(T)$, on this input $u(\cdot)$. (One may be interested instead in matching the entire output *function* $y(\cdot)$ to a desired function $y_d(\cdot)$, that is, in minimizing an error functional such as $J(\lambda) = \int_0^T \|y(t) - y_d(t)\|^2 dt$. This can be reduced to the previous case by adding a state variable $\dot{z}(t) = \|y(t) - y_d(t)\|^2$ and minimizing $z(T)$, as routinely done in optimal control. Also, one may also be dealing with a finite set of such pairs u, y_d rather than just one, but again the problem is essentially the same.) In order to perform steepest descent, it is necessary to compute the gradient of $\|y(T) - y_d(T)\|^2$ with respect to λ. Denoting by $\partial x(T)/\partial \lambda$ the differential of $x(T)$ with respect to λ, evaluated at the parameter values obtained in the previous iteration of the descent procedure, one must compute

$$\frac{\partial \|h(x(T)) - y_d(T)\|^2}{\partial x(T)} \; \frac{\partial x(T)}{\partial \lambda} \tag{3}$$

(the parameters λ are omitted, for notational convenience.) Viewing the parameters as constant states, the second term can be obtained by solving the variational or linearized equation along the trajectory corresponding to the control $u(\cdot)$, for the system obtained when using the current parameters λ; see for instance Theorem 1 in [33]. (One "old" reference is [26].) Such an approach is known in network circles as the "real time recurrent learning" algorithm, and it is often pointed out that it involves a fairly large amount of variables, as the full fundamental solution (an $n \times n$ matrix of functions) must be solved for. It has the advantage of being an "online" method, as the equations can be solved at the same time as the input u is being applied, in conjunction with the forward evolution of the system. An alternative procedure is as follows. Since the full gradient is not needed, but only the product in (3) is, one may instead propagate the first term in (3) via the adjoint equation, again as done routinely in optimal control theory. This involves two passes through the data (the adjoint equation must be solved backwards in time) but less memory requirements. It is a procedure sometimes called "backpropagation through time". In discrete time, the difference between the two procedures is nothing more than the difference between evaluating a product

$$vA_T \ldots A_2 A_1 \tag{4}$$

where $A_1, \ldots, A_T$ are matrices and v is a vector, from right to left (so a matrix must be kept at each stage, but the procedure can be started before all A_i's are known) or from left to right (so only a vector is kept). The discussion of the relative merits of each approach seems to have consumed a major amount of effort in this area. Nothing especial about neural network models seems to have been used in any papers, except for some remarks on storage of coefficients. No global convergence theorems are proved.

3 Neural Nets

Artificial neural nets give rise to a particular class of parameterized controllers and models. What is meant precisely by the term neural net varies, depending on the author. Most often, it means a linear interconnection of memory-free scalar nonlinear units, supplemented by memory elements (integrators or delay lines) when dynamical behavior is of interest. The coefficients characterizing the connections, called "weights," play a role vaguely analogous to the concentrations of neurotransmitters in biological synapses, while the nonlinear elements in this over-simplified analogy correspond to the neurons themselves. In practice, one decides first on the class of "neurons" to be used, that is, the type of nonlinearity allowed, which is typically of the "sigmoidal" type reviewed below. The weights are "programmable" parameters that are then numerically adjusted in order to model a plant or a controller. In some of the literature, see e.g. [28], each scalar nonlinear unit acts on a scalar quantity equal to the distance between the incoming signal and a reference (vector) value; this is in contrast to operating on a linear combination of the components of the incoming signal, and gives rise to "radial basis function" nets, which are not treated here. See the textbook [18] for a clear and well-written, if mathematically incomplete, introduction to neural nets.

Motivating the use of nets is the belief —still not theoretically justified— that in some sense they are an especially appropriate family of parameterized models. Typical engineering justifications range from parallelism and fault tolerance to the possibility of analog hardware implementation; numerical and statistical advantages are said to include good capabilities for learning (adaptation) and generalization.

3.1 What are Networks

As explained above, by a ("artificial neural") net one typically means a system which is built by linearly combining memory-free scalar elements, each of which performs the same nonlinear transformation $\sigma : \mathbb{R} \to \mathbb{R}$ on its input. One of the main functions σ used is sign$(x) = x/|x|$ (zero for $x = 0$), or its relative, the hardlimiter, threshold, or *Heaviside* function $\mathcal{H}(x)$, which equals 1 if $x > 0$ and 0 for $x \leq 0$ (in either case, one could define the value at zero differently; results do not change much). In order to apply various numerical techniques, one often needs a differentiable σ that somehow approximates sign(x) or $\mathcal{H}(x)$. For this, it is customary to consider the hyperbolic tangent tanh(x), which is close to the sign function when the "gain" γ is large in tanh(γx). Equivalently, up to translations and change of coordinates, one may use the *standard sigmoid* $\sigma_{\mathrm{s}}(x) = \frac{1}{1+e^{-x}}$. Also common in practice is a piecewise linear function, $\pi(x) := x$ if $|x| < 1$ and $\pi(x) = \text{sign}(x)$ otherwise; this is sometimes called a "semilinear" or "saturated linearity" function.

Whenever time behavior is of interest, one also includes dynamic elements, namely delay lines if dealing with discrete-time systems, or integrators in the continuous-time case.

In the static case, one considers nets formed by interconnections *without loops*, as otherwise the behavior may not be well-defined; these are called "feedforward" nets, in contrast to the terms "feedback," "recurrent," or "dynamic" nets used in the general case. The next Figure provides an example of a static net computing the function $y = 2\sigma[3\sigma(5u_1 - u_2) + 2\sigma(u_1 + 2u_2 + 1) + 1] + 5\sigma[-3\sigma(u_1 + 2u_2 + 1) - 1]$ and of a dynamic net representing the system $\dot{x}_1 = \sigma(2x_1 + x_2 - u_1 + u_2), \dot{x}_2 = \sigma(-x_2 + 3u_2), y = x_1$.

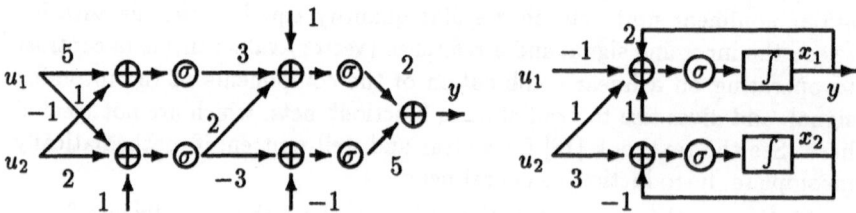

3.2   A Very Brief History

In the 1940s, McCulloch and Pitts introduced in [20:119] the idea of study-ing the computational abilities of networks composed of simple models of neurons. Hebb ([20:78]) was more interested in unsupervised learning and adaptation, and proposed the training rule known by his name, which is still at the root of much work, and which consists of reinforcing the associations between those neurons that are active at the same time.

Rosenblatt, in the late 1950s, pursued the study of "perceptrons" or, as they might be called today, "multilayer feedforward nets" such as the first one shown above, using the Heaviside activation $\mathcal{H}$. He developed various adaptation rules, including a stochastic technique. (He called the latter "backpropagation" but the term is now used instead for a gradient descent procedure. For a *very special* case, that of networks in which all nodes are either input or output nodes, a global convergence result was proved, which came to be known as the "perceptron convergence theorem." It is a widely held misconception that this latter special case was the main one that he considered.) The book [20:148] summarized achievements in the area. His goal was more the understanding of "real" brains than the design of artifi-cial pattern recognition devices, but many saw and promoted his work as a means towards the latter. Unfortunately, the popular press —in Rosen-blatt's words, "which (treated Perceptrons) with all the exhuberance and discretion of a pack of happy bloodhounds"— published hyped-up claims of perceptron-based artificial intelligence. When the expectations were not met, Rosenblatt suffered from the backlash, and the work was to a great extent abandoned. The book by Minsky and Papert [7:70] is widely credited with showing the mathematical limitations of perceptrons and contributing to the decline of the area, but Rosenblatt was already well-aware, judging from his book, of the difficulties with his models and analysis techniques.

During the mid 1970s many authors continued work on models of neural nets, notably Grossberg and his school —see for instance [20:71]— with most of this work attempting to produce differential equation models of various conditioning phenomena. Work by Kohonen and others ([20:99]) on feature extraction and clustering techniques was also prominent during this period.

The resurgence in interest during the mid 1980s can be traced to two independent events. One was the work by Hopfield [20:183] on the design of associative memories, and later the solution of optimization problems, using a special type of recurrent networks. The other was the suggestion to return to Rosenblatt's feedforward nets, but using now differentiable, "sigmoidal," activation functions σ. Differentiability makes it possible to employ steepest descent on weight (parameter) space, in order to find nets that compute a desired function or interpolate at desired values. This was emphasized in the very popular series of books and papers by the "Parallel Distributed Processing" research group —see for instance [20:150]— and came to be known under the term "backpropagation." (The term comes from the fact that in computing the gradient of an error criterion with respect to parameters, via the chain rule, one multiplies a product such as the one in equation (4) from left to right, "propagating backwards" the vector v that corresponds to the error at the output.)

The interest in Hopfield-type nets for associative storage has waned, as their limitations have been well-explored, theoretically and in applications, while the use of the "backpropagation" approach, sometimes modified in various ways, and extended to include general recurrent neural nets, has taken center stage.

It is hard to explain the popularity of the artificial neural nets from a purely mathematical point of view, given what is *currently* known about them. Some of the claimed advantages were mentioned earlier, but many of the same claims could apply in principle to several other classes of function approximators and systems models. Of course, analogies to biological structure, coupled with a substantial marketing effort by their proponents, did not hurt. Ease of use and geometric interpretability of results helps in user-friendliness, and this has been a major factor. But perhaps the best explanation is the obvious one, namely that their recent re-introduction coincided with the advent of cheap massive computational power. Compared with the mostly linear parameter fitting techniques widely in use, it is conceivable that *any* suitably rich nonlinear technique could do better, provided enough computational effort is spent. In fact, without prior knowledge that one must wait for hundreds of thousands of iterations before noticing any sort of convergence, few would have tried this approach even 20 years ago!

3.3 This Paper

The rest of this paper explores various properties of "backpropagation" networks, with an emphasis on results which are specific to them (as opposed to more general nonlinear models). The reader should not interpret these results in any manner as a justification for their use, nor should it be assumed that this is an exhaustive survey, as it concentrates on those topics which the author has found of interest. There will be a brief discussion of *some* basic theoretical results (potentially) relevant to the use of nets for identification and control. Even for the topics that are covered, most details are not included; references to the literature are given for precise definitions and proofs. There are no explanations of experimental results, solely of theory. Moreover, numerical questions and algorithms are ignored, concentrating instead on ultimate capabilities of nets.

In many papers found in the more theoretical neural nets literature, a theorem or algorithm valid for more or less arbitrary parametric families is quoted, and the contribution is to verify in a more or less straightforward manner conditions of applicability. In this category one often finds work in, for instance, the following topics (which will not be treated here):

Numerical techniques for solving neural net approximation problems. Much work has been done on using conjugate gradient and quasi-Newton algorithms.

Statistical studies. In network learning systems, all the usual issues associated to estimation arise: the tradeoff between variance and bias (or, in AI terms, between generality and generalization). Cross-validation and other techniques are often applied; see for instance the book [40] for much on this topic.

Studies of the effect of incremental (on-line) versus "all at once" (batch or off-line) learning. In actual applications that involve gradient descent, often the complete gradient is not computed at each stage, but an approximation is used, which involves only new data. This is closely related to standard issues treated in the adaptive control and identification literature, and in fact many papers have been written on the use of stochastic approximation results for choosing learning parameters.

Rather than giving a general definition and then treating particular cases, the exposition will be organized in the opposite way: first single-hidden layer nets will be considered, as these have attracted by far the largest portion of research efforts, then two-hidden layer nets, mostly in order to emphasize that such nets are more natural for control applications, and finally recurrent nets, in which feedback is allowed.

4 Feedforward Nets

Let $\sigma : \mathbb{R} \to \mathbb{R}$ be any function, and let m, n, p be positive integers. A *single-hidden layer net with m inputs, p outputs, n hidden units, and activation function σ* is specified by a pair of matrices B, C and a pair of vectors b_0, c_0, where B and C are respectively real matrices of sizes $n \times m$ and $p \times n$, and b_0 and c_0 are respectively real vectors of size n and p. Denote such a net by a 5-tuple

$$\Sigma = \Sigma(B, C, b_0, c_0, \sigma),$$

omitting σ if obvious from the context.

Let $\vec{\sigma}_n : \mathbb{R}^n \to \mathbb{R}^n$ indicate the application of σ to each coordinate of an n-vector: $\vec{\sigma}_n(x_1, \ldots, x_n) = (\sigma(x_1), \ldots, \sigma(x_n))$. (The subscript is omitted as long as its value is clear from the context.) The *behavior* of Σ is defined to be the map

$$\mathrm{beh}_\Sigma \; : \; \mathbb{R}^m \to \mathbb{R}^p \; : \; u \mapsto C\vec{\sigma}(Bu + b_0) + c_0. \tag{1HL}$$

In other words, the behavior of a network is a composition of the type $f \circ \vec{\sigma} \circ g$, where f and g are affine maps. The name "hidden layer" reflects the fact that the signals in the intermediate space $\mathbb{R}^n$ are not part of inputs or outputs. A function f is said to be *computable by a 1HL net with n hidden units and activation σ* if it is of the type beh_Σ as above. For short, "1HL" will stand from now on for "single hidden layer" when writing about nets or the functions that they compute. Due to lack of space, some notations are unavoidable: $\mathcal{F}_{n,\sigma,m}^p$ will be used for the set of 1HL functions with n hidden units, and $\mathcal{F}_{\sigma,m}^p := \bigcup_{n>0} \mathcal{F}_{n,\sigma,m}^p$. The superscripts are dropped if $p = 1$; note that one can naturally identify $\mathcal{F}_{\sigma,m}^p$ with $(\mathcal{F}_{\sigma,m})^p$.

Often is useful to allow a linear term as well, which can be used for instance to implement local linear feedback control about a desired equilibrium. Such linear terms correspond to direct links from the input to the output node. A function computable by a net *with possible direct input to output connections* (and 1HL) as any function $g : \mathbb{R}^m \to \mathbb{R}^p$ of the form $Fu + f(u)$ where F is linear and $f \in \mathcal{F}_{n,\sigma,m}^p$.

Figure illustrates the interconnection architecture of 1HL nets;
the dotted line indicates a possible direct i/o connection. (Here
$p = 1$, $b_0 = (b_{01}, \ldots, b_{0n})$, and
$C = (c_1, \ldots, c_n)$.)

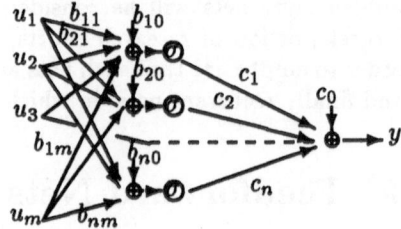

4.1 Approximation Properties

Much has been written on the topic of function approximation by nets.
Hilbert's 13th problem, on realizing functions by superpositions of scalar
ones, is sometimes cited in this context. A positive solution of Hilbert's
problem was obtained by Kolmogorov, with enhancements by Lorentz and
Sprecher. The result implies that any multivariable, real-valued continuous
function f can be *exactly* represented, on compacts, in the form $f(x) = \sum_{j=1}^{n} \sigma\left(\sum_{i=1}^{m} \mu_{ij}(u_i)\right)$ where the μ_{ij}'s are universal scalar functions (they
depend only on the input dimension, not on f). The number of "hidden
units" n can be taken to be $2m + 1$. The function σ, however, depends
on the f being represented, and, though continuous, is highly irregular.
(Essentially, one is describing a dense curve in the space of continuous
functions, parameterized by the scalar function σ.) This result can be used
as a basis of *approximation* by nets —with two hidden layers, to be defined
later— and using more regular activation functions, but it seems better to
start from scratch if one is interested in 1HL theorems.

4.2 Scalar Inputs

Given a function $\sigma : \mathbb{R} \to \mathbb{R}$, write $\mathcal{F}_\sigma$ instead of $\mathcal{F}_{\sigma,1}$; this is the affine span
of the set of all dilations and translates of σ. So the elements of $\mathcal{F}_\sigma$ are
those functions $\mathbb{R} \to \mathbb{R}$ that can be expressed as finite linear combinations

$$c_0 + \sum_{i=1}^{n} c_i \sigma(B_i u + b_i) \tag{5}$$

(denoting b_i instead of b_{0i} and letting $B_i :=$ ith entry of B).

The mapping σ is a *universal activation* if, for each $-\infty < \alpha < \beta < \infty$,
the restrictions to the interval $[\alpha, \beta]$ of the functions in $\mathcal{F}_\sigma$ constitute a
dense subset of $C^0[\alpha, \beta]$, the set of continuous functions on $[\alpha, \beta]$ endowed
with the metric of uniform convergence. Note that, at this level of gener-
ality, nothing is required besides density. In practical applications, it may

be desirable to consider special classes of functions σ, such as those used in Fourier analysis or wavelet theory, for which it is possible to provide "reconstruction" algorithms for finding, given an $f : [\alpha, \beta] \to \mathbb{R}$, a set of coefficients B_i, b_i, c_i that result in an approximation of f to within a desired tolerance.

Not every nonlinear function is universal in the above sense, of course; for instance, if σ is a fixed polynomial of degree k then $\mathcal{F}_\sigma$ is the set of all polynomials of degree $\leq k$, hence closed and not dense in any C^0. But most nonlinear functions are universal. A conclusive result has recently been obtained and is as follows; see [22]: Assume that σ is locally Riemann integrable, that is, it is continuous except at most in a set of measure zero, and it is bounded on each compact. Then, σ is a universal activation *if and only if it is not a polynomial.*

Previous results along these lines were obtained in [19], which established that any σ which is continuous, nonconstant, and bounded is universal (see also [13] and [20:(59,85)] for related older results).

The proof in [22] is based essentially on two steps: First, one reduces, by convolution, to the case in which σ is infinitely differentiable (and nonpolynomial). Locally, Taylor series expansions can be used to write σ approximately as a polynomial of arbitrary degree, and enough linearly independent polynomials can be obtained by suitable dilations and translations of σ. Now the Weierstrass Theorem completes the proof. Rather than providing details, it is instructive to briefly sketch the proof of universality in some special but still quite general cases. These special cases are also of interest since they illustrate situations in which the approximations are constructive and give rise to explicit bounds.

A *sigmoidal* function is any function with the property that both $\lim_{u \to -\infty} \sigma(u)$ and $\lim_{u \to +\infty} \sigma(u)$ exist and are distinct (without loss of generality, assume the limits are 0 and 1 respectively). If σ is also monotone, one says that it is a *squashing* function. For squashing σ, the associated "bump function" function $\overline{\sigma}(u) := \sigma(u) - \sigma(u-1)$ is (Riemann) integrable. To show that σ is universal, it is enough to show that $\overline{\sigma}$ is, so from now on assume without loss of generality that σ is integrable, nonnegative, and not identically zero. Let $\hat{\sigma}$ denote the Fourier transform of σ. As σ is nonzero, there is some ω_0 so that $\hat{\sigma}(\omega_0) \neq 0$, and one may assume $\omega_0 \neq 0$ (by continuity). Then, for each t:

$$e^{it}\hat{\sigma}(\omega_0) = \int_{-\infty}^{\infty} \sigma(x)e^{-i\omega_0 x}e^{it}\,dx = \frac{1}{|\omega_0|}\int_{-\infty}^{\infty}\sigma\left(\frac{u+t}{\omega_0}\right)e^{-iu}\,du.$$

Taking real and imaginary parts, and approximating the integral by Riemann sums, one concludes that sines and cosines can be approximated by elements of $\mathcal{F}_\sigma$. From here, density of trigonometric polynomials provides the desired result. To approximate a given function, one first needs to obtain an approximation by trigonometric polynomials, and then each sine and cosine is approximated by a 1HL net. The speed and accuracy of approximation in this manner is thus determined essentially by Fourier information about both the function to be fit and σ.

Another proof, which allows accuracy to be estimated by the local oscillation of the function being fit, is as follows. Take any continuous function f on $[\alpha, \beta]$, and assume one wants to approximate it to within error $\varepsilon > 0$. First approximate f uniformly, to error $\varepsilon/2$, by a piecewise constant function, i.e. by an element $g \in \mathcal{F}_{\mathcal{H}}$; this can be done in such a manner that all discontinuous steps be of magnitude less than $\varepsilon/4$. Assume a linear combination with k steps achieves this. Now, if σ is squashing, each term in this sum corresponding to a $c\mathcal{H}(u + b)$ can be approximated by a term of the form $c\sigma(B(u + b))$ for large enough positive B; this can be done to within error $\varepsilon/4k$ uniformly away from the discontinuity and with values everywhere between the limits of the step function. Adding all terms, there results a uniform approximation of g by an element of $\mathcal{F}_{\sigma}$, to tolerance $\varepsilon/2$, and hence also an ε-approximation of the original function.

The above proof works for any continuous function (in fact, for any "regulated" function: f must have one-sided limits at each point), but the number of terms and/or the coefficients needed may be very large, if f is fast changing. It can be proved, and this will be used below, that if f happens to be a function of bounded variation (for instance, if f has a continuous first derivative), then, letting $V(f)$ denote the total variation of f on the interval $[\alpha, \beta]$, the following holds. For each $\varepsilon > 0$ there exists a sum as in (5), with $\sigma = \mathcal{H}$, which approximates f uniformly to within error ε and so that the sum of the absolute values of the coefficients satisfies $\sum_{i>0} |c_i| \leq V$. The term c_0 can be taken to equal $f(\alpha)$. In fact, the variation V is exactly the smallest possible number so that for each ε there is an approximation with such a coefficient sum. This result holds for any function of bounded variation, even if not continuous, and characterizes the classical bounded variation property. An equivalent statement, after normalizing, is as follows. The variation of f is the smallest $V > 0$ such that $f - f(\alpha)$ is in the closure (in the uniform metric) of the convex hull of $\{\pm V\mathcal{H}(\pm u + b), b \in \mathbb{R}\}$.

4.3 Multivariable Inputs

Universality of σ implies the corresponding multivariable result. That is, for each m, p, and each compact subset K of $\mathbb{R}^m$, the restrictions to K of elements of $\mathcal{F}^p_{\sigma, m}$ (i.e., the functions $F : \mathbb{R}^m \to \mathbb{R}^p$ computable by 1HL) form a dense subset of $C^0(K, \mathbb{R}^p)$. This can be proved as follows.

Working one coordinate at a time, one may assume that $p = 1$. For any $\sigma : \mathbb{R} \to \mathbb{R}$, universal or not, a σ-*ridge* function $f : \mathbb{R}^m \to \mathbb{R}$ is one of the form $f(u) = \sigma(Bu + b)$, where $B \in R^{1\times m}$ and $b \in \mathbb{R}$. A finite sum $f(u) = \sum_{i=1}^{r} \sigma_i(B_i u + b_i)$ of functions of this type is a "multiridge" function. Such multiridge functions (not the standard terminology) are used in statistics and pattern classification, in particular when applying projection pursuit techniques, which incrementally build f by choosing the directions B_i, $i = 1,\ldots,r$ in a systematic way so as to minimize an approximation or classification error criterion; see for instance [15]. Note that $\mathcal{F}_{\sigma,m}$ is precisely the set of those multiridge functions for which there are scalars c_i such that $\sigma_i = c_i\sigma$ for all i.

To reduce to the single-input case treated in the previous section, it is enough to show that the set of all multiridge functions (with regular enough σ_i's) is dense, as one can then approximate each ridge term $\sigma_i(B_i u + b_i)$ by simply approximating each scalar function $\sigma_i(x)$ separately and then substituting $x = B_i u + b_i$. Now observe that, by the Weierstrass Theorem, polynomials are dense, so it suffices to show that each monomial in m variables can be written as sum of ridge polynomials. By induction on the number of variables, it is enough to see this for each monomial in *two* variables. Write such a monomial in the following form, with $0 < r < d$: $u^{d-r}v^r$. The claim is now that this can be written as: $\sum_{i=0}^{d} c_i(a_i u + v)^d$ for any fixed choice of distinct nonzero $a_0,\ldots,a_d$ (for instance, $a_0 = 1, a_1 = 2,\ldots,a_d = d+1$) and suitable c_i's. Dividing by u^d and letting $z := v/u$ it is equivalent to solve $\sum_{i=0}^{d} c_i(a_i + z)^d = z^r$ for the c_i's. As the polynomials $(a_i + z)^d$ are linearly independent —computing derivatives of order $0,\ldots,d$ at $z = 0$ results in a Vandermonde matrix,— they must span the set of all polynomials of degree $\leq d$, and in particular z^r is in the span, as desired.

Instead of polynomials, one can also base the reduction proof on Fourier expansions. For this, it is enough to see that trigonometric polynomials satisfy the conditions of the Stone-Weierstrass Theorem. Other proofs use various algorithms for reconstruction from projections, such as Radon transforms (see [11]). In conclusion, universality guarantees that functions of the type (5) are dense in C^0. One can also establish density results for L^q spaces, $q < \infty$, on compact sets, simply using for those spaces the density of continuous functions. Approximation results on noncompact spaces, in L^q but for finite measures, are also possible but considerably harder; see [19].

It is false, on the other hand, that one can do uniform approximation of more or less arbitrary discontinuous functions, that is, approximation in L^∞, even on compact sets. In input dimension 1, the approximated function must be regulated (see above), which is not too strong an assumption. But in higher input dimensions, there are strong constraints. For instance, the characteristic function of a unit square in the plane cannot be approximated by 1HL nets to within small error. This leads to very interesting questions for control problems, where it is often the case that the only solution to a problem is one that involves such an approximation. Later the paper deals with such issues.

4.4   Number of Units

It is interesting to ask how many units are needed in order to solve a given classification or interpolation task. This can be formalized as follows. Let $\mathcal{U}$ be a set, to be called the *input set*, and let $\mathcal{Y}$ be another set, the *output set*. In the results below, for 1HL nets, $\mathcal{U} = \mathbb{R}^m$, but the definitions are more general. To measure discrepancy in outputs, it may be assumed that $\mathcal{Y}$ is a metric space. For simplicity, assume from now on that $\mathcal{Y} = \mathbb{R}$, or $\mathcal{Y}$ is the subset $\{0,1\}$, if binary data is of interest. A *labeled sample* is a finite set $S = \{(u_1, y_1), \ldots, (u_s, y_s)\}$, where $u_1, \ldots, u_s \in \mathcal{U}$ and $y_1, \ldots, y_s \in \mathcal{Y}$. (The y_i's are the "labels;" they are *binary* if $y_i \in \{0,1\}$.) It is assumed that the sample is consistent, that is, $u_i = u_j$ implies $y_i = y_j$. A *classifier* is a function $F : \mathcal{U} \to \mathcal{Y}$. The *error* of F on the labeled sample S is defined as

$$E(F, S) := \sum_{i=1}^{s} \left(F(u_i) - y_i \right)^2 .$$

A set $\mathcal{F}$ of classifiers will be called an *architecture*. Typically, and below, $\mathcal{F} = \{F_{\vec{w}} : \mathcal{U} \to \mathcal{Y}, \vec{w} \in R^r\}$ is a set of functions parameterized by $\vec{w} \in R^r$, where $r = r(\mathcal{F})$. An example is that of nets with $m=1$ inputs, $p=1$ outputs, and n hidden units, that is, $\mathcal{F}^1_{n,\sigma,1}$; here the parameter set has dimension $r = 3n + 1$.

The sample S is *loadable into* $\mathcal{F}$ iff $\inf_{f \in \mathcal{F}} E(F, S) = 0$. Note that for a binary sample S and a binary classifier F, $E(F, S)$ just counts the number of missclassifications, so in the binary case loadability corresponds to being able to obtain exactly the values y_i by suitable choice of $f \in \mathcal{F}$. For continuous-valued y_i's, loadability means that values arbitrarily close to the desired y_i's can be obtained.

One may define the *capacity* $c(\mathcal{F})$ of $\mathcal{F}$ via the requirement that:

$$c(\mathcal{F}) \geq \kappa \text{ iff every } S \text{ of cardinality } \kappa \text{ is loadable.}$$

(Other natural definitions of capacity measures are possible, in particular the VC dimension mentioned below.) That is, $c(\mathcal{F}) = \infty$ means that all finite S are loadable, and $c(\mathcal{F}) = \kappa < \infty$ means that each S of cardinality $\leq \kappa$ is loadable but some S of cardinality $\kappa + 1$ is not.

Various relations between capacity and number of neurons are known for nets with one hidden layer and Heaviside or sigmoidal activations. It is an easy exercise to show that the results are independent of the input dimension m, for any fixed activation type σ and fixed number of units n. (Sketch of proof: the case $m = 1$ is included in the general case, by simply taking collinear points u_i. Conversely, given any set of points u_i in $\mathbb{R}^m$, there is always some vector v whose inner products with the distinct u_i's are all different, and this reduces everything to the one dimensional case.) In the case $m = 1$, parameter counts are interesting, so that case will be considered next. Observe that, for 1HL nets with one input, and n hidden units (and $p=1$), there are $3n + 1$ parameters (appearing nonlinearly) —or $3n + 2$ if allowing direct connections, that is, when there is an extra linear term $c_{n+1}u$ in (5)— though for $\mathcal{H}$, effectively only $2n + 1$ matter. (In the case of the standard sigmoid, a Jacobian computation shows that these parameters are independent.) For classification purposes, it is routine to consider just the sign of the output of a neural net, and to classify an input according to this sign. Thus one introduces the class $\mathcal{H}(\mathcal{F}^1_{n,\sigma,1})$ consisting of all $\{0,1\}$-valued functions of the form $\mathcal{H}(f(u))$ with $f \in \mathcal{F}^1_{n,\sigma,1}$.

Of interest are scaling properties as $n \rightarrow \infty$. Let $\mathrm{CLSF}(\sigma) := \underline{\lim}_{n \to \infty} c(\mathcal{F})/r(\mathcal{F})$ for $\mathcal{F} = \mathcal{H}(\mathcal{F}^1_{n,\sigma,1})$ and $\mathrm{INTP}(\sigma) := \underline{\lim}_{n \to \infty} c(\mathcal{F})/r(\mathcal{F})$ for $\mathcal{F} = \mathcal{F}^1_{n,\sigma,1}$. Define similarly $\mathrm{CLSF}^d, \mathrm{INTP}^d$ when using 1HL nets with direct connections.

Consider the property (*): $\exists c$ s.t. σ is differentiable at c and $\sigma'(c) \neq 0$. Then, for classification, the following results are given in [34]:

$$\boxed{\mathrm{CLSF}(\mathcal{H}) = 1/3, \ \mathrm{CLSF}^d(\mathcal{H}) = 2/3, \ \mathrm{CLSF}(\sigma) \geq 2/3}$$

assuming that σ is sigmoidal and (*) holds. The last bound is best possible, in the sense that for the piecewise linear π one has $\mathrm{CLSF}(\pi) = 2/3$ while it is the case that $\mathrm{CLSF}(\sigma) = \infty$ for some "nice" (even, real-analytic) sigmoidal functions σ satisfying (*). Regarding continuous-valued interpolation, these are the results:

$$\boxed{\mathrm{INTP}(\mathcal{H}) = 1/3, \ \mathrm{INTP}^d(\mathcal{H}) = 1/3, \ \mathrm{INTP}(\sigma) \geq 2/3}$$

assuming in the last case that (*) holds and σ is a continuous sigmoidal. Again here, $\mathrm{INTP}(\pi) = 2/3$, and one can also show that $2/3 \leq \mathrm{INTP}(\sigma_s) \leq 1$ for the standard sigmoid (the proof of the upper bound in this latter case involves some algebraic geometry; the value may be infinite for more general sigmoids; see also [24]). Furthermore, the inequality $\mathrm{INTP}(\sigma) \geq 1/3$ holds for any universal nonlinearity. Obtaining the precise value for σ_s is a very interesting open problem.

Note that the above discussion focused on general bounds on what can be achieved. In practice, however, one is given a particular labeled sample and the problem is to decide whether this sample can be loaded into the desired architecture. For differentiable activations, numerical techniques are used to estimate the answer. But one may ask about the abstract (in the sense of computer science) *computational complexity* of the loading problem: do there exist weights that satisfy the desired objective? For fixed input dimension and Heaviside activations, this becomes essentially a linear programming problem, but even for such activations, the problem is NP-hard when scaling with respect to the number of input or output dimensions; see [9] and [21] for much on this issue.

4.5   Learnability and VC Dimension

One of the main current approaches to defining and understanding the question of learning, which after all underlies much of the reason for the use of neural nets in control, is based on the *probably approximately correct* ("PAC") model proposed in computational learning theory by Valiant in the early 1980s. Very closely related ideas appeared in statistics even earlier, in the work of Vapnik and Chervonenkis —see the excellent book [38]— and the interactions between statistics and computer science are the subject of much current research. The next few paragraphs introduce the basic ideas, using terminology from learning theory; for more details see for instance the textbook [3].

In the PAC paradigm, a "learner" has access to data given by a labeled sample $S = (u_1, y_1), \ldots, (u_s, y_s)$. The inputs u_i are being generated at random, independently and identically distributed according to some probability measure P. There is some fixed but unknown function f so that, for each i, $y_i = f(u_i)$, and f belongs to some known class of functions $\mathcal{F}$. This class of functions is used to characterize the assumptions ("bias") being made about what is common among the observed input/output pairs. The learner knows the class $\mathcal{F}$ but not the particular f being used. For instance, in systems identification, the u_i's might correspond to inputs applied to a plant and the y_i's would be the corresponding outputs, while $\mathcal{F}$ might consist of all stable SISO systems of a certain degree. The learner's objective is to use the information gathered from the observed labeled sample in order to guess the correct f in $\mathcal{F}$. In a control environment, a learning controller might be trained in this manner to recognize certain features of the state space which are associated to a particular control action.

For simplicity, because the theory is far simpler in that case, and because of the application to pattern classification, only binary-valued functions f (and hence binary samples $y_i \in \{0, 1\}$) will be considered here. Thus instead of a class of functions $\mathcal{F}$ one could work equivalently with a class of *concepts*, that is, those subsets of the input space which are of the form $\{u \mid f(u) = 1\}$ for the various $f \in \mathcal{F}$. There are many generalizations of this basic setup, including dealing with noisy data, continuous-valued outputs, or allowing the learner to guess a function not in the original class $\mathcal{F}$, but the basic ideas are best illustrated in this simplest case.

One defines the *error* of a hypothesis $\hat{f}$ made by the learner as the probability that it will incorrectly classify a new randomly chosen example (u, y), that is, the probability that $\hat{f}(u) \neq y$ (the prediction error). It is assumed that u is picked with the same probability distribution P that was used to generate the training inputs u_i. Since only a limited number of samples are presented during training, they will in general not be sufficient to distinguish between all possible concepts (possible functions f), and this is a source of error. Another possible source of error arises from the fact that the inputs u, having been randomly chosen, might not be representative enough of future inputs. Neither of these errors need be as serious as it may seem at first sight, however. First of all, if the sample is large enough, and $\hat{f}(u_i) = y_i$ for all i, it is quite unlikely that $f \neq \hat{f}$ unless the class is too rich. It may happen, but with small probability. For the second type of error, if the probability distribution P being used to generate the inputs u_i is concentrated in a part of the input space where f and $\hat{f}$ coincide, then the new testing input u will likely come from this part of the space as well, and hence the prediction error will again be small. The term PAC learning refers to the fact that (very) "probably" the estimate will be "approximately correct."

Before giving precise definitions of learnability in the sense just discussed, it is instructive to consider very informally a case which does *not* lead to learnability and one that does. Assume that the inputs u are real numbers uniformly distributed in the interval $[0, 1]$ and that the set of functions $\mathcal{F}$ consists of the functions $f_k(u) := \mathcal{H}(\sin(ku))$, over all positive integers k. That is, $f_k(u)$ is 1 if $\sin(ku) > 0$ and zero otherwise. Now, given a random sequence $(u_1, f(u_1)), \ldots, (u_s, f(u_s))$, where $f = f_k$, there is in general no possible good prediction of $f(u)$ for a new u. Indeed, with probability one, the complete set $u_1, \ldots, u_s, u$ will consist of rationally independent real numbers, and therefore there exists some $j \neq k$ so that $f_j(u_i) = f(u_i)$ for $i = 1, \ldots, s$ but $f_j(u) \neq f(u)$. (Recall that the set of values of $(\sin(lu_1), \ldots, \sin(lu_s), \sin(lu))$, as l ranges over the positive integers, is dense in $[-1, 1]^{s+1}$.) Since the learner is not able to decide, on the basis of the observed data, if $f = f_k$ or $f = f_j$, the prediction $\hat{f}(u)$ cannot be made with any degree of reliability. Contrast this example with the following one, at the other extreme: the functions $\mathcal{F}$ are now of the type $f_a(x) := \mathcal{H}(x - a)$, with $a \in [0, 1]$. The concepts to be identified are the sets of the form $\{x > a\}$. Identifying the concept means identifying the cut point a. Given a large enough ($s \gg 1$) sample, there will be enough pairs (u_i, y_i) with u_i near a so that a good estimate of a can be obtained, for instance by estimating a as the midpoint of the interval $[a_1, a_2]$, where a_1 is the largest observed u_i so that $f(u_i) = 0$ and a_2 is the smallest one so that $f(u_i) = 1$. The only errors would be due to a bad sample (it so happened that all the u_i's were far from a, so the interval is very large) or to the test input u falling in the interval $[a_1, a_2]$, where no information about a is available. But these errors occur with low probability.

The definitions are as follows (they are standard, but the terminology is changed a bit in order to make it closer to systems and control usage). An input space $\mathcal{U}$ as well as a collection $\mathcal{F}$ of maps $\mathcal{U} \to \{0, 1\}$ are given. The set $\mathcal{U}$ is assumed to be countable, or an Euclidean space, and the maps in $\mathcal{F}$ are assumed to be Borel measurable. In addition, mild regularity assumptions are made which insure that all sets appearing below are measurable (details are omitted; see the references). Let W be the set of all sequences

$$w = (u_1, f(u_1)), \ldots, (u_s, f(u_s)) \tag{6}$$

over all $s \geq 1$, $(u_1, \ldots, u_s) \in \mathcal{U}^s$, and $f \in \mathcal{F}$. An *identifier* is a map $\varphi : W \to \mathcal{F}$. The value of φ on the sequence appearing in (6) will be denoted as φ_w. The *error* of φ with respect to a probability measure P on $\mathcal{U}$, an $f \in \mathcal{F}$, and a sequence $(u_1, \ldots, u_s) \in \mathcal{U}^s$, is $\mathrm{Err}\,(P, f, u_1, \ldots, u_s) := \mathrm{Prob}\,[\varphi_w(u) \neq f(u)]$ (where the probability is being understood with respect to P).

The class $\mathcal{F}$ is (uniformly) *learnable* if there is some identifier φ with the following property: For each $\varepsilon, \delta > 0$ there is some s so that, for every probability P and every $f \in \mathcal{F}$, $\mathrm{Prob}\,[\mathrm{Err}\,(P, f, u_1, \ldots, u_s) > \varepsilon] < \delta$ (where the probability is being understood with respect to P^s on $\mathcal{U}^s$).

In the learnable case, the function $s(\varepsilon, \delta)$ which provides, for any given ε and δ, the smallest possible s as above, is called the *sample complexity* of the class $\mathcal{F}$. It can be proved that $s(\varepsilon, \delta)$ is automatically bounded by a polynomial in $1/\varepsilon$ and $1/\delta$. In fact, it can be bounded by $-(c/\varepsilon)\log(\delta\varepsilon)$, where c is a constant that depends only on the class $\mathcal{F}$. It can also be proved that, if there is any identifier at all in the above sense, then one can always use the following naive identification procedure: pick any element f which is consistent with the observed data. This leads computationally to the loading question discussed earlier. In the statistics literature —see [38]— this "naive technique" is a particular case of what is called *empirical risk minimization*.

The above discussion corresponds to being able to learn uniformly with respect to unknown input distributions. Much research is currently taking place on the question of learning with respect to particular subfamilies of distributions on $\mathcal{U}$. It is perfectly possible for an $\mathcal{F}$ not to be learnable in the above sense but to be learnable if the inputs are assumed to be Gaussian, for instance. Notions of Kolmogorov metric entropy are used to characterize such learning. (In the case of learning with respect to particular data distributions, the naive identification procedure is no longer guaranteed to work, and more sophisticated methods must be used.) Another variation consists of studying the computational effort required in order to use the identifier φ; the definition given above is purely information-theoretic, but the original work of Valiant emphasized that aspect, essentially the loading problem, as well.

Checking learnability is in principle a difficult issue, and the introduction of the following combinatorial concept is extremely useful in that regard. A subset $S = \{u_1, \ldots, u_s\}$ of $\mathcal{U}$ is said to be *shattered* by the class of binary functions $\mathcal{F}$ if all possible binary labeled samples $\{(u_1, y_1), \ldots, (u_s, y_s)\}$ (all $y_i \in \{0, 1\}$) are loadable into $\mathcal{F}$. The *Vapnik-Chervonenkis dimension* $\text{VC}(\mathcal{F})$ is the supremum (possibly infinite) of the set of integers κ for which there is *some* set S of cardinality κ that can be shattered by $\mathcal{F}$. (Thus, $\text{VC}(\mathcal{F})$ is at least as large as the capacity $c(\mathcal{F})$ defined earlier.)

The main result, due to [10], but closely related to previous results in statistics (see [38]) is: *The class $\mathcal{F}$ is learnable if and only if* $\text{VC}(\mathcal{F}) < \infty$. So the VC dimension completely characterizes learnability with respect to unknown distributions; in fact, the constant "c" in the sample complexity bound given earlier is a simple function of the number $\text{VC}(\mathcal{F})$, independent of $\mathcal{F}$ itself.

It follows from the results in [8] that, for the class $\mathcal{H}(\mathcal{F}_{n,\sigma,m})$ of classifiers implementable by 1HL nets with with n hidden units and m inputs (n and m fixed) and activation $\sigma = \mathcal{H}$, $\text{VC}(\mathcal{F}) < dmn(1 + \log(n))$, for a small constant d. Thus 1HL nets with threshold activations are learnable in the above sense. Note that finiteness is trivial to verify in one very special case, namely $m = 1$: in this case, a function f computable by a 1HL $\mathcal{H}$-net with n hidden units is piecewise constant with at most n discontinuities. This means that, given any set with $n + 2$ elements or more, the alternate labeling $0, 1, 0, 1, \ldots$ cannot be implemented; in other words, $\text{VC}(\mathcal{F})$ is at most $n + 1$.

It is easy to construct examples of sigmoids, even extremely well-behaved ones (analytic and squashing, for instance) for which the VC dimension of the class $\mathcal{H}(\mathcal{F}_{n,\sigma,m})$ is infinite; see [34]. However, for the standard sigmoid, a recent result in [24] proves that $\text{VC}(\mathcal{F}) < \infty$, so neural nets with activation σ_s are also learnable. Thus sigmoidal nets appear to have some special properties, vis a vis other possible more general parametric classes of functions, at least from a learnability viewpoint. Other results on finiteness of learning, but from a more statistical viewpoint (nonlinear regression, estimation of joint densities), are given in [17], where metric entropy estimates are obtained for networks with bounded weights.

The results on learnability just explained extend to other classes of feedforward nets, including 2HL nets (defined below) and nets involving products of inputs ("high order nets"), but such more general results will not be reviewed here. Note also that the general question of guessing values of a function at unseen inputs is essentially also that of extrapolation, and it is classical in numerical analysis. In that context, the "prior" class of functions $\mathcal{F}$ often reflects a smoothness constraint.

4.6 Quality of Approximation

Certain recent results due to Andrew Barron and Lee Jones have been used to support the claim that 1HL neural network approximations may require less parameters than conventional techniques. What is meant by this is that, given a function f to be approximated (for instance, a pattern classifier or a controller), approximations of f to within a desired error tolerance ε can be obtained using "small" networks, while using, for instance, orthogonal polynomials, splines, or Fourier series, would require an astronomical number of terms, especially for multivariate inputs.

At least as the results currently stand, this claim represents a misunderstanding of the very nice contributions of Barron and Jones. First of all, their results would, in general terms, apply equally well to show that one can obtain efficient approximations with various types of classical basis functions, *as long as the basis elements can be chosen in a nonlinear fashion*, just as with neural networks. For instance, splines with varying (rather than fixed) nodes, or trigonometric series with adaptively selected frequencies, will have the same properties. What is important is the possibility of *selecting* terms adaptively, in contrast to the use of a large basis containing many terms and fitting these through the use of least squares. Thus, the important fact about the recent results is that they emphasize that *non*linear parameterizations may require less parameters than linear ones to achieve a guaranteed degree of approximation. (Abstractly, this is not so surprising: for an analogy, consider the fact that a *one*-parameter analytic curve, based on ergodic motions, can be used to approximate arbitrarily well every element in an Euclidean space $\mathbb{R}^d$, but no $(d-1)$-dimensional subspace can do so.) The most interesting —and as yet unresolved— issues have to do with the tradeoffs between rate of approximation and number of parameters, and the related balance between generalization capabilities and computational complexity.

The rest of this section reviews some of the basic results in question. The main tool is the following Lemma, which is often attributed to Maurey (see [29]): *Let G be a bounded subset of a Hilbert space H, with $\|g\| \leq \gamma$ for all $g \in G$, and let G_n be the subset of H consisting of all convex combinations of at most n elements of G. Then, for each f in the closed convex hull of G, and for each $n \geq 1$,*

$$\inf_{g \in G_n} \|f - g\| \leq \frac{\gamma}{\sqrt{n}}.$$

There are several different proofs, probabilistic and geometric, of this result and related ones in which the constant in the bound may be allowed to be slightly larger. Jones pointed out that one may use an incremental approximation, monotonically decreasing the approximation error while recursively adding one element of G at a time.

In applications, the set H is a space of square integrable functions and, for some fixed $\Gamma > 0$, $G = G(\Gamma)$ is the set of ridge functions $c\sigma(Bu + b)$, with $|c| \leq \Gamma$, $B \in \mathbb{R}^{1 \times m}$, and $b \in \mathbb{R}$. In that case, $G(\Gamma)_n$ is the set of 1HL σ-nets with n units and weight sums $\sum |c_i| \leq \Gamma$. The approximation result says that those functions f which can be approximated arbitrarily close by elements of the $G(\Gamma)_n$'s (fixed Γ), can in fact be also approximated with a mean square error $O(1/n)$ using 1HL nets with no more than n units. To apply all of this, one must understand what functions f are of this form. One such example was discussed earlier, namely the case of scalar functions of bounded variation on an interval $[\alpha, \beta]$. In that case, it was remarked that $f\text{-}f(\alpha)$ is in the closed convex hull of $\{\pm V\mathcal{H}(\pm u - b)\}$ (in the uniform norm, and hence also in L^2 for any finite measure on $[\alpha, \beta]$). Generalizing, Barron in [6] suggested *defining* a function $f : Q \to \mathbb{R}$, where Q is a bounded subset of $\mathbb{R}^m$, to have "bounded variation with respect to halfspaces" if this property holds: there exists a real number V so that, for some $q \in Q$, $f\text{-}f(q)$ is in the closed convex hull, in the uniform norm, of the functions $\pm V\mathcal{H}(Bu + b)$. The smallest such V he called the variation $V_{f,Q}$ of the function f on the set Q. For functions with $V_{f,Q} < \infty$, there is then a rate of approximation theorem using 1HL $\mathcal{H}$-nets, and from there also one for sigmoidal nets (just approximate each Heaviside function by a sigmoid). The conclusion is that one can approximate by 1HL nets with n units with an error as small as $V_{f,Q}/\sqrt{n}$ in $L^2(Q, \mu)$, for any probability measure μ on Q.

One source of examples of functions of bounded variation in this generalized sense is as follows. Let Q for simplicity be taken to be the unit ball. Assume that f admits a Fourier representation $f(u) = \int_{\mathbb{R}^m} e^{i\omega \cdot u} \tilde{f}(\omega) d\omega$ for each $u \in Q$, and that $C_f = \int_{\mathbb{R}^m} \|\omega\| |\tilde{f}(\omega)| d\omega < \infty$. Then, $V_{f,Q} \leq 2C_f$. (Norm of ω is standard Euclidean norm.) For the spaces $\{f \mid C_f \leq k\}$, k a fixed constant, Barron also proved (see references in [6]) that approximations by linear subspaces of dimension n would result in a worst-case error of at least $O(n^{-1/d})$, which is asymptotically worse than $O(1/\sqrt{n})$ when $d > 2$.

Note that the approximations, as they use Maurey-type arguments, hold a priori only in L^2 (or other Hilbert spaces). Indeed, these arguments are false for approximation in L^∞, if the set G is arbitrary (see [14] for this and related remarks). However, Barron in the above reference was also able to prove a similar result for the particular case in which G corresponds to characteristic functions of half-spaces, using a deeper result due to Dudley. This beautiful recent contribution is closer to being a true "neural networks" theorem, as it uses essentially a property involving VC dimension which is not true in general.

4.7   Uniqueness

As discussed, in most applications dealing with learning and pattern recognition, neural nets are employed as models whose parameters must be fit to training data characterized by a labeled sample. Gradient descent and other algorithms are used in order to minimize $E(F, S)$ over all F corresponding to a fixed network architecture. Among the many numerical complications that arise when following such a procedure are the possibilities of (1) non-global local minima, and (2) multiple global minimizers. The first issue was dealt with by many different authors —see for instance [36] and the references there— and will not be reviewed here. Regarding the second point, observe that there are obvious transformations that leave the behavior of a network invariant, such as interchanges of all incoming and outgoing weights between two neurons (mathematically, the relabeling of neurons) and, for odd activation functions, flipping the signs of all incoming and outgoing weights at any given node. Two networks differing in such a manner give the same error on the training data. When there is a net that fits perfectly the data, all nets differing from it by one of the above transformations also attain the global minimum (zero) of the error functional.

It is then natural to ask if neuron exchanges and sign flips are the *only* function-preserving transformations that can generically occur. If these are the only possible ones, then essentially all the internal structure is uniquely determined by the external behavior of the network. Moreover, the set of invariant transformations is then finite, and there is no possible dimensionality reduction in the parameter space. Such a situation is in sharp contrast to classical linear systems, where canonical forms have to be introduced in order to achieve parameter identifiability. (Seen more positively, the parameterizations provided by neural networks are then irredundant.)

For simplicity, assume from now on that $p = 1$; generalizations to the multiple-output case are not hard but they complicate notations. Also, assume that σ is an odd function. Thus, from now on, C in the definition of 1HL net is a row n-vector and c_0 is a constant. Two networks Σ and $\hat{\Sigma}$ are (input/output) *equivalent*, denoted $\Sigma \sim \hat{\Sigma}$, if $\text{beh}_\Sigma = \text{beh}_{\hat{\Sigma}}$ (equality of functions). The question to be studied, then, is: to what extent does $\Sigma \sim \hat{\Sigma}$ imply $\Sigma = \hat{\Sigma}$?

The function σ satisfies the *independence property* ("IP" from now on) if, for every positive integer l, nonzero real numbers $b_1, \ldots, b_l$, and real numbers $b_{01}, \ldots, b_{0l}$ for which the pairs (b_i, b_{0i}), $i = 1, \ldots, l$ satisfy $(b_i, b_{0i}) \neq \pm(b_j, b_{0j})\ \forall i \neq j$, it must hold that the functions $1, \sigma(b_1 u +$

$b_{01}), \ldots, \sigma(b_l u + b_{0l})$ are linearly independent. The function σ satisfies the *weak* independence property ("WIP") if the above linear independence property is only required to hold for all pairs with $b_{0i} = 0$, $i = 1, \ldots, l$.

Let $\Sigma(B, C, b_0, c_0, \sigma)$ be given, and denote by B_i the transpose of the ith row of the matrix B and by c_i and b_{0i} the ith entries of C and b_0 respectively. With these notations, $\mathrm{beh}_\Sigma(u) = c_0 + \sum_{i=1}^n c_i \sigma(B_i u + b_{0i})$. As in [37], Σ is called *irreducible* if the following properties hold: $c_i \neq 0$ for each $i = 1, \ldots, n$; $B_i \neq 0$ for each $i = 1, \ldots, n$; and $(B_i, b_{0i}) \neq \pm(B_j, b_{0j})$ for all $i \neq j$. Given $\Sigma(B, C, b_0, c_0, \sigma)$, a *sign-flip* operation consists of simultaneously reversing the signs of c_i, B_i, and b_{0i}, for some i. A *node-permutation* consists of interchanging (c_i, B_i, b_{0i}) with (c_j, B_j, b_{0j}), for some i, j. Two nets Σ and $\hat{\Sigma}$ are *sign-permutation (sp) equivalent* if $n = \hat{n}$ and (B, C, b_0, c_0) can be transformed into $(\hat{B}, \hat{C}, \hat{b}_0, \hat{c}_0)$ by means of a finite number of sign-flips and node-permutations. Of course, sp-equivalent nets have the same behavior (since σ has been assumed to be odd). With this terminology, the following holds: *Let σ be odd and satisfy property* IP. *Assume that Σ and $\hat{\Sigma}$ are both irreducible, and $\Sigma \sim \hat{\Sigma}$. Then, Σ and $\hat{\Sigma}$ are sp-equivalent.* A net Σ has *no offsets* if $b_0 = c_0 = 0$ (the terminology "biases" or "thresholds" is sometimes used instead of offsets, but these terms are used for other very different purposes as well). Then, also: *If σ is odd and satisfies* WIP, *the same statement is true for nets with no offsets.*

Characterizing WIP is especially easy, and very classical: *If σ is a polynomial,* WIP *does not hold. Conversely, if σ is odd and infinitely differentiable, and if there are an infinite number of nonzero derivatives $\sigma^{(k)}(0)$, then σ satisfies property* IP. Nets with no offsets appear naturally in signal processing and control applications, as there it is often the case that one requires that $\mathrm{beh}(0) = 0$, that is, the zero input signal causes no effect, corresponding to equilibrium initial states for a controller or filter. Results in this case are closely related to work in the 1970s by Rugh and coworkers and by Boyd and Chua in the early 1980s.

It appears to be harder to obtain elegant characterizations of the stronger property IP. For obvious examples of functions not satisfying IP, take $\sigma(x) = e^x$, any periodic function, or any polynomial. However, the most interesting activation for neural network applications is $\sigma(x) = \tanh(x)$, or equivalently after a linear transformation, the standard sigmoid $\frac{1}{1+e^{-x}}$. In this case, Sussmann showed in [37] that the IP property, and hence the desired uniqueness statement, hold. His proof was based on explicit computations for the particular function $\tanh(x)$. An alternative proof is possible, using analytic continuations and residues, and allows a more general result to be established (see [2] for details): *Assume that σ is a real-analytic function, and it extends to an analytic function $\sigma : \mathbb{C} \to \mathbb{C}$*

defined on a subset $D \subseteq \mathbb{C}$ of the form:

$$D = \{z \,|\, |\mathrm{Im}\, z| \le \lambda\} \setminus \{z_0, \bar{z}_0\}$$

for some $\lambda > 0$, where $\mathrm{Im}\, z_0 = \lambda$ and z_0 and $\bar{z}_0$ are singularities, that is, there is a sequence $z_n \to z_0$ so that $|\sigma(z_n)| \to \infty$, and similarly for $\bar{z}_0$. Then, σ satisfies property IP.

4.8   Two-Hidden Layers and Nonlinear Feedback

The previous sections dealt with 1HL nets. Next, the case of two hidden layers is treated. For the topics treated here, there is no need to define nets themselves, but just their behavior. If $\sigma : \mathbb{R} \to \mathbb{R}$ is any function, and m, p are positive integers, a function computed by a two-hidden layer ("2HL") net with m inputs and p outputs and activation function σ is by definition one of the type $f \circ \vec{\sigma}_n \circ g \circ \vec{\sigma}_l \circ h$, where f, g, and h are affine maps ($n + l$ is then called the number of "hidden units"). As earlier, a function computable by a 2HL net with direct input/output connections is one of the form $Fu + f(u)$, where F is linear and f is computable by a 2HL net.

One hidden layer networks have universal approximation properties, and rates of convergence can be estimated. However, these rates may not be as good as those achievable with 2HL nets. For instance, functions that are piecewise constant on squares approximate certain classes of functions in $\mathbb{R}^2$ very efficiently, and while it is not difficult to express them using 2HL nets, often no such 1HL net expression is possible.

Another disadvantage of 1HL vis a vis 2HL nets arises from the topologies in which the 1HL approximation theorems hold. This has serious implications in control applications, and can be illustrated with the following idea. Suppose that there is some *discontinuous* feedback law $u = k(x)$ which globally asymptotically stabilizes the planar system $\dot{x} = f(x, u)$ with respect to the origin. Assume that k is perfectly known but one wishes to restrict possible controllers to those computable by 1HL nets. It would appear that this would be easy to achieve, as one may merely approximate the given k by a 1HL function $\hat{k}$ and then use this $\hat{k}$ as the controller. The problem is that, for general discontinuous functions, the results only insure approximation in L^p norm (p finite), but it is impossible in general to approximate k *uniformly*. (Uniform approximation of functions that are continuous, or that are of "bounded variation with respect to half spaces," is possible, however, so when there is a continuous feedback k that stabilizes, this type of obstruction dissappears.) A weak type of approximation may not be enough for control purposes. For instance, it may be the case that for each approximant $\hat{k}$ there is some simple closed curve Γ encircling the origin where the approximation is bad (a set of measure zero!) and that this causes the vector field $f(x, \hat{k}(x))$ to point outward everywhere on Γ; in that case the closed loop behavior cannot be globally asymptotically stable, as trajectories cannot cross Γ.

It is possible to construct examples of systems which are otherwise stabilizable but such that every possible feedback implementable by a 1HL net (with basically any type of activation, continuous or not) must give rise to a nontrivial periodic orbit. On the other hand, it can be shown that every system that is stabilizable, by whatever k, can also be stabilized using 2HL nets with discontinuous activations (under mild technical conditions, and using sampled control). See [35] for details.

To summarize, if stabilization requires discontinuities in feedback laws, it may be the case that no possible 1HL net stabilizes. Thus the issue of stabilization by nets is closely related to the standard problem of continuous and smooth stabilization of nonlinear systems, one that has attracted much research attention in recent years. Roughly, there is a hierarchy of state-feedback stabilization problems: those that admit continuous solutions, those that don't but can still be solved using 1HL nets with discontinuous activations, and more general ones (solvable with 2HL). It can be expected that an analogous situation will be true for other control problems. The reason that most neurocontrol papers have used 1HL nets is that they almost always dealt with feedback linearizable systems, which admit continuous stabilizers.

Some Details.

Before discussing stabilization, one can understand the necessity of 2HL nets by means of a more abstract type of question. Consider the following *inversion* problem: *Given a continuous function* $f : \mathbb{R}^p \to \mathbb{R}^m$, *a compact subset* $C \subseteq \mathbb{R}^m$ *included in the image of* f, *and an* $\varepsilon > 0$, *find a function* $\phi : \mathbb{R}^m \to \mathbb{R}^p$ *so that* $\|f(\phi(x)) - x\| < \varepsilon$ *for all* $x \in C$. One wants to find a ϕ which is computable by a net, as done in global solutions of inverse kinematics problems —in which case the function f is the direct kinematics. It is trivial to see that in general discontinuous functions ϕ are needed, so nets with continuous σ cannot be used. However, and this is the interesting part, [35] establishes that nets with just one hidden layer, *even if* discontinuous σ is allowed, are *not* enough to guarantee the solution of all such problems. On the other hand, it is shown there that nets with two hidden layers (using $\mathcal{H}$ as the activation type) are sufficient, for every possible f, C, and ε. The basic obstruction is due, in essence, to the impossibility of approximating by single-hidden-layer nets the characteristic function of any bounded polytope, while for some (non one-to-one) f the only possible one-sided inverses ϕ must be close to such a characteristic function.

Consider now state-feedback controllers. The objective, given a system

$\dot{x} = f(x, u)$ with $f(0, 0)=0$, is to find a stabilizer $u=k(x)$, $k(0)=0$, making $x=0$ a globally asymptotically stable state of the closed-loop system $\dot{x}=f(x, k(x))$. The first remark is that the existence of a smooth stabilizer k is essentially equivalent to the possibility of stabilizing using 1HL nets (with smooth σ). (Thus the simple classes of systems studied in many neurocontrol papers, which are typically feedback-linearizable and hence continuously stabilizable, can be controlled using such 1HL nets.)

More precisely, assume that f is twice continuously differentiable, that k is also in C^2, that the origin is an exponentially stable point for $\dot{x}=f(x, k(x))$, and that K is a compact subset of the domain of stability. Pick any σ which has the property that twice continuously differentiable functions can be approximated uniformly, together with their derivatives, using 1HL nets (most interesting twice-differentiable scalar nonlinearities will do; see [19]). Then, one can conclude that there is also a different k, this one a 1HL net with activation σ, for which exactly the same stabilization property holds. (Sketch of proof: one only needs to show that if $k_n \rightarrow k$ in $C^2(K)$, with all $k_n(0)=0$ —this last property can always be achieved by simply considering $k_n(x) - k_n(0)$ as an approximating sequence— then $\dot{x}=f(x, k_n(x))$ has the origin as an exponentially stable point and K is in the domain of attraction, for all large n. Now, the proof of Theorem 12 in [33] shows that there is a neighborhood V of zero, independent of n, where exponential stability will hold, for all n sufficiently large, because $f(x, k_n(x))=A_n x + g_n(x)$, with $A_n \rightarrow A$ and with $g_n(x)$ being $o(x)$ *uniformly* on x (this last part uses the fact that σ approximates in $C^2(K)$). Now continuity of solutions on the right-hand side gives the result globally on K.)

In general, smooth (or even continuous) stabilizers fail to exist, as discussed for instance in [33], Section 4.8 and references there. Thus 1HLN feedback laws, with continuous σ, do not provide a rich enough class of controllers. This motivates the search for discontinuous feedback. It is easy to provide examples where 1HL $\mathcal{H}$-nets will stabilize but no net with continuous activations (hence implementing a continuous feedback) will. More surprisingly, 1HLN feedback laws, even with $\mathcal{H}$ activations, are not in general enough —intuitively, one is again trying to solve inverse problems— but two hidden layer nets using $\mathcal{H}$ (and having direct i/o connections) are always sufficient. More precisely, [35] shows that the weakest possible type of open-loop asymptotic controllability is sufficient to imply the existence of (sampled) controllers built using such two-hidden layer nets, which stabilize on compact subsets of the state space. On the other hand, an example is given there of a system satisfying the asymptotic controllability condition but for which every possible 1HL stabilizer gives rise to a nontrivial periodic orbit.

5   Recurrent Nets

A *recurrent net (or σ-system) with m inputs, p outputs, dimension n, and activation function σ* is specified by a triple of matrices A, B, C where A, B, and C are respectively real matrices of sizes $n \times n$, $n \times m$ and $p \times n$. Use the notation $\Sigma = \Sigma(A, B, C, \sigma)$, omitting σ if obvious from the context. One interprets the above data (A, B, C) as defining a controlled and observed dynamical system evolving in $\mathbb{R}^n$ (in the standard sense of control theory; see e.g. [33]) by means of a differential equation $\dot{x} = \vec{\sigma}(Ax + Bu)$, $y = Cx$ in continuous-time (dot indicates time derivative), or a difference equation $x^+ = \vec{\sigma}(Ax + Bu)$, $y = Cx$ in discrete-time ("+" indicates a unit time shift). See the block diagram in the next Figure, where $\Delta = x^+$ or $= \dot{x}$ in discrete or continuous time respectively.

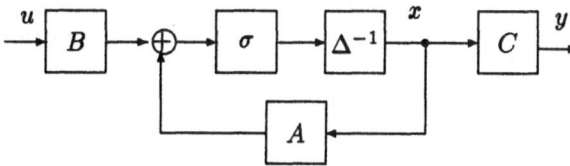

Other systems models are possible; for instance, "Hopfield nets" have dynamics of the form $\dot{x} = -Dx + \vec{\sigma}(Ax + Bu)$ (with D a diagonal matrix and often A symmetric); results analogous to those to be described can be obtained for these more general models as well.

Depending on the interpretation (discrete or continuous time), one defines an appropriate behavior beh$_\Sigma$, mapping suitable spaces of input functions into spaces of output functions, again in the standard sense of control theory, for any fixed initial state. For instance, in continuous time, one proceeds as follows: For any measurable essentially bounded $u(\cdot) : [0, T] \to \mathbb{R}^m$, denote by $\phi(t, \xi, u)$ the solution at time t with initial state $x(0) = \xi$; this is defined at least on a small enough interval $[0, \varepsilon)$, $\varepsilon > 0$. (The maps σ of interest in neural network theory are usually globally Lipschitz, in which case $\varepsilon = T$.) For each input, let beh$_\Sigma(u)$ be the output function corresponding to the initial state $x(0) = 0$, that is, beh$_\Sigma(u)(t) := C(\phi(t, 0, u))$, defined at least on some interval $[0, \varepsilon)$. Two recurrent nets Σ and $\hat{\Sigma}$ (necessarily with the same numbers of input and output channels, i.e. with $p = \hat{p}$ and $m = \hat{m}$) are *equivalent* (in discrete or continuous time, depending on the context) if it holds that beh$_\Sigma$ = beh$_{\hat{\Sigma}}$; as before, denote $\Sigma \sim \hat{\Sigma}$. (To be more precise, in continuous-time, one requires that for each u the domains of definitions of beh$_\Sigma(u)$ and beh$_{\hat{\Sigma}}(u)$ coincide, and their values be equal for all t in the common domain.)

Electrical circuit implementations of recurrent nets, employing resistive-ly connected networks of n identical nonlinear amplifiers, with the resistor characteristics used to reflect the desired weights, have been suggested as analog computers, in particular for solving constrained optimization prob-lems and for implementing content-addressable memories. In speech pro-cessing applications and language induction, as well as in signal processing ([25]) and control ([30]), recurrent nets are used as identification model-s or as prototype dynamic controllers (for partially observed systems or systems given in input/output form); they are often fit to experimental da-ta by means of the gradient-descent optimization (the so-called "dynamic backpropagation" procedure) of some cost criterion.

5.1   Approximation

Recurrent nets provide universal identification models, in a suitable sense. Consider a continuous- or discrete-time, time-invariant, control system Σ:

$$\dot{x}\,[\text{ or } x^+\,] \;=\; f(x,u) \qquad\qquad (7)$$
$$y \;=\; h(x)$$

under standard smoothness assumptions. (For instance, $x(t) \in \mathbb{R}^n$, $u(t) \in \mathbb{R}^m$, and $y(t) \in \mathbb{R}^p$ for all t, and f and h are continuously differentiable.) For any measurable essentially bounded control $u(\cdot) : [0,T] \to \mathbb{R}^m$, denote by $\phi(t,x_0,u)$ the solution at time t of (7) with initial state $x(0) = x_0$; this is defined at least on a small enough interval $[0,\varepsilon)$, $\varepsilon > 0$. For recurrent networks, when σ is bounded or globally Lipschitz with respect to x, it holds that $\varepsilon = T$; so assume here that the controls being considered are so that solutions exist globally, at least for initial states on some compact set of interest. It is not hard to prove that, on compacts and for finite time intervals, the behavior of Σ can be approximated by the behavior of a recurrent σ-net, if σ is universal.

Assume given two systems Σ and $\widetilde{\Sigma}$, as in (7), where tildes denote data associated to the second system, and with same number of inputs and outputs (but possibly $\widetilde{n} \neq n$). Suppose also given compact subsets $K_1 \subseteq \mathbb{R}^n$ and $K_2 \subseteq \mathbb{R}^m$, as well as an $\varepsilon > 0$ and a $T > 0$. Suppose further (this simplifies definitions, but can be relaxed) that for each initial state $x_0 \in K_1$ and each control $u(\cdot) : [0,T] \to K_2$ the solution $\phi(t,x_0,u)$ is defined for all $t \in [0,T]$. The system $\widetilde{\Sigma}$ *simulates* Σ *on the sets* K_1, K_2 *in time* T *and up to accuracy* ε if there exist two continuous mappings $\alpha : \mathbb{R}^{\widetilde{n}} \to \mathbb{R}^n$ and $\beta : \mathbb{R}^n \to \mathbb{R}^{\widetilde{n}}$ so that the following property holds: For each $x_0 \in K_1$ and each $u(\cdot) : [0,T] \to K_2$, denote $x(t) := \phi(t,x_0,u)$ and $\widetilde{x}(t) := \widetilde{\phi}(t,\beta(x_0),u)$; then this second function is defined for all $t \in [0,T]$, and

$$\|x(t) - \alpha(\widetilde{x}(t))\| < \varepsilon, \quad \|h(x(t)) - \widetilde{h}(\widetilde{x}(t))\| < \varepsilon$$

for all such t. One may ask for more regularity properties of the maps α and β as part of the definition.

Assume that σ is a universal activation, in the sense defined earlier. Then, *for each system Σ and for each K_1, K_2, ε, T as above, there is a σ-system $\widetilde{\Sigma}$ that simulates Σ on the sets K_1, K_2 in time T and up to accuracy ε.* The proof if not hard, and it involves first simply using universality in order to approximate the right-hand side of the original equation, and then introducing dynamics for the "hidden units" consistently with the equations. This second part requires a little care; for details, see [31].

Thus, recurrent nets approximate a wide class of nonlinear plants. Note, however, that approximations are only valid on compact subsets of the state space and for finite time, so that many interesting dynamical characteristics are not reflected. This is analogous to the role of bilinear systems, which had been proposed previously (work by Fliess and Sussmann in the mid-1970s) as universal models. As with bilinear systems, it is obvious that if one imposes extra stability assumptions ("fading memory" type) it will be possible to obtain global approximations, but this is probably not very useful, as stability is often a goal of control rather than an assumption.

5.2 Computation

The paper [31], and the references given there, dealt with computational capabilities of recurrent networks, seen from the point of view of classical formal language theory. This work studied discrete-time recurrent networks, focusing on the activation $\sigma = \pi$. (Though more general nonlinearities as well as continuous-time systems are of interest, note that using $\sigma = $ sign would give no more computational power than finite automata.) The main results —after precise definitions— are: (1) with rational matrices A, B, and C, recurrent networks are computationally equivalent, up to polynomial time, to Turing machines; (2) with real matrices, all possible binary functions, recursive or not, are "computable" (in exponential time), but when imposing polynomial-time constraints, an interesting class results. Computational universality, both in the rational and real cases, is due to the unbounded precision of state variables, in analogy to the potentially infinite tape of a Turing machine.

To state precisely the simulation results, consider then recurrent networks with $\sigma = \pi$, the piecewise-linear saturation, and having just one input and output channel ($m = p = 1$). A pair consisting of a recurrent network Σ and an initial state $\xi \in \mathbb{R}^n$ is *admissible* if the following property holds: Given any input of the special form $u(\cdot) = \alpha_1, \ldots, \alpha_k, 0, 0, \ldots,$

where each $\alpha_i = \pm 1$ and $1 \leq k < \infty$, the output that results with $x(0) = \xi$ is either $y \equiv 0$ or y is a sequence of the form

$$y(\cdot) \ = \ \underbrace{0, 0, \ldots, 0}_{s}, \beta_1, \ldots, \beta_l, 0, 0, \ldots , \qquad (8)$$

where each $\beta_i = \pm 1$ and $1 \leq l < \infty$. The pair (Σ, ξ) will be called *rational* if the matrices defining Σ as well as the initial ξ all have rational entries; in that case, for rational inputs all ensuing states and outputs remain rational.

Each admissible pair (Σ, ξ) defines a partial function

$$\phi : \{-1, 1\}^+ \rightarrow \{-1, 1\}^+ ,$$

where $\{-1, 1\}^+$ is the free semigroup in the two symbols ± 1, via the following interpretation: Given a sequence $w = \alpha_1, \ldots, \alpha_k$, consider an input as above, and the corresponding output, which is either identically zero or has the form in Equation (8). If $y \equiv 0$, then $\phi(w)$ is undefined; otherwise, if Equation (8) holds, then $\phi(w)$ is defined as the sequence $\beta_1, \ldots, \beta_l$, and one says that the response to the input sequence w was computed *in time* $s + l$. The (partial) function ϕ is *realized* by (Σ, ξ).

In order to be fully compatible with standard recursive function theory, the possibility is allowed that a decision is never made, corresponding to a partially defined behavior. On the other hand, if for each input sequence w there is a well-defined $\phi(w)$, and if there is a function on positive integers $T : \mathbb{N} \rightarrow \mathbb{N}$ so that the response to each sequence w is computed in time at most $T(|w|)$, where $|\alpha_1, \ldots, \alpha_k| = k$, then (Σ, ξ) *computes in time* T.

In the special case when ϕ is everywhere defined and $\phi : \{-1, 1\}^+ \rightarrow \{-1, 1\}$, that is, the length of the output is always one, one can think of ϕ as the characteristic function of a subset L of $\{-1, 1\}^+$, that is, a *language* over the alphabet $\{-1, 1\}$.

Disregarding computation time, some of the main results can be summarized as follows: *Let $\phi : \{-1, 1\}^+ \rightarrow \{-1, 1\}^+$ be any partial function. Then ϕ can be realized by some admissible pair. Furthermore, ϕ can be realized by some rational admissible pair if and only if ϕ is a partial recursive function.*

Given $T : \mathbb{N} \rightarrow \mathbb{N}$, the language L is *computed in time* T if the corresponding characteristic function is, for some admissible pair that computes in time T. It can be proved that languages recognizable in polynomial time by rational admissible pairs are exacly those in the class P of polynomial-time recursive languages. Using real weights, a new class, "analog P," arises. This class includes many languages not in P, but a theorem shows that NP is most likely not included in analog P.

For the rational case, one shows how to simulate an arbitrary Turing machine. In fact, the proof shows how to do so in linear time, and tracing the construction results in a simulation of a universal Turing machine by a recurrent network of dimension roughly 1000. The main idea of the proof in the real case relies in storing all information about ϕ in one weight, by a suitable encoding of an infinite binary tree. Then, π-operations are employed, simulating a chaotic mapping, to search this tree. In both the real and rational cases, the critical part of the construction is to be able to write everything up in terms of π, and the use of a Cantor set representation for storage of activation values. Cantor sets permit making binary decisions with finite precision, taking advantage of the fact that no values may appear in the "middle" range.

It is of course much more interesting to impose resource constraints, in particular in terms of computation time. Restrict to language recognition, for simplicity of exposition, but similar results can be given for computation of more general functions. The main result is that the class of languages recognized in polynomial time using recurrent nets with real weights, that is, "analog P," is exactly the same as a class also studied in computer science, namely the class of languages recognized in polynomial time by Turing machines which consult oracles, where the oracles are sparse sets. This gives a precise characterization of the power of recurrent nets in terms of a known complexity class. In summary, even though networks, as analog devices, can "compute" far more than digital computers, they still give rise to a rich theory of computation, in the same manner as the latter.

5.3   Identifiability

Finally, there are analogs for recurrent nets of the uniqueness questions discussed earlier. Assume from now on that σ is infinitely differentiable, and that it satisfies the following assumptions:

$$\sigma(0) = 0, \quad \sigma'(0) \neq 0, \quad \sigma''(0) = 0, \quad \sigma^{(q)}(0) \neq 0 \text{ for some } q > 2. \qquad (*)$$

Let $S(n, m, p)$ denote the set of all recurrent nets $\Sigma(A, B, C, \sigma)$ with fixed n, m, p. Two nets Σ and $\hat{\Sigma}$ in $S(n, m, p)$ are *sign-permutation equivalent* if there exists a nonsingular matrix T such that $T^{-1}AT = \hat{A}, T^{-1}B = \hat{B}, CT = \hat{C}$, and T has the special form: $T = PD$, where P is a permutation matrix and $D = \text{diag}(\lambda_1, \ldots, \lambda_n)$, with each $\lambda_i = \pm 1$. The nets Σ and $\hat{\Sigma}$ are just *permutation equivalent* if the above holds with $D = I$, that is, T is a permutation matrix.

Let $\mathbf{B}^{n,m}$ be the class of $n \times m$ real matrices B for which: $b_{i,j} \neq 0$ for all i, j, and for each $i \neq j$, there exists some k such that $|b_{i,k}| \neq |b_{j,k}|$. For any choice of positive integers n, m, p, denote by $S^c_{n,m,p}$ the set of all triples of matrices (A, B, C), $A \in R^{n \times n}$, $B \in R^{n \times m}$, $C \in R^{p \times n}$ which are "canonical" (observable and controllable, as in [33], section 5.5). This is a generic set of triples, in the sense that the entries of the ones that do not satisfy the property are zeroes of certain nontrivial multivariable polynomials. Finally, let:

$$\tilde{\mathcal{S}}(n,m,p) = \left\{ \Sigma(A,B,C,\sigma) \ \middle|\ B \in \mathbf{B}^{n,m} \text{ and } (A,B,C) \in S^c_{n,m,p} \right\}.$$

Then, in [1], the following result was proved: *Assume that σ is odd and satisfies property (*). Then $\Sigma \sim \hat{\Sigma}$ if and only if Σ and $\hat{\Sigma}$ are sign-permutation equivalent.* An analogous result can be proved when σ is not odd, resulting in simply permutation equivalence. Also, discrete-time results are available.

6 Acknowledgements

The author acknowledges extremely useful suggestions made by H.L. Trentelman and M. Vidyasagar, based on a preliminary version of this paper, as well as many constructive comments by C. Darken, S. Hanson, and G. Lafferriere.

References

[1] Albertini, F., and E.D. Sontag, "For neural networks, function determines form," *Neural Networks*, to appear. See also *Proc. IEEE Conf. Decision and Control, Tucson, Dec. 1992*, IEEE Publications, 1992, pp. 26-31.

[2] Albertini, F., E.D. Sontag, and V. Maillot, "Uniqueness of weights for neural networks," in *Artificial Neural Networks with Applications in Speech and Vision* (R. Mammone, ed.), Chapman and Hall, London, 1993, to appear.

[3] Anthony, M., and N.L. Biggs, *Computational Learning Theory: An Introduction*, Cambridge U. Press, 1992.

[4] Aubin, J.-P., *Mathematical Methods of Artificial Intelligence*, to appear.

[5] Baker, W.L., and J.A. Farrell, "An introduction to connectionist learning control systems," in [39].

[6] Barron, A.R., "Neural net approximation," in *Proc. Seventh Yale Workshop on Adaptive and Learning Systems*, Yale University, 1992, pp. 69-72.

[7] Barto, A.G., "Connectionist learning for control: An overview," in [27].

[8] Baum, E.B., and D. Haussler, "What size net gives valid generalization?," *Neural Computation* 1(1989): 151-160.

[9] Blum, A., and R.L. Rivest, "Training a 3-node neural network is NP-complete," in *Advances in Neural Information Processing Systems 2* (D.S. Touretzky, ed), Morgan Kaufmann, San Mateo, CA, 1990, pp. 9-18.

[10] Blumer, A., A. Ehrenfeucht, D. Haussler, and M. Warmuth, "Classifying learnable geometric concepts with the Vapnik-Chervonenkis dimension," in *Proc. 18th. Annual ACM Symposium on Theory of Computing*, pp. 273-282, ACM, Salem, 1986.

[11] Carroll, S.M., and B.W. Dickinson, B.W., "Construction of neural nets using the Radon transform," in *Proc. 1989 Int. Joint Conf. Neural Networks*, pp. I: 607-611.

[12] Chen, F.C., and H.K. Khalil, "Adaptive control of nonlinear systems using neural networks," *Proc. IEEE Conf. Decision and Control, Hawaii, Dec. 1990*, IEEE Publications, 1990.

[13] Cybenko, G., "Approximation by superpositions of a sigmoidal function," *Math. Control, Signals, and Systems* 2(1989): 303-314.

[14] Darken, C., M. Donahue, L. Gurvits, and E. Sontag, "Rate of approximation results motivated by robust neural network learning," submitted.

[15] Flick, T.E., L.K. Jones, R.G. Priest, and C. Herman, "Pattern classification using projection pursuit," *Pattern Recognition* 23(1990): 1367-1376.

[16] Franklin, J.A., "Historical perspective and state of the art in connectionist learning control," *Proc. IEEE Conf. Decision and Control, Tampa, Dec. 1989*, IEEE Publications, 1989.

[17] Haussler, D., "Decision theoretic generalizations of the PAC model for neural net and other learning applications," *Information and Computation* 100(1992): 78-150.

[18] Hertz, J., A. Krogh, and R.G. Palmer, *Introduction to the Theory of Neural Computation*, Addison-Wesley, Redwood City, 1991.

[19] Hornik, K., "Approximation capabilities of multilayer feedforward networks," *Neural Networks* 4(1991): 251-257.

[20] Hunt, K.J., D. Sbarbaro, R. Zbikowski, and P.J. Gawthrop, "Neural networks for control systems: A survey," *Automatica* 28(1992): 1083-1122.

[21] Judd, J.S., *Neural Network Design and the Complexity of Learning*, MIT Press, Cambridge, MA, 1990.

[22] Leshno, M., V.Ya. Lin, A. Pinkus, and S. Schocken, "Multilayer feedforward networks with a non-polynomial activation function can approximate any function," *Neural Networks*, 1993, to appear.

[23] Livstone, M.M., J.A. Farrell, and W.L. Baker, "A computationally efficient algorithm for training recurrent connectionist networks," in *Proc. Amer. Automatic Control Conference*, Chicago, June 1992.

[24] Macintyre, M., and E.D. Sontag, "Finiteness results for sigmoidal 'neural' networks," in *Proc. 25th Annual Symp. Theory Computing*, San Diego, May 1993, to appear.

[25] Matthews, M., "A state-space approach to adaptive nonlinear filtering using recurrent neural networks," *Proc. 1990 IASTED Symp. on Artificial Intelligence Applications and Neural Networks*, Zürich, pp. 197-200, July 1990.

[26] McBride, L.E., and K.S. Narendra, "Optimization of time-varying systems," *IEEE Trans. Autom. Control*, 10(1965): 289-294.

[27] Miller, T., R.S. Sutton, and P.J. Werbos (eds.), *Neural networks For Control*, MIT Press, Cambridge, 1990.

[28] Niranjan, M. and F. Fallside, "Neural networks and radial basis functions in classifying static speech patterns," *Computer Speech and Language* 4 (1990): 275–289.

[29] Pisier, G., "Remarques sur un resultat non publiè de B. Maurey," in *Seminaire d'analyse fonctionelle 1980-1981*, Ecole Polytechnique, Palaiseau, 1981.

[30] Polycarpou, M.M., and P.A. Ioannou, "Neural networks and on-line approximators for adaptive control," in *Proc. Seventh Yale Workshop on Adaptive and Learning Systems*, pp. 93-798, Yale University, 1992.

[31] Siegelmann, H.T., and E.D. Sontag, "Some results on computing with 'neural nets'," *Proc. IEEE Conf. Decision and Control, Tucson, Dec. 1992*, IEEE Publications, 1992, pp. 1476-1481.

[32] Slotine, J.-J., "Neural networks for control: Part 2," this volume.

[33] Sontag, E.D., *Mathematical Control Theory: Deterministic Finite Dimensional Systems*, Springer, New York, 1990.

[34] Sontag, E.D., "Feedforward nets for interpolation and classification," *J. Comp. Syst. Sci.* 45(1992): 20-48.

[35] Sontag, E.D., "Feedback stabilization using two-hidden-layer nets," *IEEE Trans. Neural Networks* 3 (1992): 981-990.

[36] Sontag, E.D., and H.J. Sussmann, "Backpropagation separates where perceptrons do," *Neural Networks*, 4(1991): 243-249.

[37] Sussmann, H.J., "Uniqueness of the weights for minimal feedforward nets with a given input-output map," *Neural Networks* 5(1992): 589-593.

[38] Vapnik, V.N., *Estimation of Dependencies Based on Empirical Data*, Springer, Berlin, 1982.

[39] White, D.A., and D.A. Sofge (eds.), *Handbook of Intelligent Control: Neural, Fuzzy and Adaptive Approaches*, Van Nostrand Reinhold, NY, 1992.

[40] Weiss, S.M., and C.A. Kulikowski, *Computer Systems That Learn*, Morgan Kaufmann, San Mateo, CA, 1991.

11 Neural Networks for Adaptive Control and Recursive Identification: A Theoretical Framework

J.-J.E. Slotine and R.M. Sanner*

Abstract

Massively parallel arrays of simple processing elements, the so-called "neural" network models, can be used to greatly enhance and extend techniques for identification and control of complex, nonlinear dynamic systems. However, the design of practical algorithms capable of ensuring prespecified performance levels requires a comprehensive, cross-disciplinary treatment, drawing techniques and insights from such diverse fields as machine learning, constructive approximation, nonlinear dynamic stability, and robust systems analysis. The goal of this paper is to review techniques for assembling all of these elements into an integrated and systematic framework, highlighting both *constructive* neural network analysis and synthesis methods, as well as algorithms for adaptive prediction and control with guaranteed levels of performance. These designs also contain supervised network training methods as a special case, and hence provide tools for evaluating the quality of a training set, avoiding overtraining, and maximizing the speed of learning.

1 Introduction

1.1 Historical Overview

The period from the early 1950s to approximately the mid-1960s was an era of explosive growth in learning systems research: tremendous resources were directed toward imbuing computing machinery with the speed, flexibility, and adaptability of biological computing structures. The ability to learn from mistakes, and to use newly acquired knowledge to anticipate and react correctly to new situations, was (and still is) seen as the next major threshold in the evolution of "artificial intelligence" [103].

---

*Department of Mechanical Engineering, Massachusetts Institute of Technology, Cambridge MA 02139, USA

Among the many groups exploring these methods, two stand out as precursors to the research undertaken herein: the so-called "neural network" research groups, and the "adaptive control" research groups. While neural network researchers used biologically inspired computing and learning models in an attempt to create machines with these adaptive features [33, 53, 78], the control system community concentrated on the task of designing control laws whose adjustable parameters could be continually tuned so as to elicit an "optimal" performance from a dynamic system [9, 46, 97].

Both fields developed mathematical models for their training algorithms, based partly upon heuristics, and partly upon mathematical optimization theory [101, 78, 19, 51, 97]. While proofs of the stability and convergence of these various approaches to the learning problem were scarce or lacking altogether, the great proliferation of successful heuristic applications of these models drove the research. The most celebrated proof from this era was the *perceptron convergence theorem* (PCT) [78] which demonstrated that one (biologically inspired) training algorithm for the perceptron model of neural computation could in fact be guaranteed to produce successful solutions, and in finite time, for a class of pattern recognition problems. In most cases however, successful applications of the learning procedures in each field required substantial amounts of trial-and-error tuning of various free parameters in the designs. The importance of a convergence theorem such as PCT should thus not be underestimated, and much of this paper is devoted to developing similar guarantees for the new neural network models and training procedures.

The two research groups were not long kept separate, of course. Control theorists realized early on that the learning and pattern recognition capabilities of neural networks could also be very useful in solving difficult control tasks. By learning to associate measurable properties of a dynamic system, presented to the network as a "pattern" of signals, with the control actions required to keep these measurements at specified levels, neural networks could be used to develop control systems for devices too complicated to model and control by conventional techniques. Many of the early methods used to train networks for control tasks also drew inspiration from biology, often not from the low level, intercellular models of learning, but rather from observation of higher cognitive functions. Learning techniques based upon Pavlovian stimulus/response and Skinner's operant conditioning and positive reinforcement ideas were used [54, 24], as were methods of entraining precomputed "optimal" control strategies into a network representations [102]. Unfortunately, as with the learning models used by control theorists, it was not possible to develop prior conditions which would ensure the success of these methods; as a result, "neural" control system design was also very much an *ad hoc*, trial and error process.

Retrenchment in both neural network and adaptive control research occurred in the mid- and late 1960s, as field experiments and more rigorous analysis showed the potential instability mechanisms underlying the adaptive methods developed by the control community [4, 69], while [59] rigorously demonstrated that there was no theoretical basis for the universal computing properties implicitly assumed to underlie the perceptron learning model.

The adaptive control community responded to this challenge by shifting their aims to more gradual goals. Starting first with single-input, single-output, linear, time invariant dynamic systems, and carefully progressing to more complex multivariable and nonlinear systems, control researchers developed over the next two decades the analysis techniques and algorithms required to ensure the stability and convergence of their recursive identification and adaptive tracking control designs [50, 63, 64, 68, 22, 60, 92, 61, 89, 47].

Neural network researchers, meanwhile, developed progressively more complex models which finally appeared to be capable of overcoming at least some of the previous criticisms [31, 32, 35, 36, 37, 79, 48, 72]. Driven especially by the apparent ease with which biological computing structures seem to be capable of solving computationally "hard" problems in speech and visual processing, the new *multilayer, sigmoidal* networks in particular have been extensively applied to a wide range of such problems in recent years, usually using *backpropagation* or similar gradient descent training techniques. The local nature of signal processing performed in these models, which facilitates a high degree of parallelism in the computation, is thought to embody a key reason for the success of biological solutions to these problems, and the extremely simple input-output behavior of the individual processing elements makes the model particularly amenable to implementation in analog or digital hardware, which further increases their popularity.

The new, multilayer network models have met with a frustrating pattern of successes and failures, however; there seems often to be no clear indication how complex a network is required for a given task, nor how well a trained network will provide correct responses for inputs not encountered during training. In order to analyze these problems rigorously, several researchers began to re-evaluate from first principles the learning problems being posed to their networks, seeking a mathematically formal framework in which to address the matter. Making a complete break from biologically inspired neural models, the *regularization* approach [72] views the common supervised learning application of neural networks as an ill-posed problem in interpolatory function theory. Perhaps surprisingly, when the necessary constraints are added to the problem statement, by including information about the nature of the functions assumed to underlie the example data, the formal solutions often have a very biological "feel" to them. The layered, parallel arrangements of neuron-like elements, and the local nature of the required signal processing both naturally emerge from this analysis, although the input-output behavior of the individual elements is somewhat different from sigmoidal or perceptron models [72].

Often, however, applications demand more than simple interpolation of the training data; one is willing to sacrifice an exact fit to the training set for the more desirable feature of ensuring a prespecified worst case error over a range of possible network inputs, Even in this extended framework, the principle underlying the regularization formulation, that of using assumed constraints on the function to be learned as a guide for developing a network architecture, is still quite sound. By replacing the idea of function interpolation with that of function *approximation*, one can draw upon the formidable, and still rapidly developing, field of approximation theory to develop architectures suitable for representing entire classes of functions to a specified level of uniform accuracy. Provided then that the function generating the training set belongs to the assumed function class, the network can be guaranteed to have sufficient flexibility to correctly learn and generalize the mapping implicit in the data.

1.2   Overview of the Paper

At this point, however, the march away from biology in search of a rigorous formulation has perhaps come too far; in abstracting the learning problem to this level we may have sacrificed the features which make neural network designs so attractive as a computing model. And yet as in the regularization analysis, it is possible to identify within formal approximation theory certain representations which still seem to capture many of the essential features of the biological model. Section 4.1 will attempt to strike a balance between these approximation theoretic and engineering requirements, by proposing a more general definition of "neural network" than commonly found in the literature. This definition encompasses not only the more popular (static) neural network models, but a great number of additional models as well including several powerful representations provided by approximation theory.

Inevitably, the new learning models have again attracted the attention of control engineers, and the past several years have seen development of a great variety of adaptive control and system identification algorithms which employ the newer network architectures and gradient training methods. Section 2 briefly reviews the salient features of these designs, and discusses some of their strengths and weaknesses. Unfortunately, despite the great number of reported successful applications, these methods suffer from same drawback as their predecessors: there is no method of guaranteeing when the training will be successful, or what the quality of the resulting control will be. In fact, Section 3.1 demonstrates that, viewed as nonlinear dynamic systems, gradient network training algorithms such as backpropagation actually contain potential instability mechanisms. At least one manifestation of this instability is the well known *overtraining* phenomenon, which has long plagued neural network applications. Section 3.3 then shows how similar instability mechanisms may be present when a dynamic system is controlled by a trained neural network, even when no additional learning occurs.

A careful analysis of the nature of these problems shows that the measures necessary to circumvent them require a *quantitative* characterization of the ability of the chosen network architecture to approximate the mappings necessary for successful identification or control. This discussion reveals the intimate connection between the stability and the approximation theoretic aspects of the overall analysis; a connection which will arise often in the sequel as more sophisticated identification and control applications are considered. It also demonstrates that stability considerations are not merely theoretical technicalities, but are rather vitally important for understanding the circumstances under which a training procedure will be successful, and for explaining observed pathologies of existing network training procedures.

It is only quite recently, however, that the kind of rigorous stability analyses which have marked the past 25 years of development in adaptive control have been extended to identification and control applications of neural networks. Thus, starting with Section 4, the remainder of this paper reviews methods for accomplishing a formal merger of the current state of the art in both neural network and adaptive nonlinear control theory. The resulting network training procedures are as far removed from the Pavlovian conditioning ideas of the original learning control methods as the approximation theoretic networks are from their biological counterparts, but, as in these latter, there are surprising similarities between the formal mathematical solutions and the biologically inspired heuristics.

To provide the quantitative characterization of network approximation capabilities required in the stability analysis, Section 4 reviews a number of *constructive* analysis techniques for single hidden layer networks. The analysis begins with consideration of the celebrated *sampling theorem*, which itself provides an excellent example of an approximation theoretic expansion that can be mapped onto a neural network architecture, providing a parallel arrangement of neuron-like elements each of whose input-output behavior is that of an ideal low-pass filter. Moreover, the construction inspired by sampling theory permits precise relations to be established between the components of this network and the properties of the function it approximates, providing a powerful synthesis tool.

The classical sampling theorem, however, has some undesirable numerical properties which imply that quite large networks would be required in order to ensure good approximations over a given range of network inputs. However, by modifying the processing elements in our networks to have better *spatial localization* properties, while preserving to the extent possible the frequency localization (low-pass) properties of the elements in the classical model, representations can be devised which employ much smaller networks to achieve the same level of accuracy. A formal analysis shows that the *Gaussian radial basis function* is the optimum choice in this trade-off, and that by arranging the centers of these elements to lie on a regular grid the components of the resulting networks admit the same interpretation as those of the classical sampling model, allowing the specification of explicit design procedures. Moreover, adopting a distributional view of this sampling procedure, it is possible to establish a strict correspondence between radial Gaussian approximation and multivariate spline techniques, which permits specification of constructive Gaussian network designs for approximations to polynomial and trigonometric functions as well.

Armed with a constructive theory of network computation, Section 5.1 re-examines applications of these devices in the framework of formal stability theory, starting with recursive identification. Algorithms for stable, convergent identification of continuous- and discrete-time nonlinear systems are reviewed, and include supervised learning applications as a special case. This analysis demonstrates the importance of using *robust* adaptive techniques for the training algorithms, and also again highlights the interplay between stability and approximation theory, as the quantitative measures of the degree and extent of network approximation capabilities are vital components of the robust designs which ensure stable and convergent learning.

Many of the tracking control methods discussed in Section 2 require use of an identification model of the process to be controlled. Satisfactory control of this type then demands this model be of high fidelity, and to guarantee this it is necessary that certain excitation conditions be satisfied by the signals used to train the networks during the identification phase. An extremely attractive feature of the Gaussian network identification models developed in Section 5.1 is that one can exactly determine the conditions a sequence of network inputs must satisfy to guarantee that the signals used in the learning process will satisfy these conditions. Section 5.2 reviews these excitation conditions, and discusses the consequences for the stability and effectiveness of indirect control algorithms.

Unfortunately, as Section 2.3 shows, regardless of how accurate a network identification model is, it is still possible for a control law developed from this model to drive the actual process unstable unless careful attention is given to the interaction between the structure of the control law, the approximation capabilities of the network, and the actual process dynamics. Moreover, the assumption of a persistently exciting identification phase is somewhat constraining and may be difficult to ensure in practice. These considerations lead to the exploration, in Section 5.3, of *direct* adaptive tracking control algorithms which permit learning and control to proceed simultaneously, and which do not require the development of accurate models of the process. By combining the robust adaptation methods developed in Section 5.1 with techniques of robust nonlinear control, it is possible to specify direct adaptive tracking control algorithms which ensure the global stability of the adaptive system and the asymptotic convergence of the model tracking errors. While current state of the art in nonlinear control system design and the complexity of the stability analysis somewhat limits the class of processes for which these structures can be reliably developed, very satisfying convergence results can be give for important classes of nonlinear systems.

2   Indirect Control Algorithms and Gradient Training

This section reviews some very common, and potentially quite powerful, methods for using neural networks to control unknown or partially known dynamic systems. The methods reviewed are both *indirect* and (generally) *off-line*; that is, they involve developing explicit, accurate models of the process dynamics during an initial training phase which uses either pre-collected input-output data or (assuming the process is stable) limited on-line, *open loop* interaction, in which the required input-output data is generated as the plant evolves in response to probing inputs. Once the quality of the model thus developed is assured, a control network is either trained using, or created directly from, the identification model. The trained control network is then brought on-line, and its outputs used to drive the actual process.

The papers [76, 44, 45, 65, 66, 67] and the book [58] provide a comprehensive discussion of these approaches; this section will only briefly review some salient features, so as to contrast them with the on-line, direct methods developed in later sections. Limited space precludes a comprehensive discussion of the many other methods proposed to use neural networks in control systems; the reader is referred to the recent survey [39] for a more complete overview. The identification aspect of the problem is also treated at length in [14, 15, 16].

2.1   Identification Methods and Neural Networks

The first step in an indirect approach to the control problem is to develop an accurate model of the process: either its forward dynamics, that is, the relation between the process outputs in response to its inputs, or its inverse dynamics, that is, the relation which describes the process inputs required to produce a particular set of outputs. Since techniques for accomplishing this with a great variety of processes are well known in the system identification literature, it is necessary to understand the conditions under which a neural network would ever be needed to solve the problem.

Creating an identifier requires first introducing some prior knowledge about the process so that an appropriate model can be selected. For example, in discrete time a general model would be

$$y[t] = f(y[t-1], \ldots, y[t-N], u[t], \ldots, u[t-M+1]). \qquad (2.1)$$

Most physical systems usually have at least a one step delay between the control and the output; the instantaneous dependence on $u[t]$ is included here so as to subsume supervised learning models into the same architecture. In continuous time a similarly general description would be

$$\dot{z}(t) = f(z(t), u(t)), \quad y(t) = h(z(t), u(t)), \qquad (2.2)$$

where $z(t) \in \mathcal{R}^n$. In both models y is the process output, and u is the process input, both assumed scalar for convenience. For simplicity the nonlinear functions f and h will be assumed to be at least continuously differentiable on $\mathcal{R}^{N+M}$ or $\mathcal{R}^{n+1}$.

Of course, several assumptions have already been made about the behavior of the process at this level. First, in the discrete time case, it is assumed that upper bounds are known on the extent of the dependence of the process output on its past outputs and the past values of the control input; equivalently, in the continuous time description, appropriate signals describing the state, z, of the system are assumed known, and will in the sequel be assumed available for measurement. Second, these models assume that the process is autonomous, that is, there is no explicit dependence on time in its equations of motion.

For convenience in discussing these models, define the signal vector x to be $x[t] = [y[t-1], \ldots, y[t-N], u[t], \ldots, u[t-M+1]]$ in discrete time models, and $x^T(t) = [z^T(t) \ u(t)]$ in continuous time. To simplify the ensuing discussion, and to mirror the historical development, the discrete time model will be used to illustrate the techniques described in this section; similar ideas can applied to the continuous model, which may appear more naturally in nonlinear physical systems, and which will be analyzed in detail in later sections.

With these assumptions about the process behavior, two common structures are utilized to model it. The first is the so-called *parallel* identification model,

$$\hat{y}[t] = \hat{f}(\hat{\mathbf{x}}[t], \hat{\mathbf{c}}[t-1]),$$

where $\hat{\mathbf{x}}$ is defined similarly to $\mathbf{x}$ but using the past estimates $\hat{y}[t-i]$ in place of $y[t-i]$. The function $\hat{f}$ is parameterized by the adjustable values $\hat{\mathbf{c}} \in \mathcal{R}^p$, and in the classical identification algorithms is chosen such that, for at least one choice of parameters, $\mathbf{c} \in \mathcal{R}^p$, $\hat{f}(\hat{\mathbf{x}}, \mathbf{c}) = f(\hat{\mathbf{x}})$ for any $\hat{\mathbf{x}} \in \mathcal{R}^{N+M}$.

Notice that the parallel model receives no direct information from the process it attempts to model, but rather feeds its own estimates of the process behavior back to itself to predict future behavior. This is in contrast to the so-called *series-parallel* model, which has the form

$$\hat{y}[t] = \hat{f}(\mathbf{x}[t], \hat{\mathbf{c}}[t-1]),$$

the chief difference being that predictions of future process behavior are here made based on the observed input-output history of the process itself, not that of the model.

In both identifier models, different functional forms are possible for $\hat{f}$ depending upon the amount of additional prior information available about the process. For example, if the process is known to be linear an appropriate structure is

$$\hat{f}(\mathbf{x}[t], \hat{\mathbf{c}}) = \sum_{i=1}^{N} \hat{c}_i y[t-i] + \sum_{i=0}^{M-1} \hat{c}_{i+N} u[t-i] = \sum_{i=1}^{N} \hat{c}_i x_i[t]$$

or, if the nonlinearity f is known to lie in the span of a finite set of basis functions Y_i, one can choose

$$\hat{f}(\mathbf{x}[t], \hat{\mathbf{c}}) = \sum_{i=1}^{N} \hat{c}_i \, Y_i(\mathbf{x}[t]).$$

It is now clear how neural networks may be used to extend this paradigm; networks provide parameterized expansions for entire *classes* of nonlinear functions [20, 38, 25, 28]. By using an appropriately designed neural network, $\mathcal{N}(\mathbf{x}, \mathbf{c})$, to implement the function $\hat{f}$, the resulting structure can theoretically recover any f belonging to an appropriate function class (for example, any continuous function), instead of the quite restricted forms which can be recovered by the above models. The symbol $\mathcal{N}(\mathbf{x}, \mathbf{c})$ here stands for a neural network structure with inputs $\mathbf{x}$, parameterized by the weights in $\mathbf{c}$, and its exact composition and configuration may depend upon the specific function class to which f belongs. In the discussion which follows, any network with differentiable components can be assumed; later sections of this paper will focus on a specific class of networks models.

However, these theoretical results show that a finite size neural network can only *approximate* a given function class, and this only on closed bounded subsets of the input space. As a result, the precise equality between $\hat{f}(\mathbf{x}, \mathbf{c})$ and $f(\mathbf{x})$ exploited in the classical algorithms no longer holds, but is replaced by the weaker condition that, given an $\epsilon_f > 0$, it is possible to choose a neural network structure such that, for some $\mathbf{c} \in \mathcal{R}^p$

$$\hat{f}(\mathbf{x}, \mathbf{c}) = \mathcal{N}(\mathbf{x}, \mathbf{c}) = f(\mathbf{x}) + d_f(\mathbf{x}) \tag{2.3}$$

with $|d_f| \le \epsilon_f, \forall \mathbf{x} \in A$, where A is a closed, bounded subset of $\mathcal{R}^{N+M}$. That is, in contrast to classical identification algorithms, which provide globally exact parameterizations of the process dynamics, neural networks provide parameterizations which are both *local* (i.e., accurate only in a bounded region) and *approximate*.

Additional prior information about the process may be introduced into the structure of the identifier model; this additional information can often substantially facilitate the design of control laws. For example, if

$$y[t] = f(y[t-1], \ldots, y[t-N]) + \sum_{i=0}^{M-1} c_i u[t-i]$$

the nonlinear function f can be identified by an appropriate neural network, and a more conventional linear identifier used to determine the influence of the control terms $u[t-i]$.

2.2   Gradient Network Training

Training neural networks typically requires selection of an appropriate cost function to minimize. In supervised learning applications (which indeed are easily seen to be a special case of the above identification problem by assuming a memoryless process with $N = 0$ and $M = 1$), the cost is the sum of the squares of the errors $\mathcal{N}(\mathbf{x}, \hat{\mathbf{c}}) - f(\mathbf{x})$ at a finite set of test points [79, 49]. The network weights are then adjusted to minimize this cost. For more general identification problems, one cannot assume the availability of a finite, prior set of input-output responses, but rather must be prepared for any possible input-output sequence. One generalization of the supervised learning cost function to this setting would be to compute the cost over a moving window into the evolving time series,

$$J(\hat{\mathbf{c}}) = \sum_{\tau=t}^{t+T-1} e^2[\tau]$$

where $e[\tau] = \hat{y}[\tau] - y[\tau]$ is the prediction error at time τ, and hence implicitly depends upon the parameters $\hat{\mathbf{c}}$ being used in the model.

Commonly (even in supervised learning applications), T is chosen to be 1, so that the cost to be minimized is simply the instantaneous squared error. In this case, the gradient descent law for parameter adjustment gives:

$$\hat{c}[t] = \hat{c}[t-1] - k_a e[t] \frac{\partial \hat{y}[t]}{\partial \hat{c}}.$$

For the series-parallel identifier model, this training law is exactly equivalent to back-propagation (omitting the common "momentum" term [79]); that is

$$\hat{c}[t] = \hat{c}[t-1] - k_a e[t] \frac{\partial \mathcal{N}}{\partial \hat{c}},$$

where the network partial derivatives are evaluated at $(\mathbf{x}[t], \hat{c}[t-1])$, i.e. using the "old" weights and the node activations produced by the current network inputs.

For the parallel model, however, the feedback structure of the identifier requires use of the sensitivity function method [19], producing a more complicated mechanism for computing the gradient of a squared prediction error metric. The sensitivity method adjusts the network weights using the following system of equations [66]:

$$\begin{aligned} \gamma_i[t] &= (D_1\mathcal{N})[t]\gamma_i[t-1] + \cdots + (D_N\mathcal{N})[t]\gamma_i[t-N] + (D_{N+M+i}\mathcal{N})[t] \\ \hat{c}_i[t] &= \hat{c}_i[t-1] - k_a e[t]\gamma_i[t] \end{aligned}$$

Where the notation $(D_i\mathcal{N})[t]$ is used to indicate the partial derivative of the function $\mathcal{N}$, with respect to its ith argument, evaluated as above at $(\mathbf{x}[t], \hat{c}[t-1])$. Notice that unlike the adjustment mechanism for the series-parallel model, which directly uses the derivatives of the network output with respect to each weight, the sensitivity method instead passes this signal through a time-varying filter. A filter of this form is associated with each weight of the network, creating a substantial computational burden. In principle, however, this filtering could be accomplished in hardware with the same degree of parallelism which characterizes the forward operation of the network.

These sensitivity methods give only an approximation to gradient descent in the chosen cost functional; the use of partial derivatives in this fashion implicitly assumes that the adjustable parameters are constant, while in fact they are continually changing. To ensure that these methods provide a reasonable approximation to the gradient, the network parameters should change only quite slowly, requiring very small adaptation gains, or should change only after several time steps, instead of changing at each instant. A complete discussion of the tradeoffs and limitations can be found in [66]. Choosing a large adaptation rate, so that the parameters change rapidly, may result in very erratic learning, and even instability. In fact, it was the difficulty in determining the exact conditions under which the stability of sensitivity methods could be assured that led control researchers to re-design adaptation mechanisms from first principles using stability methods [62, 69]. Methods for accomplishing a comparable "Lyapunov redesign" for identification and control algorithms which use neural network models will be explored in Section 5.

2.3   Indirect Controller Structures and Training

The design of an effective control law for a dynamic system is often ham-pered by lack of complete knowledge about the structure of its equations of motion. The goal of the above identification algorithms has been to recon-struct the missing information by observing the input-output behavior of the process over a period of time. Assuming that a sufficiently slow adap-tation rate has permitted the development of an identification model which adequately describes the observed behavior of the process, for example such that $|\hat{y}[t] - y[t]| \le \epsilon$ for a sufficiently long period of time, two possibilities exist for utilizing the information contained in the models. If, given exact prior information about the process, a nonlinear control design methodo-logy exists which will produce the desired behavior, the components of the identification model may be used in place of the actual plant information in the design. Often, however, the design of an effective control law based upon the identification model is more difficult; in this case, neural net-works and gradient training techniques can again be helpful in developing a representation of the required control law.

To illustrate the first possibility, suppose that it is known that the true process is modeled by

$$y[t] = u[t-1] - f(y[t-1]).$$

Given f, the control law

$$u[t-1] = f(y[t-1]) + \alpha y[t-1] + (1-\alpha)r[t-1],$$

where r is the reference input the plant must follow, and $|\alpha| < 1$, will cause the process to asymptotically follow the model response

$$y_m[t] = \alpha y_m[t-1] + (1-\alpha)r[t-1],$$

since the tracking error $e[t] = y[t] - y_m[t]$ satisfies $e[t] = \alpha e[t-1]$ and hence decays asymptotically to zero. In the absence of prior knowledge about f, one could identify this function using a neural network as in the previous section, and instead use for the control

$$u[t-1] = \mathcal{N}(y[t-1], \hat{c}) + \alpha y[t-1] + (1-\alpha)r[t-1],$$

where $\hat{c}$ is the parameter choice learned by the network during the identi-fication phase; note that this is not necessarily the same as the parameter

choice c which minimizes the term d_f in (2.3). Even if the optimal parameters are learned, however, $\mathcal{N}$ will not exactly cancel f, and generally the
ideal asymptotic decay of the error dynamics provided by the exact control
law will be perturbed by the approximation error $\mathcal{N}(\mathbf{x}[t], \hat{\mathbf{c}}) - f(\mathbf{x}[t])$. Given
the structure of the error dynamics in this example, provided the approxi-
mation error is uniformly small in magnitude, its effects on the tracking
error will also be small. The conditions under which $|\mathcal{N} - f|$ can be gu-
aranteed to be small, and the general effects of such perturbations, are
clearly an important factors in evaluating this design, and will be discussed
at length in the following sections.

When the process assumes a more general form, it may be quite difficult
to directly specify the form of the control signal required to asymptotically
track the output of a particular model. One possible approach in this
situation would be to place a neural network in series with the process,
closing the feedback loop by presenting the process outputs as inputs to the
network, and using the network output as the control input to the process.
The network parameters would then be adjusted so as to continually reduce
the model tracking error. For example, if it is known that there exists a
continuous control law of the form,

$$u[t-1] = \theta(y[t-1], \ldots, y[t-N], r[t-1], u[t-2], \ldots, u[t-M])$$

such that $y[t]$ asymptotically tracks $y_m[t]$, a (recurrent) neural network with
the same inputs could be employed to learn the continuous function θ,

$$\hat{u}[t-1] = \mathcal{N}_c(y[t-1], \ldots, y[t-N], r[t-1], \hat{u}[t-2], \ldots, \hat{u}[t-M], \hat{\mathbf{c}}_c[t-1])$$

(the c subscripts are used to distinguish the control network and its parameters from those of the network used in the identification model). With
this control strategy the closed-loop process output can be written as

$$y[t] = f(y[t-1], \ldots, y[t-N], \hat{u}[t-1], \ldots, \hat{u}[t-1-M])),$$

but unfortunately, a gradient descent procedure to reduce the error $y[t] -
y_m[t]$ cannot be implemented in this setting, since evaluation of the closed
loop dependence of the process output on parameter variations in the con-
trol network, $\frac{\partial y[t]}{\partial \hat{\mathbf{c}}_c}$, would require exact knowledge of the process dynamics,
which by assumption are unknown.

To avoid this complication, the control network is instead placed in
series with a trained parallel identification model of the process and the
feedback loop closed as before. The parameters of the control network are
then adjusted to generate the inputs required to force the *identifier* output
to follow a model response, i.e. so that $|\hat{y}[t] - y_m[t]| \leq \epsilon$ for sufficiently
large t. The known dynamics of the identifier can then be used to compute
the sensitivity functions needed to train the control network [45, 66, 67].

When acceptable tracking has been obtained by using the control network with the identification model, the control network can be brought on-line with the actual process as described above.

Often in these applications, the model response is actually the identity mapping, $y_m[t] = r[t]$. In this case the control network is being trained to precisely invert the identifier model (and hopefully therefore also the process itself) [57, 58]. Often, this inverse can be learned in a more direct fashion, by using the identification phase to instead develop a model of the inverse dynamics from the available input-output data. That is, instead of predicting the process outputs in response to its inputs, the network is used to make predictions of the control input to the process based upon observations of the current and past process outputs, and the past control inputs, i.e.

$$\hat{u}[t-1] = \mathcal{N}_c(y[t], \ldots, y[t-N], u[t-2], \ldots, u[t-M], \hat{c}_c[t-1]).$$

In this case, the cost to be minimized would be the squared control prediction error, $e[t] = \hat{u}[t-1] - u[t-1]$ (or possibly a summation of these terms over a moving window) for the range of signals seen during the identification phase. Once this network has accurately identified the inverse dynamics, it can be used to control the process by replacing the input signal $y[t]$ with the desired process behavior $r[t]$, and using the network output as the control input to the process [76]. Of course, identifying the inverse in this fashion assumes that a functional relation exists in the input-output data, that is, that the relation between $u[t-1]$ and the signals input to the network is not one-to-many; if this is not the case, inverting an identified forward model may provide better results [45].

In these latter methods of training and utilizing networks for control, it is very difficult to assess *a priori* what happens when the forward or inverse estimates are not precisely equal to the actual functions required to force the process to follow the model response. Unlike the simpler control models considered at the beginning of this section, the inevitable differences between the ideal control signal and that produced by a network trained in the fashion above may *not* produce small discrepancies in the process output compared to the model. That is, training a control network such that $|\hat{y}[t] - y_m[t]| \le \epsilon$ for sufficient large t does not necessarily guarantee that, when the actual plant is controlled by the resulting network, $|y[t] - y_m[t]|$ will be comparably small. Similarly, learning the inverse dynamics so that $|\hat{u}[t-1] - u[t-1]|$ is as small as required, does not necessarily guarantee good tracking when the inverse model is used to control the process as indicated above.

Further, the required inverses may not even exist, and when they do exist may not be representable by continuous functions. The arguments justifying the use of differentiable network models are no longer applicable in this latter case, and more complex network designs, employing discontinuous elements, are required in order to ensure the required inverse can be learned to a specified tolerance [95]. Even assuming the existence of a continuous inverse, the process may not be *stably invertible*, that is, the control signals required to force the process to follow the given model may grow without bound [65]. The methods presented above may indeed learn to implement such a control, but clearly this would not represent a desirable solution.

There are thus many questions regarding the efficacy of the control methods presented in this section which require answers. Before attempting to address these, however, there is are much more fundamental concerns which must be addressed regarding the stability and convergence properties of the gradient training process itself. The next section examines these questions in detail, before returning to the question of the closed-loop stability of these control algorithms.

3 Performance of Indirect, Gradient Techniques

Despite the possible problems, there are have been many successful demonstrations of the above algorithms applied to practical problems in identification and control, although often heuristic tuning of certain of the training procedures and controller parameters is required to obtain satisfactory results. If, however, neural networks are to be truly useful in control applications, guesswork and heuristics should be replaced by precise criteria delineating a successful design from a failure. Indeed, the intention of this section is to show that, even in ideal conditions, it is really not so difficult to obtain poor results from gradient training procedures and indirect control algorithms, and to identify the factors which influence the success of these methods.

3.1 Parameter Drift Instability

Consider, for example, a special case of the above discrete-time identification model where the process is described by the memoryless input-output relation $y[t] = f(u[t])$, and $f(u) = \sin(2\pi u)/(\pi u)$. To identify the function f describing the process, a series-parallel identifier structure is employed, using a neural network whose input is $u[t]$, and whose output is the prediction, $\hat{y}[t]$ of the process value at time t. In this application, the network is constructed using a single hidden layer of Gaussian radial basis function nodes [11, 72]. There are 17 nodes in the hidden layer of this network, and their centers lie on a regular mesh of size $\Delta = 4/9$, so that the network prediction at time step t of the required value $f(u[t])$ can be written as:

$$\hat{y}[t] = \mathcal{N}(u[t], \hat{c}[t-1]) = \sum_{k=-8}^{8} \hat{c}_k[t-1]\exp(-\pi^2(u[t]-k\Delta)^2). \qquad (3.4)$$

To train this network, the training data **t** shown in Figure 3.1 is used. Since there is only a finite data record, and since a very large amount of

Example	1	2	3	4	5	6	7	8
u	-1.250	-1.083	-0.833	-0.750	-0.583	-0.333	-0.250	-0.083
$f(u)$	0.255	0.146	-0.331	-0.424	-0.272	0.829	1.273	1.911

Example	9	10	11	12	13	14	15	16	17
u	0.000	0.167	0.250	0.417	0.500	0.667	0.917	1.167	1.500
$f(u)$	2.000	1.653	1.273	0.380	0.000	-0.414	-0.173	0.237	0.

Figure 3.1: Data set used to train the network.

training time may be required before satisfactory results are obtained, the inputs $u[t]$ in the identification model are chosen to cycle through the data in the training set; i.e. $u[t] = \mathbf{t}_{1+(t \bmod 17)}$.

The error metric to be minimized is the instantaneous squared prediction error, and to give a gradient algorithm the best possible opportunity for success, the input weights of this network are held fixed and only the output weights are adjusted, so that the squared prediction error is quadratic in the parameter error. The gradient descent algorithm for each output weight is then

$$\hat{c}_k[t] = \hat{c}_k[t-1] - k_a e[t]\exp(-\pi^2(u[t]-k\Delta)^2)$$

where $e[t]$ is the prediction error, $e[t] = \hat{y}(t) - y[t]$. In this simulation the adaptation gain is chosen as $k_a = 0.2$.

Figure 6.1 shows the worst case and total square error over the training set as a function of the time index t, while Figure 6.2 shows both the worst case error over the compact set $A = [-2.0, 2.0]$, during the first 5000 training steps. These plots demonstrate the classic *overtraining* phenomenon; as the network constructs a better fit to the available data, the error generalizing to points outside the training set worsens. The actual reason for this behavior, however, is cause for alarm: the steady worsening of the generalization error is caused by a monotonic growth in one of the network output weights, as shown in Figure 6.3. Increasing the number of training iterations produces minimal improvement in the training set error, but virtually linear growth in this weight and in the corresponding generalization error. Though it appears from the training errors in Figure 6.1 that the learning has "converged", in fact every additional training cycle serves to *worsen* the quality of the identification model, as Figure 6.4 shows.

Insight into this phenomenon can be obtained by recasting the identification problem posed to the network as an interpolation problem. Removed of its iterative structure, the network is being asked to find a weight vector c such that

$$\mathcal{G}\,\mathbf{c} = \mathbf{t}$$

where $\mathcal{G}_{ij} = \exp(-(\mathbf{t}_i - (j-9)\Delta)^2/\pi^2)$.

In this case, there are exactly as many training examples as network nodes, but the square matrix $\mathcal{G}$ is very nearly singular (with a condition number exceeding 10^{50}) and part of the training data in $\mathbf{t}$ lies in its "effective" nullspace. Solving the interpolation problem exactly thus requires effectively unbounded weights, which the gradient algorithm merrily attempts to recover, oblivious to the catastrophic effect this has on the overall quality of the identification model.

If one were solving this interpolation problem directly, processing the observed data in an off-line, "batch" fashion, the magnitude of the weights used could be restricted by computing a pseudo-inverse of $\mathcal{G}$ which omits singular values smaller than a particular threshold, thus making a tradeoff between the accuracy of the fit to the data and the size of the weights used. However, nothing in a pure gradient descent solution procedure is capable of performing this tradeoff; moreover, there is no prior method for determining when such techniques might be necessary, aside from monitoring in real time the evolution of the singular values of the evolving interpolation matrix (in fact this is a viable strategy, and is related to the idea of *persistency monitoring* [5] in adaptive systems theory, as the discussion in Section 5.2 will illustrate).

There are a number of heuristics which can be used to address the growth problem observed above. For example, by adjusting the centers of the Gaussian network to correspond to the unique points in the process input, the interpolation matrix $\mathcal{G}$ is guaranteed to be nonsingular [55, 72]. Similarly, by halting adaptation when the "true" minimum is reached, that is, when $|e[t]|$ is as small as the best possible network approximation to f on A would allow, the network will not fruitlessly strain to eliminate the last bit of error from the training set, and unbounded growth in the parameters can again be avoided [22, 71, 40]. But each of these heuristics merely replaces one problem with another. Adjusting the network centers is only effective if there is a finite number of distinct points in the input-output stream, and if the network is of comparable size. This will rarely be the case in practical applications. On the other hand, deciding when to halt adaptation requires a method of *a priori* characterizing the representational capabilities of the chosen network design.

More quantitative insight into this problem can be gained by instead considering the dynamic aspects of the training. Performing a Taylor series expansion of the prediction error in terms of the network parameters, and recalling the limitations of network approximations discussed in Section 2.1,

$$
\begin{aligned}
e[t] = \hat{y}[t] - y[t] &= \mathcal{N}(u[t], \hat{c}[t-1]) - \mathcal{N}(u[t], c) + d_f(u[t]) \\
&= \frac{\partial \mathcal{N}}{\partial \hat{c}}\tilde{c}[t-1] + d_g(u[t], \hat{c}[t-1]) + d_f(u[t]), \quad (3.5)
\end{aligned}
$$

where as usual, the indicated gradient is evaluated using the instantaneous values of the network inputs and parameter values. The term d_f represents the inevitable error incurred approximating a continuous function using a finite extent neural network and $\tilde{c}[t-1] = \hat{c}[t-1] - c$ is then the *mistuning* between the parameters being used at time $t-1$ and the parameter values, c, which provide the smallest possible ϵ_f such that $|d_f(u)| \leq \epsilon_f$ for all $u \in [-2., 2.]$. The term d_g then contains the terms of the expansion higher than first order in the weight mistuning.

In the first two identifier/process models considered in Section 2.1, for which stability based analyses are well developed, the terms d_f and d_g vanish identically, leaving only the gradient term of the expansion. The vanishing of d_f indicates that the chosen basis is capable of a globally exact representation of the process dynamics for an appropriate choice of parameters, while the vanishing of d_g indicates that the error approximating f with $\hat{f}$ can be *linearly parameterized* in terms of the mistuning of the parameter values. It is these two features which have been crucial in determining conditions which ensure the stability and convergence of identification algorithms employing these models [30, 62].

When a neural network is used to estimate the unknown function, however, these terms are generally nonzero. Yet, the training algorithms considered above account only for the gradient term in their attempts to minimize the squared error, neglecting completely the effects of d_f and d_g. In fact, analysis of the linearly parameterized models show that when such perturbations to the error equation are ignored by the adaptation mechanism, it is no longer possible to assure stability and convergence [22, 40]. The adjustable parameters may wander away from their optimal values in response to the perturbations, and possibly even diverge to infinity; this is known as *parameter drift instability* [40].

In the specific example considered above, keeping the input weights fixed renders d_g identically zero but, as will be shown below, d_f is nonnegligible for the chosen network architecture. The overtraining phenomenon witnessed, and the monotonic growth of one of the network weights, can thus be seen as a manifestation of the parameter drift problem. Fortunately, methods exist in the adaptive systems literature for modifying the gradient algorithm to guarantee its stability. For example, by placing a small *deadzone* into the error signal used for adaptation, so that no adaptation occurs when $e[t]$ is smaller than the worst expected magnitude of d_f, no parameter drift occurs, and overtraining is avoided. The above figures and Figure 6.5

illustrate this idea with a deadzone of 0.26 units; Section 5 will formalize this notion of using *robust* adaptation techniques to accommodate the use of inexact expansions of the functions to be learned.

3.2   Generalization, Persistency of Excitation, and Network Design

Even assuming that the pathological behavior described above can be avoided, the success of the indirect control methods described in Section 2.3 depend critically upon the fidelity of the identification model. Unfortunately, as is well demonstrated by the previous (unmodified) example, *convergence of the prediction error does not necessarily imply a high fidelity identification model*; that is, asymptotic satisfaction of the inequality $|\hat{y}[t] - y[t]| \leq \epsilon$ does not imply that the quantity $|\hat{f}(\mathbf{x}) - f(\mathbf{x})|$ is necessarily small for every choice of $\mathbf{x}$. This is well known in the neural network literature as the generalization problem–convergence to minimal error on a training set does *not* guarantee good predictions of the required values for points outside the training set. Just how acute this fidelity problem can be is graphically illustrated by Figure 6.4, which shows the approximation implemented by the network for the previous example over the range of inputs $[-2, 2]$. Clearly this approximation deviates in places substantially from the worst case 0.2 units observed during training.

This phenomenon is well known in the recursive identification literature as well, which has actually developed criteria, known as the *persistency of excitation* (PE) conditions, which can guarantee when convergence of the prediction error to zero will ensure asymptotically perfect fidelity in the identification model, at least for the linearly parameterized identifier structures discussed in Section 2.1. In the absence of persistency of excitation conditions for neural networks, researchers have again used heuristics to attempt to ensure high fidelity identification. A typical procedure is to train the identification model by injecting random inputs into the process, in the hope that the resulting training will be sufficiently exciting. Often, however, models are trained using physical data collected by passively observing a process over a period of time, and there may not be an easy way to obtain new data, or to directly inject stochastic perturbations so as to obtain more "interesting" input-output examples. Ideally, what is required is either a precise specification of the PE conditions for neural networks, permitting a designer to judge *a priori* the quality of a particular input-output record, or else the development of control algorithms which are not so critically dependent on a precise identification model. Both of these options will be explored in Section 5.

These convergence questions are moot, however, if the networks used in the control law are incapable of providing an approximation of sufficient quality. While the oft-cited network approximation theorems [20, 38, 25, 28] provide a justification for using these devices to learn the functions required for control applications, they provide only general *existence* statements, and do not specify the size, topology, or composition of a network required to achieve a given accuracy for a particular range of input signals. Once again, heuristics have arisen to fill the void: usually a particular network structure is assumed, for example, a one hidden-layer, sigmoidal network, and if, after a sufficiently long and (assumed) exciting training period, the results are unsatisfactory, the number of hidden layer nodes is increased, and the training is continued (or restarted). This procedure continues until the prediction errors are within the required tolerance.

In addition to the above generalization and stability issues, there is a more subtle problem underlying this strategy. A neural network with a finite number of components can provide an approximation with a particular accuracy only for a limited range of input signals; after the training phase this range is implicitly fixed by the size of the network and the weights it has learned. As noted at the end of Section 2.3, even the small errors incurred approximating the ideal control law over this range may cause large deviations between the true process output and the model response. If these deviations are such that the outputs of the process leave the set for which network accuracy can be assured, the difference between the network output and the required control will become quite large, potentially forcing the process outputs even further from the accuracy range, and instability may result. The following section examines in greater detail the possible instabilities which can occur when using networks to control dynamic systems.

3.3   Stability of Indirect Network Control Laws

Assuming a successful and persistently exciting identification phase, the resulting trained networks in the identifier, which accurately model the dynamics of the process, can either be used directly as the basis of nonlinear control laws, or used to train separate control networks as detailed in Section 2.3. There remain, however, still the questions raised at the end of that section about the ultimate effectiveness of these strategies. In particular, assuming no additional learning occurs when these networks are used to control the actual process, under what circumstances can guarantees of stability and convergence be given for the resulting closed loop behavior?

To approach this problem in its most general form, several intermediate questions must be answered on the basis of the available prior information. Given the process dynamics, does there exist a control law capable of achieving the desired objectives (i.e. tracking a model response) using bounded control inputs? Can the chosen neural network be used to accurately approximate this required control law? What is the effect of the inevitable small perturbations between the network approximation and the ideal control law?

The simplest situation to analyze is when the structure of a control law is known which permits direct utilization of components of the identification model. For example, the continuous time system $\dot{y}(t) = -f(y(t)) + u(t)$ can be forced to follow the output of a reference model, $y_m(t)$, by exactly canceling the nonlinearity f. Indeed, a linearizing control law of the form $u(t) = f(y(t)) + \dot{y}_m(t)) - k_D e(t)$. forces the tracking error $e(t) = y(t) - y_m(t)$ to converge exponentially to zero from any initial condition, since with this control law $\dot{e}(t) = -k_D e(t)$. As discussed in Section 2.3, this ideal control law can instead be approximated by using a network which has accurately identified f over a range of states.

However, by using a network approximation in place of f in the control law, the convergence properties of the ideal control may not be preserved. As an example, consider a linear process with $f(y) = cy$, and suppose an identification model of the process has been developed using the "network" approximation $\mathcal{N}(y, \hat{c}) = \hat{c}y$, such that $|\mathcal{N}(y, \hat{c}) - f(y)| \leq \epsilon_f$ everywhere on the set $A = [-1, 1]$. If the output of the model response is identically zero, i.e. it is desired to regulate the process about the origin, the closed loop output of the process evolves according to the differential equation

$$
\begin{aligned}
\dot{y}(t) &= \mathcal{N}(y(t), \hat{c}) - cy(t) + \dot{y}_d(t) - k_D(y(t) - y_d(t)) \\
&= (\tilde{c} - k_D)y(t),
\end{aligned}
$$

where $\tilde{c} = \hat{c} - c$. Since $\tilde{c}$ can be as large as ϵ_f and still provide the uniform bound given by the identification model, the feedback gain must at least satisfy $k_D > \epsilon_f$ to guarantee global stability of the closed loop system. The parameters of the controller required to ensure stable operation thus depend upon the approximation capability of the structure chosen to estimate the function f.

The situation is more complex if f is nonlinear, for example $f(y) = cy^2$. Using a "network" in the above regulation law of the form $\mathcal{N}(y, \hat{c}) = \hat{c}y^2$, trained to approximate f with the same range and uniform bounds as above, results in the closed loop differential equation $\dot{y}(t) = \tilde{c}y^2(t) - k_D y(t)$, and it is easy to show that the origin is a stable equilibrium point of this differential equation, whose basin of attraction is $(-\infty, k_D/\tilde{c})$ if $\tilde{c} > 0$, and $(k_D/\tilde{c}, \infty)$ if $\tilde{c} < 0$. The point $y^* = k_D/\tilde{c}$ is an unstable equilibrium, and for initial conditions $y(0) > k_D/\tilde{c}$, $\tilde{c} > 0$, or $y(0) < k_D/\tilde{c}$, $\tilde{c} < 0$, the process output will diverge to infinity. Thus, provided that k_D is large enough that the set A lies within the basin of attraction of the origin, and provided that the initial process condition is contained within this set, the closed loop output will converge to the origin, and the process output will remain within the set A for all $t \geq 0$.

When tracking a time varying trajectory, however, it may not be possible to ensure that the state remains within A, even if it initially begins inside this set. The tracking errors caused by the mismatch between c and $\hat{c}$ in the control law can easily carry the process state outside A, especially if the desired trajectory comes close the the boundary of this set. Suppose then that the set $A = [-k_D/|\tilde{c}|, k_D/|\tilde{c}|]$, for example, and that at some time, t_0, the output $y(t_0)$ lies a certain distance outside A as it attempts to follow a model trajectory which lies near its border. If the model trajectory suddenly becomes $y_m(t) = \dot{y}_m(t) = 0$ for $t \geq t_0$, from the above analysis the state may be "stranded" outside the basin of attraction of the origin, and, depending upon the sign of $\tilde{c}$ may diverge to infinity using the above control law. Even if continuity and differentiability constraints are placed on the model trajectory, it is still possible to devise a similar situation given a sufficiently large mistuning $\tilde{c}$.

To prevent this situation from occurring, one could either choose k_D to make A very large with respect to the desired trajectories commanded, so that the state has no chance to ever "escape" this set, or else introduce additional modifications to the *structure* of the control law designed to provide additional robustness to the perturbations caused by the network mistuning. In this situation, for example, if the linearizing control law is augmented with the term $-\bar{c}y^2(t)\,\mathrm{sat}(y(t))$, with $\bar{c} > |\tilde{c}|$, the regulation will again be successful, since the Lyapunov function $V(t) = y^2(t)$, has time derivative,

$$\dot{V}(t) = -k_D y^2(t) + y^3(t)(\tilde{c} - \bar{c}\,\mathrm{sat}(y(t)) \leq -k_D V(t),$$

and y converges exponentially to the origin from any initial condition.

In the two examples above, the "networks" used *did* have the capacity to develop globally accurate models of the process dynamics, and would produce stable, convergent closed-loop dynamics if $\tilde{c} = 0$, without further constraints on k_D or need for additional robustness. In actual network designs, however, even when the adjustable parameters assume their best possible values, there will still be discrepancies between the network output and the required linearizing control, and these discrepancies will become pronounced outside the set A, necessitating the more careful stability analysis above.

Even in these relatively simple applications of networks for control, one again observes the delicate interplay between the approximation capabilities of the neural network and designs which ensure stability and convergence. In addition to the modifications required for the training mechanism, a control law which employs a trained network must also be robust to the perturbations introduced by use of neural network approximations. Similar closed loop stability analyses for more complex nonlinear processes depend crucially upon the availability of appropriate tools from nonlinear control theory (some of which will be introduced below), but will not change the demonstrated importance of the *robustness* of the control and adaptation mechanisms.

It is hence crucial to be able to quantify the ability of a particular network architecture to approximate the functions required for identification and control applications, both for the success of the initial training phase and for the success of the ultimate control application. This analysis is so basic to the success of these designs that it must form an integral part of any neural network control system design. Accordingly, the next section examines in detail the network analysis and synthesis techniques before turning, in Section 5, to identification and control structures explicitly designed using stability theory.

4 Network Construction

The discussion in the preceding section demonstrates the intimate connection between the approximation abilities of the chosen network architecture, and the design of effective training and control algorithms. The ability to quantify the impact of the nonlinear perturbation terms d_f and d_g on the evolution of the tracking and prediction error, as well as their effects on the training process; the ability to ensure a specified level of fidelity in the trained identification model; and the ability to assure the stability of the controlled process, even without on-line learning, all require an explicit characterization of how well, and over what range of inputs, the network can learn the functions required to solve the control problem posed to it. This section examines in detail the network construction problem, and presents an analysis which permits the exact specification of networks capable of providing a specified degree of approximation to certain classes of functions. For a much more complete discussion of the ideas in this section, the reader is referred to [81, 83].

4.1 Networks and Approximation Theory

In all of the applications considered above, a neural network is being used as a method for carrying out nonlinear function approximation strategies; even when a recurrent network is used, the "interesting" part of the computation is the nonlinear processing accomplished by the feedforward components of the network, the remainder can be modeled and analyzed by placing the network in feedback with linear dynamic systems [65]. In this approximation theoretic sense, network computation is related to Fourier series, spline, and even wavelet expansions. In a network with one hidden layer, for example, selection of a set of input weights for the approximation is comparable to choosing a set of frequencies for a Fourier series expansion, or knots for a spline expansion, or dilation and translation parameters for a wavelet expansion. Similarly, choice of the output weights in a (one hidden layer) network is analogous to, in each of these three cases, determining the degree to which each resulting basis function contributes to the approximation of f.

Given this equivalence, what makes the approximations implemented by a neural network so interesting? At least part of the answer lies with the architectural simplicity of its construction: a neural network is particularly well suited to carrying out these approximations in an efficient manner. A few prototype building blocks, henceforth called *nodes*, are utilized in a massively parallel fashion,; each type of node performs exactly the same kind of computation, simultaneous with and independent of the other nodes in the network.

In this paper, each node is considered to be composed of two structures: the *input function*, $\varphi : \mathcal{R}^n \times \mathcal{R}^{p_1} \to \mathcal{R}$, and the *activation function*, $g : \mathcal{R} \to \mathcal{R}$. The input function is parameterized by a set of adjustable local weights, $\xi_k \in \mathcal{R}^{p_1}$, and serves to reduce the signals $\mathbf{x} \in \mathcal{R}^n$ incident on node k to a scalar "activation energy", $r_k = \varphi(\mathbf{x}, \xi_k)$; the scalar output of node k is then simply $g(r_k)$. The activation function is usually a nonlinear function of r_k, but this is not required. For example, a weighted summing junction fits easily into this model by taking $p_1 = n$, choosing the input function φ to be the inner product of its arguments, and using a linear activation function, $g(r) = r$. Figure 6.6 provides a more detailed view of a one hidden layer network illustrating these ideas.

As stated, however, this model is too general: any parameterized nonlinear mapping from $\mathcal{R}^n$ to $\mathcal{R}$ could be expressed by taking $g(r) = r$, and using an appropriate φ. To keep closer to the kind of local, weighted signal processing which has provided the inspirations for these models, and which indeed is one of their principal attractions, some sort of restriction is needed. Thus, φ is assumed to be such that, for each n and any $\xi \in \mathcal{R}^{p_1}$, $\varphi(\mathbf{x}, \xi)$ requires only a fixed number of multiplications and summations to compute. This allows, for example, biased linear combinations of the afferent signals to be used as the activation function, as is common in sigmoidal models, and also admits weighted and translated norms on $\mathcal{R}^n$, which arise as the activation functions in some regularized models. Multidimensional Fourier series expansions and radial wavelet expansions can thus also be mapped onto this parallel model, but choices in which $g \circ \varphi$ is a tensor product of univariate nonlinear mappings, such as tensor product spline expansions and multivariate polynomial expansions, are disallowed.

Since the general existence theorems have demonstrated the sufficiency of networks with one hidden layer to approximate continuous functions, using a variety of different types of hidden layer nodes, the remainder of this paper will restrict attention to these architectures. It is possible that additional hidden layers may enable a reduction in the number of required nodes [17], but much more work is required before a constructive theory of network approximation is available for these models. For one hidden layer architectures of the type depicted above, however, a number of construction techniques are immediately applicable, as will be shown below. The approximation implemented by such a network can be represented mathematically as;

$$\mathcal{N}(\mathbf{x}, \mathbf{c}) = \sum_{k \in \mathcal{K}} c_k \, g(\varphi(\mathbf{x}, \xi_k)),$$

where $\mathcal{K}$ is an arbitrary index set, and the size of the network is $K = \mathrm{Card}(\mathcal{K})$. It is usual practice to identify the parameters in this expansion relative to their position in the hidden layer, thus the ξ_k are called the *input weights*, while the c_k are the *output weights* (these latter could also be considered as input weights to a summing junction node which performs the indicated superposition of hidden layer outputs). Such networks are thus parameterized by their collection of input and output weights, so that $\mathbf{c} \in \mathcal{R}^p$ with $p = K(p_1 + 1)$.

It is the simplicity and potential power of this structure that holds such promise for engineering applications. The ability to compute approximations to nonlinear functions in a massively parallel fashion, using arrays of extremely simple processing elements, is very attractive for real time nonlinear control and identification applications. It must be stressed that these particular features are *independent* of the actual models φ and g used for each of the building blocks. Indeed, the above structure seeks to capture what is, possibly, an important component of the information processing mechanism underlying biological nervous systems. To the extent it succeeds at this goal, the actual neural structures which are observed by neurophysiology can be viewed as just specific instances of this paradigm, constrained by the capabilities of the biological building blocks available for their construction. In human engineered implementations the construction constraints are quite different, and practical applications are free to, and in fact should, exploit this flexibility and chose those models which are easiest to analyze and construct.

Thus, while this computing structure has been *inspired* by neurobiology, the connection with biological nervous systems ends at the abstract level described above. The chosen forms for φ and g need have nothing to do with the input-output behavior of "actual" neurons, but should rather be driven by engineering and mathematical convenience and the demands of the specific applications.

4.2   Sampling Theory and Networks

To quantitatively characterize the approximation capabilities of a particular network architecture, therefore, it is necessary to understand when the table of numbers, $\{\xi_k\}$, $\{c_k\}$, together with the functional form of g and φ serve to characterize a particular class of functions. That is, to what extent can more "interesting" functions be synthesized from the fundamental building blocks g and φ? The simplest such representation theorem available in approximation theory, and indeed one which has deep connections to more recent developments in the field, is well known to most engineers as the *sampling theorem* [52, 12]. That is, provided $f : \mathcal{R} \to \mathcal{R}$ is square-integrable and has a Fourier transform, $F(\nu)$, which vanishes outside the interval $[-\beta, \beta]$, then at each $x \in \mathcal{R}$,

$$f(x) = \sum_{k=-\infty}^{\infty} f(k\Delta) \frac{\sin(\Delta^{-1}\pi(x - k\Delta))}{\Delta^{-1}\pi(x - k\Delta)}$$

for any $\Delta^{-1} > 2\beta$. The function $\frac{\sin(\pi r)}{\pi r}$ (often referred to as the *sinc* function, and denoted $\mathrm{sinc}(r)$) is the impulse response of an ideal low-pass filter, corresponding to a "door" function in the frequency domain. The elegance of this discovery, that there are classes of functions which can be *exactly* reconstructed using just the *discrete* translates of a *single* function, so moved the mathematician E.T.Whittaker that he proclaimed sinc to be "the function of royal blood in the family of entire functions" [100].

This representation also has great significance in terms of the parallel, analog networks outlined in the previous section, for not only does it map exactly onto these structures, it does so in such a manner as to allow a precise interpretation of every component of the resulting network. That is, if the input function is chosen to be the dilated translation, $\varphi(x, \xi) = \Delta^{-1}(x - \xi)$, and the node activation function is taken as the sinc function, $g(r) = \mathrm{sinc}(r)$, then the network implements the expansion

$$\mathcal{N}(x, \mathbf{c}) = \sum_{k \in \mathcal{Z}} c_k \, \mathrm{sinc}(\Delta^{-1}(x - \xi_k)).$$

By identifying the input weights of this network with the sample points $\xi_k = k\Delta$, and the output weights with the samples of the function being approximated, $c_k = f(k\Delta)$, this structure exactly implements the univariate sampling theorem, and hence can exactly represent any function band-limited to the interval $[-\beta, \beta]$ with $2\beta < \Delta^{-1}$.

There are two drawbacks to using this insight as the basis for a constructive theory of networks. The first is that the canonical reconstructing function in higher dimensions is the tensor product of univariate sinc functions [70], and hence does not take the form $g(\varphi(\mathbf{x}, \boldsymbol{\xi}))$ required to map the representation onto the neural networks considered above. The second problem is well known to approximation theorists: while this cardinal series has an elegantly simple structure, it has very poor convergence properties. As noted by I.M. Schoenberg, who attempted to use this expansion to approximate aerodynamic functions, "its excessively slow rate of damping ... makes the classic cardinal series inadequate for numerical purposes" [90].

That is, to be practical, and indeed implementable on a network, only a finite number of neighboring sample points should be required to compute an approximation to $f(x)$ accurate to a chosen tolerance,

$$|f(x) - \sum_{|x - \Delta k| \leq \rho} f(\Delta k) g(\varphi(x, \Delta k))| \leq \epsilon_f.$$

In the classical sampling theorem, however, very large values of ρ, and hence a great number of terms in the series (or nodes in the network), are required to achieve reasonably small values of ϵ_f. The central problem lies with the decay properties of the reconstructing function $g(r) = \mathrm{sinc}(r)$; this function is not well localized, indeed it is not even absolutely integrable, and thus in most cases the series converges very slowly as a function of ρ. However, this analysis suggests a method for addressing the convergence and network equivalence problem. Perhaps it is possible to identify multi-dimensional low-pass filters which still conform to the network constraints in higher dimensions, and which are also well localized in space so as to assure rapid convergence of the corresponding cardinal series. These new filters may introduce some error into the exact representation otherwise provided by sampling theory, which holds only for strict low-pass filters, but since approximation errors will occur in any case by truncating the series expansion, these new error sources can be tolerated so long as they are small compared to the reduction of truncation error afforded by the increased spatial localization.

4.3 Space-Frequency Localization and Gaussian Networks

The localization of a filter, G, and of its impulse response, g, can be quantified using the expressions (assuming, for clarity, that the filter and impulse response are even functions)

$$w(g)^2 \triangleq \|g\|^{-2} \int_{\mathcal{R}} x^2 |g(x)|^2 dx$$

$$w(G)^2 \triangleq \|G\|^{-2} \int_{\mathcal{R}} \nu^2 |G(\nu)|^2 d\nu,$$

and it is assumed that each of the above L^2 norms are well-defined. To be an effective low-pass filter $w(G)$ must be small as possible, while to ensure rapid series convergence $w(g)$ must also be minimized. However, these two quantities are linked by the uncertainty relation [18]

$$w(g)w(G) \geq \frac{1}{4\pi}.$$

It is thus impossible to independently reduce both quantities to arbitrarily small levels: good spatial concentration implies a slowly decaying spectrum, and conversely a concentrated spectrum implies a slowly decaying impulse response.

Among all possible functions for which these localization quantities are well defined, the lower bound in the uncertainty relation is attained if and only if $G(\nu) = e^{-\pi\nu^2/\sigma^2}$, so that $g(x) = \sigma e^{-\pi\sigma^2 x^2}$ [18, 26]. That is, of all low-pass filters, a Gaussian achieves the best possible space-frequency localization tradeoff. Taking tensor products, similar arguments can be made in multiple dimensions, in which case g is the radial Gaussian [21]. These are quite attractive results, especially given the widespread popularity of the radial Gaussian function in neural network models [11, 72].

A rigorous analysis of the effect of replacing the sinc function with the Gaussian in the sampling theorem shows that the additional sources of error introduced are quite small indeed, and that the resulting series has very favorable convergence properties. Taking the input function as the Euclidean norm $r = \varphi(\mathbf{x}, \boldsymbol{\xi}) = \|\mathbf{x} - \boldsymbol{\xi}\|$, and the node function as the Gaussian $g(r) = \exp(-\pi\sigma^2 r^2)$, this expansion can be mapped onto the one hidden layer network,

$$\mathcal{N}(\mathbf{x}, c) = \sum_{\mathrm{dist}(A, \mathbf{k}\Delta) \leq \rho} c_{\mathbf{k}} \exp(-\pi\sigma^2 \|\mathbf{x} - \mathbf{k}\Delta\|^2).$$

where $\mathbf{k} \in \mathcal{Z}^n$ and here $\mathrm{dist}(A, \mathbf{k}\Delta) \triangleq \inf_{\mathbf{z} \in A} \|\mathbf{z} - \mathbf{k}\Delta\|_\infty$. The interpretation of each component of these Gaussian networks is exactly the same as in the sampling theorem. The input weights encode a regular mesh of sampling points, and the output weights are the samples of the continuous function . $c_{\mathbf{k}} = c(\mathbf{k}\Delta)$, where the function c is related to f through a simple convolution.

Thus, provided that the function is bandlimited, given any prespecified tolerance ϵ_f and any compact set A, one can choose the construction parameters $\Delta, \sigma, \rho, c_{\mathbf{k}}$ such that a Gaussian network approximation satisfies $|\mathcal{N}(\mathbf{x}, c) - f(\mathbf{x})| < \epsilon_f$, $\forall \mathbf{x} \in A$. Most importantly, this analysis gives precise expressions relating the magnitude of ϵ_f and the size of the set A to the function class to which f belongs, expressed in terms of its bandwidth and an upper bound on its total energy, and the network construction parameters (Δ, σ, ρ). Exact expressions for the required output weights, and formulae relating the remaining parameter choices to the degree and extent of the network's accuracy, can be found in [81, 83].

Since very few functions are ideally bandlimited, a more precise and useful statement of this result is: assuming that f can be represented on a compact set A to an arbitrary uniform tolerance, ϵ_1, by a function with an absolutely integrable Fourier transform, then it is possible to choose a set of Gaussian network parameters so that the resulting network can approximate f on A with any uniform accuracy $\epsilon_f > \epsilon_1$.

As an example, consider the function used to generate the training data for the identification problem examined in Section 3.1, $f(x) = \sin(2\pi x)/(\pi x)$. This function is ideally bandlimited to the interval $[-1, 1]$, and hence can be well approximated by a Gaussian network. The formulae in [83] suggest that with the choices $\sigma^2 = \pi$, $\Delta = 4/9$, and $\rho = 5\Delta$ the magnitude of the error approximating f should be no worse than 0.26 everywhere on the set $A = [-2, 2]$. Indeed, by explicitly computing the required output weights, the error between the Gaussian expansion and the actual function can be computed and is plotted in Figure 6.7, from which the predicted bound on ϵ_f is seen to be correct. Note especially, however, the rapid decrease in accuracy outside of the set A; this is the problem, discussed at the end of Section 2.3, encountered using a finite sized network. As demonstrated above, and further explored below, special care must be taken in identification and control applications to ensure that the network inputs remain within the set for which accuracy can be assured.

This example was deliberately constructed to have a large error on A, primarily so that the divergence observed in Section 3.1 would be dramatic. Small increases in the oversampling bring a great improvement in the uniform accuracy the network can provide. For example, reducing the mesh slightly to $\Delta = 4/10$ improves the uniform accuracy on A to .12, a factor of two improvement, without requiring an increase in the size of the network.

But do these Gaussian networks really achieve better accuracy with fewer components than classical sampling theory would provide? To answer this question, Figure 6.8 shows the results of a 25 node approximation to the function $f(x - .33)$ using both sinc and Gaussian nodes a zero-centered grid of mesh $\Delta = 0.25$. The Gaussian approximation is approximately 60 times more accurate on the interval $[-2, 2]$, although it degrades much more rapidly outside this set. Notice that the translation does not affect the frequency content of the function, and was performed since otherwise f is exactly a multiple of one of the basis functions in a sampling theory expansion, which would hardly provide a fair comparison.

Interestingly, sampling theoretic arguments can also be used to predict the ability of these Gaussian networks to approximate polynomials. In a distributional sense, polynomials are bandlimited since their transforms are supported only at the origin. This insight can be used to show that polynomials can be exactly reconstructed using the regular translates of a sufficiently localized function, g, whose Fourier transform, G, and an appropriate number of its derivatives vanish at the reciprocal lattice points $\Delta^{-1}\mathbf{k}$, save at the origin where the derivatives must also vanish, but $G(0) = \Delta^n$. In fact, these are precisely the Strang-Fix conditions [96] for polynomial reconstruction, well known in the analysis of finite element methods, and recently used to great advantage to characterize the approximation abilities of radial basis functions [75].

Just as the Gaussian is only, strictly speaking, an approximate low-pass filter, so too it only approximately satisfies the Strang-Fix conditions. However, the errors introduced by its nonzero derivatives at reciprocal lattice points away from the origin can be made as small as desired by appropriate mesh size. The effects of its nonzero derivatives at the origin can be offset by using in the expansion the samples of a slightly different polynomial than that being approximated, analogous to the convolution which relates c to f in the bandlimited case [81]. This analysis thus allows specification of a construction procedure for polynomial approximation using Gaussian networks, employing exactly the same synthesis methods as the sampling theoretic construction.

5   Stable Identification and Control with Networks

One problem with the training algorithm used in Section 3.1 is that the cost function being optimized does not penalize the magnitude of the weights used. A gradient descent method is thus free to follow a potential minimizing solution along a direction in which the weights increase effectively without bound in order to try to squeeze the last bit of error out of the estimates to the values in a training set. A stability theoretic approach to the learning problem creates a positive definite cost function which contains *all* the time varying parameters of the identification or control problem. The adaptation mechanism is then designed analytically to ensure that this cost function is non-increasing at every instant in time.

To extend stability methods to identification and control algorithms which use neural network models, one possible approach is to try to develop stable adaptation algorithms which directly address the nonlinear effect of parameter mistuning represented by the term d_g. While some preliminary work in this direction has begun [8], a complete solution which could be applied to the present application is still lacking. A second alternative is to treat the terms d_f and d_g in (3.5) above as *disturbances* to existing linearly parameterized adaptation algorithms, whose stability properties have been exhaustively explored, and to call upon *robust* adaptation techniques to preserve stability in the face of these perturbations. This latter is the most promising approach, and has been extensively explored in algorithms developed in [82, 83, 84, 86, 13, 74, 80, 98], several of which will be reviewed below.

In order to implement each of these robust algorithms, however, and to predict the ultimate capabilities of neural identification and control structures, it is necessary to quantify the magnitude of these disturbances. Fortunately, the analysis and synthesis techniques for Gaussian networks summarized above provide the tools required to evaluate these terms for a large class of functions. Moreover, the design procedure exactly specifies the input weights, variances, and number of nodes required to approximate with a chosen accuracy any function from a particular class, for example, any function bandlimited to $[-\beta, \beta]^n$ whose transform is bounded in magnitude by F_{max}. The assumed function class (measured by β and F_{max}) and the target accuracy (measured by ϵ_f and A) thus determine *a priori* a correct set of input weights and variances; only the correct set of output weights must be learned to develop an accurate representation of a particular function in this class.

This prior analysis thus renders the network estimate *linear* in the adjustable output weights, making d_g identically zero. Provided the actual function underlying the observed data lies in the assumed function class, the magnitude of the term d_f is then by design bounded by ϵ_f for every network input in the set A. Compared with the network expansions discussed in Section 4.1, the input weights $\xi_k = k\Delta$ are considered a fixed part of this architecture, and only the output weights are adjusted, s o that $\hat{c} \in \mathcal{R}^p$ where here $p = \text{Card}(\{k \mid \text{dist}(A, k\Delta) \le \rho\})$.

With the magnitude of the disturbances in the resulting linearly parameterized expansion thus quantified, robust adaptation theory provides a number of techniques to modify classical adaptation algorithms to prevent the kind of (possibly unbounded) parametric drift observed in Section 3.1. Not surprisingly, some of these strongly resemble heuristics independently discovered by neural network researchers. For example, weight decay methods, which add a term to the adaptation law biasing the parameter estimates toward zero, thus preventing unbounded growth, are equivalent to the *sigma modification* technique in the robust adaptation literature [41]. Similarly, explicitly limiting parameter magnitudes, for example limiting $\|\hat{c}\| \le C_2$ or $|\hat{c}_k| \le C_\infty$, is equivalent to a *parameter projection* robustness technique [22, 40]; such limits are often implicitly imposed by hardware implementations of neural networks. Note that the Gaussian network construction also provides the necessary bounds on the magnitudes of the true parameters $\|c\|$ or $|c_k|$ needed to implement this latter modification.

A more common heuristic, demonstrated at the end of Section 3.1, involves stopping adaptation when the measured errors fall below a particular threshold. This is equivalent to a *deadzone* robustness modification [22, 71], and much of this section details applications of this technique, starting first with the stable identification algorithms developed in Section 5.1. However, as discussed in Section 3.2, even a stable and convergent identifier will not necessarily produce high fidelity models of the plant dynamics unless the persistency of excitation conditions are satisfied. Section 5.2 thus shows how the Gaussian network constructions of the previous section also permit precise characterization of the sequence of network inputs necessary to satisfy the persistency conditions.

As Section 3.3 has shown, regardless of how good a network identification model is, it is still possible for a control law developed from this model to drive the actual process unstable unless careful attention is given to the interaction between the structure of the control law and the approximation capabilities of the network. Moreover, the assumption of a persistently exciting identification phase is somewhat constraining and may be difficult to ensure in practice. These considerations lead to the detailed exploration, in Section 5.3, of on-line, direct adaptive control algorithms which permit learning and control to proceed simultaneously, and which ensure the global stability of the adaptive system and the asymptotic convergence of the model tracking error. While current state of the art in nonlinear control system design and the complexity of the stability analysis somewhat limits the class of processes for which these structures can be reliably developed, very satisfying convergence results are available for important classes of nonlinear systems.

5.1   Stable Identification

The design of a stable identification model using Gaussian networks thus requires a slightly different set of prior knowledge than classical designs. The functions driving the process dynamics should belong to one of the function classes considered in the previous section, and bounds must be known for the parameters of the function class to which the process dynamics are assumed to belong. Furthermore, prior bounds are required for the magnitude of control signals input to the process, and on the magnitude of the resulting process states and outputs. These bounds on the input-output behavior of the process define the boundaries of the set A on which the network approximation will be required; together with the function class information, they completely specify the structure of a Gaussian network required in the identification model. Only the output weights must be recursively learned during the training procedure, and this section presents the identifier structures and adaptation mechanisms which guarantee that this learning will be both stable and convergent.

For example, if the process can be adequately modeled by $y[t] = f(\mathbf{x}[t])$ where $\mathbf{x}[t] \in A \subset \mathcal{R}^{N+M}$ for all $t \geq 0$, then a series-parallel identification structure

$$\hat{y}[t] = \sum_{\text{dist}(A,\mathbf{k}\Delta)\leq\rho} \hat{c}_{\mathbf{k}}[t-1]\exp(-\pi\sigma^2\|\mathbf{x}[t]-\mathbf{k}\Delta\|^2),$$

can be used to predict future outputs of the process. Provided the function f belongs to one of the function classes discussed in Section 4, the variance, σ, mesh size, Δ, truncation radius, ρ, can be chosen such that the Gaussian network can represent any function in the assumed function class to a target uniform accuracy, ϵ_f everywhere on the set A.

To learn the specific function from the available data, the output weights of this network are adjusted according to the gradient adaptation law,

$$\hat{c}_{\mathbf{k}}[t] = \hat{c}_{\mathbf{k}}[t-1] - k_a e_\Delta[t]\exp(-\pi\sigma^2\|\mathbf{x}[t]-\mathbf{k}\Delta\|^2).$$

where the adaptation signal $e_\Delta[t]$ incorporates a deadzone of size Φ into the prediction error $e[t] = \hat{y}[t] - y[t]$, i.e.

$$e_\Delta[t] = e[t] - \Phi\,\text{sat}(e[t]/\Phi),$$

and sat is the saturation function ($\text{sat}(y) = y$ if $|y| < 1$, and $\text{sat}(y) = \text{sign}(y)$ otherwise). If the adaptation gain is chosen to satisfy

$$0 < \delta = 2 - k_a \sup_{\mathbf{x}\in A} \sum_{\text{dist}(A,\mathbf{k}\Delta)\leq\rho} \exp(-2\pi\sigma^2\|\mathbf{x}-\mathbf{k}\Delta\|^2), \qquad (5.6)$$

and the deadzone is chosen so that $\Phi \geq \epsilon_f$, then each of the parameter estimates $\hat{c}_{\mathbf{k}}$ will remain bounded, and further, the prediction error will converge asymptotically to $|e[t]| \leq \Phi$ as $t \to \infty$. To prove these claims one proceeds by demonstrating that a quadratic function of the prediction error and the output weight mistuning is nonincreasing during the adaptation process, and further that the signal e_Δ tends asymptotically toward zero. The specific details can be found in [83].

These results exactly quantify the relations between the approximation capabilities of the network and the adaptation parameters required for stable, convergent operation. The allowable adaptation gains are restricted by (5.6), while the required deadzone is determined by the bound ϵ_f. The reason for the success of the modified supervised learning example at the end of Section 3.1 are now clear: for the given network the adaptation gain $k_a = 0.2$ is easily seen to satisfy the above constraint, and from the construction example in Section 4.3, the deadzone size $\Phi = 0.26$ is larger than the ϵ_f provided by this Gaussian network.

Consider now the continuous time process model (2.2) introduced in Section 2.1, with measured states $\mathbf{z}(t) \in \mathcal{R}^n$. Again assuming the functions f_j and h belong to classes which can be approximated by Gaussian networks, and that information about the respective function classes is available to allow the proper choice of construction parameters, a series-parallel identifier can again be used

$$\dot{\hat{z}}_j(t) = -k_D e_j(t) + \sum_{\text{dist}(A, \mathbf{k}\Delta_f) \leq \rho_f} \hat{c}^f_{j,\mathbf{k}}(t) \exp(-\pi\sigma_f^2 \|\mathbf{x}(t) - \mathbf{k}\Delta_f\|^2)$$

$$\hat{y}(t) = \sum_{\text{dist}(A, \mathbf{k}\Delta_h) \leq \rho_h} \hat{c}^h_{\mathbf{k}}(t) \exp(-\pi\sigma_h^2 \|\mathbf{x}(t) - \mathbf{k}\Delta_h\|^2),$$

for $j = 1 \ldots n$, and recall that here $\mathbf{x}(t)^T = [\mathbf{z}^T(t) \ u(t)] \in \mathcal{R}^{n+1}$. Once again, the set A, which together with the function class information determines the size of the require networks, is determined by the assumed bounds on the control input and resulting state magnitudes.

Note that this identification structure effectively attempts to identify each of the nonlinear functions f_j and h simultaneously; $n + 1$ different networks can be used for this task, allowing different choices for the network design parameters, or, if $\mathbf{f}$ and h belong to similar function classes, the same network with $n + 1$ outputs can be used. The structure above uses a single network with n outputs to estimate each component of the dynamics $\mathbf{f}$, and a different network to estimate the output mapping h.

To stably train the networks in this example, the adaptation laws

$$\dot{\hat{c}}^f_{j,\mathbf{k}}(t) = -k_1 \ e_{\Delta j}(t) \ \exp(-\pi\sigma_f^2 \|\mathbf{x}(t) - \mathbf{k}\Delta_f\|^2)$$

$$\dot{\hat{c}}^h_{\mathbf{k}}(t) = -k_2 \ s_\Delta(t) \ \exp(-\pi\sigma_h^2 \|\mathbf{x}(t) - \mathbf{k}\Delta_h\|^2),$$

can be used, where $e_{\Delta j}(t)$ is formed by incorporating a deadzone into each component of the state estimation error $\mathbf{e}(t) = \hat{\mathbf{z}}(t) - \mathbf{z}(t)$, so that

$$e_{\Delta j}(t) = e_j(t) - \Phi_f \ \text{sat}(e_j(t)/\Phi_f).$$

A deadzone can similarly be incorporated into the output estimation error $s(t) = \hat{y}(t) - y(t)$, resulting in a signal $s_\Delta(t)$ defined similarly to the $e_{\Delta j}(t)$, using instead a deadzone of width Φ_h.

If the deadzone Φ_f is chosen such that $k_D\Phi_f \geq \max_j \epsilon_{f,j}$, and $\Phi_h \geq \epsilon_h$ then the states of the identifier and the output weights of the network remain bounded, and further the state and output estimation errors asymptotically converge to their respective deadzones. For proof, one can consider the quadratic function of the estimation errors and weight mistunings

$$V(t) = \frac{1}{2}\mathbf{e}_\Delta^T(t)\mathbf{e}_\Delta(t) + \frac{1}{2k_1} \sum_{\text{dist}(A, \mathbf{k}\Delta_f) \leq \rho_f} \sum_{j=1}^{n} (\tilde{c}^f_{j,\mathbf{k}})^2 + \frac{1}{2k_2} \sum_{\text{dist}(A, \mathbf{k}\Delta_h) \leq \rho_h} (\tilde{c}^h_{\mathbf{k}})^2$$

where the outer sums are over the lattice indices corresponding to centers of the nodes in each of the networks. To show that $V(t)$ has a nonpositive time derivative and that $e_\Delta(t)$ and $s(t)$ converge to zero, the analysis is identical to that in [86], with the addition of the nonpositive terms $-s_\Delta^2(t) - |s_\Delta(t)|(\Phi_h - \epsilon_h)$ to the inequality for $\dot{V}(t)$.

Once again, this analysis serves to quantify the adaptation parameters required for stable, convergent operation. Unlike the discrete time case, there are no limits on the allowable adaptation gains (which can, in fact, be different for each output weight), while the deadzones required for the estimates of the functions f_j are actually smaller than those required in the previous section, since the filtering provided by each error equation helps to attenuate the effects of the network approximation errors.

5.2 Persistency of Excitation

With a perturbed linearly parameterized error system, it is well known [30, 62] that, if the adaptive system is sufficiently excited, there is no need for robust modification such as a deadzone in the face of uniformly bounded disturbances; the excitation is itself a form of robustness which prevents the parameters from drifting too far from their correct values. Moreover if this parameterization is globally exact, persistently exciting signals in the adaptation mechanism guarantee asymptotically perfect identification of the components of the process dynamics. This section identifies particular sequences of network inputs which ensure satisfaction of the PE conditions in the above adaptive algorithms, and bounds the asymptotic fidelity of the identification models which result when these conditions are satisfied.

The conditions for a vector of regressors, $\mathbf{g}(\mathbf{x}[t])$ used in an adaptation algorithm to be PE are, in discrete time,

$$P[t] := \sum_{\tau=t}^{t+T-1} \mathbf{g}(\mathbf{x}[\tau]) \, \mathbf{g}(\mathbf{x}[\tau])^T \geq \alpha \mathbf{I},$$

and in continuous time,

$$P(t) := \int_t^{t+T} \mathbf{g}(\mathbf{x}(\tau)) \mathbf{g}(\mathbf{x}(\tau))^T d\tau \geq \alpha \mathbf{I}$$

for some $T > 0, \alpha > 0$ and all $t > t_0$; see e.g., [62, 30] for a more complete discussion. Let $\mathcal{K} = \{\mathbf{k} \in \mathcal{Z}^n \,|\, \mathrm{dist}(A, \mathbf{k}\Delta) \leq \rho\}$, let K be the size of the network, so that $K = \mathrm{Card}(\mathcal{K})$, and define $\gamma : \{1, \ldots, K\} \to \mathcal{K}$ to be a one-to-one mapping of the integers from 1 to K to the indices contained in $\mathcal{K}$. With this notation, the relevant signals for a persistency analysis of the preceding algorithms are the outputs of the Gaussian nodes, and hence $g_i(\mathbf{x}) = \exp(-\pi\sigma^2\|\mathbf{x} - \gamma(i)\Delta\|^2)$, $i = 1 \ldots K$.

The Gaussian satisfies the hypotheses of Micchelli's theorem [55] on completely monotone functions, and thus the matrix, $\mathcal{M}$, whose (i, j)th entry is $\exp(-\pi\sigma^2\Delta^2\|\gamma(i)-\gamma(j)\|^2)$, has a determinant bounded away from zero. Suppose then that the input sequence steps through the centers of the nodes in the network, that is $\mathbf{x}[t] = \gamma(1+(t \bmod K))\Delta$, for all $t \geq 0$. Under these conditions [83] shows that with $T = K$, $P[t] = (\mathcal{M}^T\mathcal{M}) \geq \alpha I$, $\alpha > 0$ for all $t \geq 0$. Thus, more generally, if some subset of the network inputs correspond to the centers of each Gaussian node at least once during every time interval of length $T \geq K$, the outputs of the hidden layer nodes in the network are persistently exciting.

This result is easily extended to the continuous time case [87]. In this setting sufficient conditions for PE are that there exists a $T > 0$ such that, for every $t > 0$, there exist K constants $\alpha_i \geq \delta > 0$ and a collection of $K + 1$ points $\{t_i\}$ in each interval $[t, t + T]$ with $t_0 = t$ and $t_K = t + T$, so that the following equalities hold for each $i = 1 \ldots K$

$$\int_{t_{i-1}}^{t_i} \mathbf{g}(\mathbf{x}(\tau))\mathbf{g}(\mathbf{x}(\tau))^T\, d\tau = \alpha_i\, \mathbf{g}(\gamma(i))\, \mathbf{g}(\gamma(i))^T,$$

Under these conditions $P(t) \geq \delta\mathcal{M}^T\mathcal{M} \geq \alpha I$ for some $\alpha > 0$ and all $t \geq 0$ [86]. The interpretation of these conditions in terms of the required network inputs is the same as in the discrete time case, but in a time averaged fashion. Certainly they will be satisfied if the network inputs dwell at each center for a nonzero length of time during every interval of length T, but in practice this situation will rarely occur, and the above conditions give a more general formulation.

When the persistency conditions are satisfied it can be shown that, using the above adaptation algorithms (perhaps omitting the deadzone, since persistency also ensures the required robustness), the parameter estimates will converge to a ball about their tuned values whose size depends continuously on the magnitude of the disturbance ϵ_f, i.e. $\lim_{t\to\infty} \|\tilde{c}(t)\| \leq \tilde{c}_\infty(\epsilon_f)$ in both the continuous and discrete time identification models [81, 80]. Thus, asymptotically

$$|\mathcal{N}(\mathbf{x}, \hat{c}[t]) - f(\mathbf{x})| \leq \left| \sum_{\text{dist}(A,\mathbf{k}\Delta)<\rho} \tilde{c}_{\mathbf{k}}[t] \exp(-\pi\sigma^2\|\mathbf{x} - \mathbf{k}\Delta\|^2) \right| + |d_f(\mathbf{x})|$$
$$\leq \kappa\tilde{c}_\infty(\epsilon_f) + \epsilon_f$$

for any input $\mathbf{x}$ in the set A, where the constant κ can be easily evaluated from the network architecture. Hence, if the training is persistently exciting the resulting identifier will develop a model of the process dynamics whose fidelity on A is asymptotically limited only by the magnitude of ϵ_f.

5.3   Stable Online, Direct Adaptive Control

The uses of networks in the control laws discussed in Section 2.3 assume that good identification models have been obtained. Despite the explicit characterization given above of the conditions which ensure high fidelity identification, it may not always be possible to maintain the prerequisite persistency conditions during the identification phase. An indirect control law may thus be forced to use approximations which may be quite poor in certain parts of the state space. It may be possible to compensate for the use of these low quality approximations by appropriate modifications to the control law, however this could result in the generation of undesirably large control signals (e.g, large k_D and $\bar{c}$ in the examples of Section 3.3). Moreover, the separation of the algorithms discussed thus far into explicit identification and control stages seems artificial; ideally learning and control should occur simultaneously, allowing low quality approximations to be improved as needed. In contrast to the off-line, indirect algorithms, one would thus prefer on-line algorithms which directly estimate the required control signal and are less sensitive to the persistency requirements; these are the subject of the current section.

Of course, it is possible to continue training the identifier and controller networks in the indirect algorithms even after the networks have been brought on-line. Indeed, provided that the off-line training phase has produced network estimates sufficiently close to the actual functions required, and that the network parameters are changed sufficiently slowly, this strategy may result in a stable, closed loop adaptive system capable of providing quite satisfactory results [65] . Once again, the difficulty lies in quantifying exactly how accurate the models developed off-line must be, and just how slowly adaptation must proceed, in order to guarantee stability.

Carrying out learning and control simultaneously introduces a level of complexity into the stability analysis which goes far beyond that used to design the identifier adaptation laws above. Whereas in an identification model the network merely passively observes the (assumed stable) process and adjusts its parameters accordingly, in an on-line, direct adaptive control setting the network interacts immediately with the process it observes, and must react quickly enough to stabilize even a potentially unstable process. The adaptation mechanism then couples the evolution of the time-varying parameters in the network to the dynamics of the system being controlled, producing a high-dimensional, nonlinear, closed-loop system, whose states are those of the original process together with the adjustable parameters of the network.

Generalities aside, the ability to carry the analysis further depends crucially upon the availability of suitable direct adaptive control structures for nonlinear systems. The class of nonlinear systems for which such techniques are known has grown substantially in the past few years [47, 2], but still represents only a limited subset of possible nonlinear process dynamics. Nonetheless, these special cases have wide applicability to a variety of practical applications. To clearly illustrate the central ideas, the remainder of this section reviews a design technique developed in [83] for nonlinear systems in the *canonical form* [42, 94]:

$$y^{(n)}(t) + f(\mathbf{x}(t)) = b(\mathbf{x}(t))u(t) \tag{5.7}$$

where b is assumed globally bounded away from zero. The control objective is to have the plant output and all of its derivatives, $\mathbf{x}^T = [y, \dot{y}, \ldots, y^{(n-1)}] \in \mathcal{R}^n$, assumed to be available for measurement, asymptotically follow a specified model trajectory, $\mathbf{x}_m^T = [y_m, \dot{y}_m, \ldots, y_m^{(n-1)}]$. (Note that this definition of $\mathbf{x}(t)$ omits the input $u(t)$ compared with the definition used for the identification models). Defining a tracking error vector, $\tilde{\mathbf{x}}(t) = \mathbf{x}(t) - \mathbf{x}_m(t)$, the problem is thus to design a control law $u(t)$ which ensures that $\tilde{\mathbf{x}}(t) \to 0$ as $t \to \infty$.

For nonlinear processes governed by these dynamics, the so-called *feedback linearizing* control law

$$u(t) = b^{-1}(\mathbf{x}(t))[y_m^{(n)}(t) + f(\mathbf{x}(t)) - \boldsymbol{\alpha}^T \tilde{\mathbf{x}}(t)]$$

would accomplish the control objective if the functions f and b were known exactly [94, 42], since, with proper choice of feedback gains, α_i, this control law causes the tracking error to evolve as the zero-input response of a stable, linear time-invariant system

$$\tilde{y}^{(n)}(t) + \alpha_n \tilde{y}^{(n-1)}(t) + \cdots + \alpha_1 \tilde{y}(t) = 0.$$

If the functions f and b are continuous, they can be estimated on compact subsets of the state space by neural networks whose inputs are the measured values of the states of the process. These estimates can then used immediately in place of the actual functions in the feedback linearizing control law; this is known as the *certainty equivalence* principle [5]. The actual control signal input to the process is then

$$\hat{u}(t) = \frac{1}{\mathcal{N}_b(\mathbf{x}(t), \hat{\mathbf{c}}^b(t))} \left[y_m^{(n)}(t) - \boldsymbol{\alpha}^T \tilde{\mathbf{x}}(t) + \mathcal{N}_f(\mathbf{x}(t), \hat{\mathbf{c}}^f(t)) \right].$$

For the resulting closed loop system, a convenient measure of the model tracking error is

$$s(t) = (\frac{d}{dt} + \lambda)^{n-1} \tilde{y}(t) = \boldsymbol{\lambda}^T \tilde{\mathbf{x}}(t)$$

for some $\lambda > 0$ so that $\lambda_i = \binom{n-1}{i-1} \lambda^{n-i}$. Choosing the feedback gains as $\alpha_i = k_D \lambda_i + \lambda_{i-1}$ with $\lambda_0 \triangleq 0$ and $k_D > 0$, this measure evolves as

$$\dot{s}(t) = -k_D s(t) + \left[b(\mathbf{x}(t)) - \mathcal{N}_b(\mathbf{x}(t), \hat{\mathbf{c}}^b(t)) \right] \hat{u}(t) + \left[\mathcal{N}_f(\mathbf{x}(t), \hat{\mathbf{c}}^f(t)) - f(\mathbf{x}(t)) \right].$$

A first order relation of this type (or, more generally, a *strictly positive real* condition; see e.g. [62, 94]) is crucial to the design of a stable adaptive mechanism, for it ensures that there is not too much lag in the impact of the mistuning in the network parameters on the tracking error measure. This property was also exploited in the continuous time identification algorithm in Section 5.1, since each of the $e_i(t)$ evolved in this first order fashion. In the identification setting this was easy to arrange, since the "obvious" error measure, i.e. comparing each identifier state with the corresponding measured state, immediately produced a stable, first order differential equation relating the measured error to the function estimation error. In the direct adaptive control setting, however, the "obvious" metric with which to train the network would be the discrepancy between the ideal linearizing control signal and that generated using the network estimates; since this is not a measurable quantity, more care is required in constructing a suitable error measure from the available measurements.

Most importantly, naively using the model tracking error $e(t) = y(t) - y_m(t)$ as the network training signal will generally *not* produce stable, closed-loop adaptive algorithms, since this signal will usually not satisfy the required first order relation with the estimation errors. The measure $s(t)$, however, both satisfies the required relation, and provides a suitable indication of how well the controller is satisfying the tracking requirements: driving $s(t)$ to zero asymptotically ensures that $e(t)$ also approaches zero [93].

Assuming a properly chosen Gaussian network architecture, permitting a linear expansion of the functions f and b in Gaussian basis functions to uniform accuracy ϵ_f and ϵ_b respectively on a chosen subset of the state space, $A \subset \mathcal{R}^n$, the evolution of the error metric can be further expanded as

$$
\begin{aligned}
\dot{s}(t) &= -k_D s(t) + d_f(\mathbf{x}(t)) - d_b(\mathbf{x}(t))\hat{u}(t) \\
&\quad + \sum_{\text{dist}(A,\mathbf{k}\Delta_f)\le\rho_f} \tilde{c}^f_{j,\mathbf{k}}(t) \exp(-\pi\sigma_f^2\|\mathbf{x}(t) - \mathbf{k}\Delta_f\|^2) \\
&\quad -\hat{u}(t) \sum_{\text{dist}(A,\mathbf{k}\Delta_b)\le\rho_b} \tilde{c}^b_{j,\mathbf{k}}(t) \exp(-\pi\sigma_f^2\|\mathbf{x}(t) - \mathbf{k}\Delta_b\|^2))
\end{aligned}
$$

where $\tilde{c}^f_{\mathbf{k}}(t) = \hat{c}^f_{\mathbf{k}}(t) - c^f_{\mathbf{k}}$ and similarly for $\tilde{c}^b_{\mathbf{k}}(t)$. The disturbances are given by $d_f(\mathbf{x}) = \mathcal{N}_f(\mathbf{x}, \mathbf{c}^f) - f(\mathbf{x})$ and similarly for d_b so that $|d_f(\mathbf{x})| \le \epsilon_f$ and $|d_b(\mathbf{x})| \le \epsilon_b$ for all $\mathbf{x} \in A$.

The effects of these disturbances while the state is within the set A can be accommodated by the adaptation mechanism using the robust adaptive techniques discussed above, and would produce only asymptotically small tracking errors when the networks are correctly tuned. No component of the proposed control law, however, is capable of offsetting the effects of these disturbances outside the set A, and since the amount of initial parameter mistuning is not known, it is not possible to make an *a priori* choice of A which the state will never leave. The degradation of the ability of the networks to approximate the required functions thus requires a structural modification to the linearizing control law in these regions.

Given the first order relation which exists between $s(t)$ and the uncertainty on f, a nonlinear control methodology known as *sliding control* [94, 99] can be also used to "overpower" large disturbances and drive $s(t)$ toward zero, given only crude upper bounds on the magnitudes of d_f and d_b. The sliding control component usually employs the signum function, as in the example in Section 3.3, however it is possible to smooth out the resulting discontinuity by instead using the saturating control input

$$u_{sl}(t) = - \operatorname{sat}(s(t)/\Phi)\underline{b}(\mathbf{x}(t))^{-1} \left[\overline{d}_b(\mathbf{x}(t))|\hat{u}(t)| + \overline{d}_f(\mathbf{x}(t))\right]$$

where $0 < \underline{b}(\mathbf{x}) \le b(\mathbf{x})$, $\overline{d}_f(\mathbf{x}) \ge |d_f(\mathbf{x})|$, and $\overline{d}_b(\mathbf{x}) \ge |d_b(\mathbf{x})|$. The constant Φ describes the width of a *boundary layer*, which can be taken to be the same as the deadzone width in the adaptation mechanism.

This sliding component can be added to the adaptive linearizing control law as needed to ensure stability. However, the definitions of the required gains include the (inaccurate) outputs of the tuned neural networks when the state lies outside of A; this means that the sliding component must actually fight against the outputs of the networks in these regions. To prevent this contention, it is useful to simply shut off the adaptive components as the state moves beyond the boundaries of A, relying only upon the sliding controller in this region. Similarly, since the adaptation mechanism is robust to small disturbances and small tracking errors can be tolerated, the sliding component can be omitted from the control when the state is within A.

To illustrate the above ideas with a specific design, consider the case where it is known that $b(\mathbf{x}) = 1$ in (5.7), obviating the need for a network to approximate this function. The more general case requires a special treatment to ensure that the network estimates N_b can be inverted (see e.g. [83, 74, 98]) and would obscure the main features of the algorithm. Let A_d be a subset of the state space which defines the "nominal operating range" for the process, that is, A_d contains all the model trajectories the process will be required to follow. Let A be a compact subset of the state space be chosen such that A_d is a proper subset. Let $m(\mathbf{x}(t))$ be a continuous modulation function chosen such that

$$\begin{aligned} m(t) &= 0 & &\text{if } \mathbf{x}(t) \in A_d \\ 0 < m(t) &< 1 & &\text{if } \mathbf{x}(t) \in A - A_d \\ m(t) &= 1 & &\text{if } \mathbf{x}(t) \in A^c. \end{aligned}$$

Assume that the function class to which f belongs is known, and allows a choice of Gaussian network parameters (σ, Δ, ρ) such that the network can represent functions in this class to a uniform accuracy of ϵ_f everywhere on A. Finally, assume that an upper bound, possible state dependent, is known for f for points outside the set A_d, i.e. $|f(\mathbf{x})| \leq \overline{f}(\mathbf{x}), \forall \mathbf{x} \in A_d^c$.

For this process, the control law

$$
\begin{aligned}
u(t) \quad = \quad & x_m^{(n)}(t) - \boldsymbol{\alpha}^T \tilde{\mathbf{x}}(t) - m(t)\,\overline{f}(\mathbf{x}(t))\,\mathrm{sat}(s(t)/\Phi) \\
& +(1-m(t)) \sum_{\mathrm{dist}(A,\mathbf{k}\Delta)\leq\rho} \hat{c}_{\mathbf{k}}(t)\exp(-\pi\sigma^2\|\mathbf{x}(t) - \mathbf{k}\Delta\|^2),
\end{aligned}
$$

where $\boldsymbol{\alpha}$ is defined as above, together with the adaptation mechanism

$$
\dot{\hat{c}}_{\mathbf{k}}(t) = -k_a(1 - m(t))\,s_{\Delta}(t)\exp(-\pi\sigma^2\|\mathbf{x}(t) - \mathbf{k}\Delta\|^2)
$$

where the deadzone, boundary layer, and feedback gains are chosen to satisfy $k_D\Phi > \epsilon_f$, will ensure the global stability of the adaptive system, and further will guarantee that asymptotically, for each $i = 0, \ldots, n-1$,

$$
\lim_{t\to\infty} |\tilde{x}^{(i)}(t)| \leq \frac{2^i\Phi}{\lambda^{n-1-i}}.
$$

Here, as above, $s_{\Delta}(t)$ is the result of continuously incorporating a deadzone of width Φ into the error measure $s(t)$.

A detailed proof of these claims is given in [83], together with extensions designed to handle the non-unity gain case. Note again how these results quantify the relations between the approximation capabilities of the network and the control laws and adaptation mechanisms which guarantee stability and convergence. The required deadzone size again reflects the attenuation of the effects of the worst case network approximation errors on A through the dynamics of the measure $s(t)$, and the new structure which must be added to the control law to ensure global stability requires knowledge of the set A on which the network can maintain its accurate approximation to f.

Several additional features of this control and adaptation law deserve mention. First, adaptation is stopped both when the measured error is less than that which would be expected using the best possible network approximation in the control law, and also when the state moves outside of the region on which the quality of the best network approximation can be guaranteed to the specified tolerance. Second, as the state moves outside this region, the network is taken off line and the sliding controller becomes active, smoothly pushing the process back into the nominal operating region, whereupon network learning and control are resumed. Finally, since the tracking errors converge to a vicinity of zero, implying that all of the process states are within or very close to the set A_d, the control is eventually dominated by the feedforward term and the output of the network. The sliding control component is thus eventually inactive, and indeed it may never become active at all if the initial parameter error is small enough.

Results similar to these can be obtained using parameter projection or weight decay (sigma-modification) robustness methods in the adaptation mechanism for the network output weights [74, 40]. A possible drawback to these approaches is that, while stability is guaranteed, convergence is specified in terms of asymptotic energy bounds on the error measure, not the pointwise results given above. This time averaged convergence criterion allows for periodic, short episodes of very bad tracking, caused by the (bounded) drift of the network weights from their tuned values in response to the small disturbances; this is sometimes referred to as a "bursting" phenomenon [1]. On the other hand, these robustness methods may produce superior tracking if the disturbances are smaller than expected and the system is sufficiently excited.

For discretized nonlinear processes, [80] shows how many of these general control and adaptation robustness considerations, together with linearly parameterized network designs, can be translated into effective model tracking algorithms which use networks for stable adaptive *feedforward* compensation of a limited amount of uncertainty in the required control law. The robustness in these algorithms assumes that the outputs of the networks are persistently exciting, making the results of Section 5.2 quite useful for practical application of this method.

6   Concluding Remarks

This paper has presented, in an integrated fashion, the outlines of a theory for using neural networks as elements of recursive identification and adaptive control algorithms for nonlinear dynamic systems. By combining theories of robust adaptive and nonlinear control together with constructive neural network design procedures developed from approximation theory, we have been able to specify adaptive algorithms with guaranteed stability and convergence properties. The synthesis effected reflects our belief that the approximation and the stability aspects of the analysis should be treated simultaneously, as part of the unified whole.

This unified viewpoint has many beneficial side effects. For example, by noting that the accuracy of a Gaussian network approximation at a point $x \in A$ depends only upon nodes in the sampling mesh within the truncation distance ρ of this point, new nodes of the network can be brought into existence only as needed, i.e. only when the input comes within ρ of a particular center. The sampling mesh thus defines a lattice of *latent* node positions, and the approximation theoretic analysis hence naturally specifies a mechanism for minimally expanding the network in response to a particular sequence of inputs, while still ensuring that the required mapping is learned to the specified tolerance [85]. Similarly, by replacing the sum-squared output weight mistuning terms in the Lyapunov functions used above with more general positive definite forms, lateral inhibition and excitation can be stably added to the learning algorithm, permitting the learning of the output weight for a particular node to be influenced by the outputs of neighboring hidden layer nodes [81].

Finally, consider the problem of stable, on-line adjustment of the centers and variances in a Gaussian network. The nonlinear impact perturbations to these parameters have on the approximation being implemented makes it very difficult to develop an effective adaptive algorithm, apart from treating the nonlinear effects as disturbances and adapting only on the first order information, as briefly discussed above. If, however, the changes to these parameters are *coupled* appropriately, it is possible to find a measure of the aggregate mistuning which appears linearly in the expansion of the resulting approximation error. In fact, the sampling theoretic analysis provides just such a coupling mechanism via the assumed bandwidth of the function being approximated; higher bandwidths produce more tightly packed centers and larger variances, and vice-versa. If the bandwidth is made an estimated parameter, and this estimate used in a certainty equivalence network design procedure, the bandwidth mistuning appears linearly when the resulting estimation errors are bounded, and hence stable mechanisms can be developed for simultaneously adjusting both the output weights and the bandwidth estimate [85].

We expect this unified framework to continue to yield valuable insights as more complex network architectures and adaptation mechanisms are pursued, and as the above control algorithms are extended to more complicated dynamical system structures. Of course, in so doing one must be careful not to overlook existing, non-neural solutions for the problems under consideration, which may indeed more effectively exploit available prior information to achieve the required performance. A neural network should thus be viewed not as a panacea, but rather as yet another utility in an engineer's toolbox.

References

[1] Anderson, B. D. O., "Adaptive systems, lack of persistency of excitation and bursting phenomena", *Automatica*, vol. 21, 247-258, 1985.

[2] Annaswamy, A., and Seto, D., "Adaptive control of a class of nonlinear systems", *Proc. 7th Yale Workshop Adaptive and Learning Systems*, New Haven, CT, May 1992.

[3] Arsac, J., *Fourier Transforms and the Theory of Distributions*, Prentice-Hall, Englewood Cliffs, N.J., 1966.

[4] Astrom, K. J., "Theory and applications of adaptive control: a survey", *Automatica*, vol. 19, 471-486, 1983.

[5] Astrom, K. J., and Wittenmark, B., *Adaptive Control*, Addison-Wesley, Reading, MA, 1989.

[6] Atkeson, C. G., "Memory-based approaches to approximating continuous functions", *Proc. 6th Yale Workshop on Adaptive and Learning Systems*, 212-217, Yale, New Haven, CT, 1990.

[7] Atkeson, C. G., and Reinkensmeyer, D. J., "Using associative content-addressable memories to control robots", in *Neural Networks for Control*,T.W. Miller, R.S. Sutton, and P.J. Werbos, eds., MIT Press, Cambridge, MA, 1990.

[8] Bastin, G., Bitmead, R.R., Campion, G., and Gevers, M., "Identification of linearly overparameterized nonlinear systems", *IEEE Trans. Autom. Control*, vol. 37, 1073-1078, July 1992.

[9] Bellman, R., *Adaptive Control Processes–A Guided Tour*, Princeton University Press, Princeton, 1961.

[10] Bracewell, R., *The Fourier Transform and Its Applications*, McGraw-Hill, New York, NY, 1965.

[11] Broomhead, D. S., and Lowe, D., "Multivariable functional interpolation and adaptive networks", *Complex Systems*, 2, 321-355, 1988.

[12] Butzer, P.L., Splettstosser and Stens, R.L., "The sampling theorem and linear prediction in signal analysis", *Jber. d.. Dt. Math.-Verein*, vol. 90, 1-70, 1988.

[13] Chen, F.-C., and Khalil, H.K, "Adaptive control of nonlinear discrete-time systems using neural networks", *preprint* Electrical Engineering Dept, Michigan State Univ., East Lansing, Mich, 1991.

[14] Chen, S., Billings, S.A, and Grant, P.M., "Non-linear system identification using neural networks", *Int. J. Control*, vol. 51, 1191-1214, 1990.

[15] Chen, S., Cowan, C. F. N., Billings, S.A., and Grant, P.M., "Parallel recursive error prediction algorithm for training layered neural networks", *Intl. J. Control*, vol. 51, 1215-1228, 1990.

[16] Chen, S., Cowan, C. F. N., Billings, S.A., and Grant, P.M., "Practical identification of NARMAX models using radial basis functions", *Intl. J. Control*, vol. 52, 1327-1350, 1990.

[17] Chester, D., "Why two hidden layers are better than one", *IEEE Intl. Conf. on Neural Nets.*, I, 265-268, Washington, DC, 1990.

[18] Chui, C. K., *An Introduction to Wavelets*, Academic Press, San Diego, 1992.

[19] Cruz, Jr. J. B., *System Sensitivity Analysis*, Dowden,Hutchinson and Ross, Inc, Stroudsburg, PA, 1973.

[20] Cybenko, G., "Approximations by superposition of a sigmoidal function", *Math. Cont., Sig, and Sys.*, vol. 2, 303-314, 1989.

[21] Daugman, J. G., "Complete discrete 2D Gabor transforms by neural networks for image analysis and compression", *IEEE Trans. ASSP*, 36, 1169-1179, 1988.

[22] Egardt, B., *Stability of adaptive controllers*, Springer-Verlag, Berlin, 1979.

[23] Fu, K. S., "Learning control systems", in *Computer and Information Sciences*, Julius Tou and Richard Wilcox, ed., Spartan Books, Washington, pp. 318-343, 1964.

[24] Fu, K. S., and Waltz, M.D., "A heuristic approach to reinforcement-learning control systems",*IEEE Trans. Autom. Cont.*, vol. 10, 390-398, 1965.

[25] Funahashi, K., "On the approximate realization of continuous mappings by neural networks", *Neural Networks*, 2, 183-192, 1989.

[26] Gabor, D., "Theory of communication", *Jour. IEE*, 93, 429-457, 1946.

[27] Girosi, F. and Anzellotti, G. " Rates of convergence of approximation by translates", *Artificial Intelligence Lab. Memo*, No. 1288, MIT, Cambridge, MA, (in preparation), 1992.

[28] Girosi, F., and Poggio, T., "Networks and the best approximation property", *Biological Cybernetics*, vol 63, 169-176, 1990.

[29] Girosi, F., and Poggio, T., "Representation properties of networks: Kolmogorov's theorem is irrelevant", *Neural Computation*, vol. 1, 465-469, 1989.

[30] Goodwin, G. C., and Sin K. S., *Adaptive Filtering, Prediction, and Control*, Prentice-Hall, Englewood Cliffs, NJ, 1984.

[31] Grossberg, S., *Neural Networks and Natural Intelligence*, MIT Press, Cambridge, MA, 1988.

[32] Grossberg, S., "Nonlinear neural networks: principles, mechanisms, and architectures", *Neural Networks*, vol. 1, 17-61, 1988.

[33] Hebb, D.O., *The Organization of Behavior*, New York, Wiley, 1949.

[34] Higgins, J.R., "Five short stories about the cardinal series", *Bull. Am. Math. Soc.*, vol. 12, 45-89, 1985.

[35] Hopfield, J. J., "Neural networks and physical systems with emergent collective computational abilities", *Proc. Natl. Acad. Sci., USA*, vol. 79, pp. 2554-2558, 1982.

[36] Hopfield, J. J., "Neurons with graded response have collective computational properties like those of two-state neurons", *Proc. Natl. Acad. Sci., USA*, vol. 81, pp. 3088-3092, 1984.

[37] Hopfield, J. J., and Tank, D. W. "'Neural' computation of decisions in optimization problems", *Bio. Cybern.*, vol. 52, pp. 141-152, 1985.

[38] Hornik, K., Stinchcombe, M., and White, H., "Multilayer feedforward networks are universal approximators", *Neural Networks*, 2, 359-366, 1989.

[39] Hunt, K.J., Sbarbaro, D., Zbikowski, R., and Gawthrop, P.J., "Neural networks for control systems-a survey", *Automatica*, vol. 28, 1083-1112, 1992.

[40] Ioannou, P., and Datta, A., "Robust adaptive control: a unified approach", *Proc. IEEE*, 79, 12, pp. 1736-1768, Dec. 1991.

[41] Ioannou, P., and Kokotovic, P.V., "Instability analysis and the improvement of robustness of adaptive control", *Automatica*, 20,5, pp.583-594, Sept. 1984.

[42] Isidori, A., *Nonlinear Control Systems*, Springer-Verlag, Berlin, 1989.

[43] Jerri, A.J., "The Shannon sampling theorem-its various extensions and applications: a tutorial review", *Proc. IEEE*, vol. 65, 1565-1596, 1977.

[44] Jordan, M.I., "Learning inverse mappings using forward models", *Proc. 6th Yale Workshop on Adaptive and Learning Systems*, 146-151, Aug. 1990.

[45] Jordan, M. I., and Rumelhart, D.E., "Forward models: supervised learning with a distal teacher", *Cognitive Science*, vol. 16, 307-354, 1992.

[46] Kalman, R. E., "Design of a self-optimizing control system", *Trans. ASME*, vol. 80, 468-478, 1958.

[47] Kanellakopoulos, I., Kokotovic, P. V., Morse, A.S., "Systematic design of adaptive controllers for feedback linearizable systems", *IEEE Trans. Autom. Cont.*, vol. 36, 1241-1253, 1991.

[48] Kohonen, T., *Self-Organization and Associative Memory*, Springer-Verlag, Berlin, 1984.

[49] Kosko, B., *Neural Networks and Fuzzy Systems*, Prentice Hall, Englewood Cliffs, NJ, 1992.

[50] Kudva, P. and Narendra, K. S., "Synthesis of an adaptive observer using Lyapunov's direct method", *Intl. J. Cont.*, vol. 18, 1201-1210, 1973.

[51] Margolis, M. and Leondes, C. T., "On the theory of adaptive control systems: Method of teaching models", *Proc. 1st IFAC Congress*, vol. 2, 1961.

[52] Marks, R., J., *Introduction to Shannon Sampling and Interpolation Theory*, Springer-Verlag, New York, 1991.

[53] McCulloch, W.S. and Pitts, W., "A logical calculus of ideas immanent in nervous activity", *Bull. Math. Biophys.*, vol. 5, pp.1115-1133, 1943.

[54] Mendel,J.M., and McLaren, R.W., "Reinforcement-learning control and pattern recognition systems", in *Computer and Information Sciences*, Julius Tou and Richard Wilcox, ed., Spartan Books, Washington, pp. 318-343, 1964.

[55] Micchelli, C., "Interpolation of scattered data: distance matrices and conditionally positive definite functions", *Const. Approx.*, 2, pp. 11-22, 1986.

[56] Miller, W. T., An, E., and Glanz, F., "The design of CMAC neural networks for control", *Proc. 6th Yale Workshop on Adaptive and Learning Systems*, 218-224, Yale, New Haven, CT, 1990.

[57] Miller, W. T., Glanz, F. H., and Kraft, L.G., "Application of a general learning algorithm to the control of robotic manipulators", *Intl. Jour. Rob. Res.*, 6, 84-98, 1987.

[58] T.W. Miller, R.S. Sutton, and P.J. Werbos, eds., *Neural Networks for Control*, MIT Press, Cambridge, MA, 1990.

[59] Minsky, M., and Papert, S., *Perceptrons*, MIT Press, Cambridge, MA, 1969.

[60] Morse, A.S., "Global stablity of parameter adaptive systems", *IEEE Trans. Autom. Cont.*, vol. 25, 433-439, 1980.

[61] Nam, K. and Arapostathis, "A model reference adaptive control scheme for pure-feedback nonlinear systems", *IEEE Trans. Autom. Control*, 33, 9, 803-811, 1988.

[62] Narendra, K. S., and Annaswamy, A., *Stable Adaptive Systems*, Prentice Hall, Englewood Cliffs, NJ, 1989.

[63] Narendra, K.S., and Kudva, P., "Stable adaptive schemes for system identification and control–Parts I and II", *IEEE Trans. Sys., Man, and Cyber.*, vol. SMC-4, 542-560, 1974.

[64] Narendra, K.S., Lin, Y.H., and Valavani, L.S., "Stable adaptive controller design–Part II: proof of stability",*IEEE Trans. Autom. Cont.*, vol. 25, 440-448, 1980.

[65] Narendra, K. S., and Parthasarathy, K., "Identification and control of dynamical systems using neural networks", *IEEE Trans. on Neural Networks*, vol. 1, 4-27, 1990.

[66] Narendra, K. S., and Parthasarathy, K., "Gradient methods for the optimization of dynamical systems containing neural networks", *IEEE Trans. on Neural Networks*, vol. 2, 252-262, 1991.

[67] Narendra, K. S., and Parthasarathy, K., "Identification and control of dynamical systems using neural networks–Part II", *Proc. 6th Yale Workshop on Adaptive and Learning Systems*, 164-172, Aug. 1990.

[68] Narendra, K.S., and Valavani, L.S., "Stable adaptive controller design–Direct control", *IEEE Trans. Autom. Cont.*, vol. 23, 570-583, 1978.

[69] Parks, P. C., "Lyapunov redesign of model reference adaptive control systems", *IEEE Trans. Autom. Control.*, vol. 11, 362-367, 1966.

[70] Petersen, D. P., and Middleton, D., "Sampling and reconstruction of wave number limited functions in n-dimensional Euclidean spaces", *Information and Control*, 5, 279-323, 1962.

[71] Peterson, B. B., and Narendra, K. S., "Bounded Error Adaptive Control", *IEEE Trans. Autom. Cont.*, 27,6, 1161-1168, 1982.

[72] Poggio, T., and Girosi, F., "Networks for approximation and learning", *Proc. IEEE*, 78, 9, 1481-1497, 1990.

[73] Poggio, T., and Girosi, F., "Extensions of a theory of networks for approximation and learning: dimensionality reduction and clustering", *Artificial Intelligence Lab. Memo*, No. 1167, MIT, Cambridge, MA, April 1990.

[74] Polycarpou, M., and Ioannou, P., "Identification and control of nonlinear systems using neural network models: design and stability analysis", TR No. 91-09-01, USC Dept. EE-Systems, Sept. 1991.

[75] Powell, M. J. D., "The theory of radial basis function approximation in 1990", *DAMTP Report*, No. NA11, Univ. of Cambridge, Cambridge, England, Dec. 1990.

[76] Psaltis, D., Sideris, A., and Yamamura, A., "Neural Controllers", *Proc. 1st IEEE Conf. on Neural Networks*, San Diego, CA, vol. IV, 551-557, June 1987

[77] Rohrs, C., Valavani, L., Athans, M. and Stein, G., "Robustness of adaptive control algorithms in the presence of unmodelled dynamics", *Proc. 21st Conf. Dec. Control*, 3-11, 1982.

[78] Rosenblatt, F., *Principles of Neurodynamics*, Washington, Spartan Books, 1962.

[79] Rumelhart, D. E., and McClelland, J. L., *Parallel Distributed Processing, Explorations in the Microstructure of Cognition*, vol. 1, MIT Press, Cambridge, MA, 1986.

[80] Sadegh, N., "Nonlinear identification and control via neural networks", ASME Winter Annual Meeting, pp. 45-56, 1991.

[81] Sanner, R.M., "Stable adaptive control and recursive identification of nonlinear systems using radial gaussian networks", *Ph.D. Thesis*, MIT Dept. of Aeronautics and Astronautics, May, 1993.

[82] Sanner, R. M., and Slotine, J.-J. E., "Gaussian networks for direct adaptive control", *Proc. 1991 Automatic Control Conference*, vol. 3, 2153-2159, Boston MA, June 1991.

[83] Sanner, R. M., and Slotine, J.-J. E., "Gaussian networks for direct adaptive control", *IEEE Trans. Neural Networks*, 3(6), pp 837-863, November 1992.

[84] Sanner, R. M., and Slotine, J.-J. E., "Stable adaptive control and recursive identification using radial Gaussian networks", *IEEE Conf. Decision and Control*, Brighton, England, December 1991.

[85] Sanner, R. M., and Slotine, J.-J. E., "Enhanced algorithms for adaptive control using radial gaussian networks", *Proc. 1992 Automatic Control Conference*, Chicago,IL, June 1992.

[86] Sanner, R. M., and Slotine, J.-J. E., "Stable recursive identification using radial basis function networks", *Proc. 1992 Automatic Control Conference*, Chicago,IL, June 1992.

[87] Sanner, R. M., and Slotine, J.-J. E., "Approximation error bounds for scattered-center Gaussian networks', *Proc. 1992 Conf. Dec. and Cont.*, Tuscon, AZ, Dec 1992.

[88] Sastry, S., and Bodson, M., *Adaptive Control: Stability, Convergence, and Robustness*, Prentice-Hall, Englewood Cliffs, NJ, 1989.

[89] Sastry, S. , and Isidori, A. "Adaptive control of linearizable systems", *IEEE Trans. Autom. Control*, 34, 1123-1131, 1989.

[90] Schoenberg, I.J., "Metric spaces and positive definite functions", *Trans. Am. Math. Soc.*, 44, pp. 522-536, 1938.

[91] Slotine, J.-J. E., "Sliding controller design for nonlinear systems", *Intl. Journal of Control*, 40, 421, 1984.

[92] Slotine, J.-J. E. and Coetsee, J.A., "Adaptive sliding controller synthesis for nonlinear systems", *Intl. Journal of Control*, 1986.

[93] Slotine, J.-J. E. and Li, W., "On the adaptive control of robotic manipulators", *Intl. Journal of Rob. Res.*, 6, 3, 1987.

[94] Slotine, J.-J. E. and Li, W., *Applied Nonlinear Control*, Prentice Hall, Englewood Cliffs, NJ, 1991.

[95] Sontag, E. D., "Feedback stabilization using two-hidden-layer networks", *IEEE Trans. Neural Networks*, vol. 3, 981-990, 1992.

[96] Strang, G., and Fix, G., "A Fourier analysis of the finite element variational method", in *Constructive Aspects of Functional Analysis*, G. Geymonet, ed., 793-840, 1973.

[97] Tsypkin, Ya. Z., *Adaptation and Learning in Automatic Systems*, Academic Press, New York, 1971.

[98] Tzirkel-Hancock, E., and Fallside, F., "Stable control of nonlinear systems using neural networks", *Intl. J. Robust and Nonlinear Control*, 2, 63-86, May 1992.

[99] Utkin, V. I., "Variable structure systems with sliding mode: a survey", *IEEE Trans. Auto. Contr.*, 22, 212, 1977.

[100] Whittaker, E.T., "On the functions which are represented by the expansion of the interpolation theory", *Proc. Royal Soc. Edinburgh*, vol. 35, 181-194, 1915.

[101] Widrow, B., and Hoff, M. E., "Adaptive switching circuits", *Inst. Rad. Eng., WestCon Convention Record*, IV, 96-104, 1960.

[102] Widrow, B. and Smith, F. W, "Pattern-recognizing control systems", in *Computer and Information Sciences*, Julius Tou and Richard Wilcox, ed., Spartan Books, Washington, pp. 318-343, 1964.

[103] Winston, P. H., *Artificial Intelligence*, 3rd. ed., Addison-Wesley, Reading, MA, 1992.

[104] Zaanen, A. C., *Continuity, Integration, and Fourier Theory*, Springer-Verlag, Berlin, 1989.

[105] Zemanian, A. H., *Distribution Theory and Transform Analysis*, McGraw-Hill, New York, 1965.

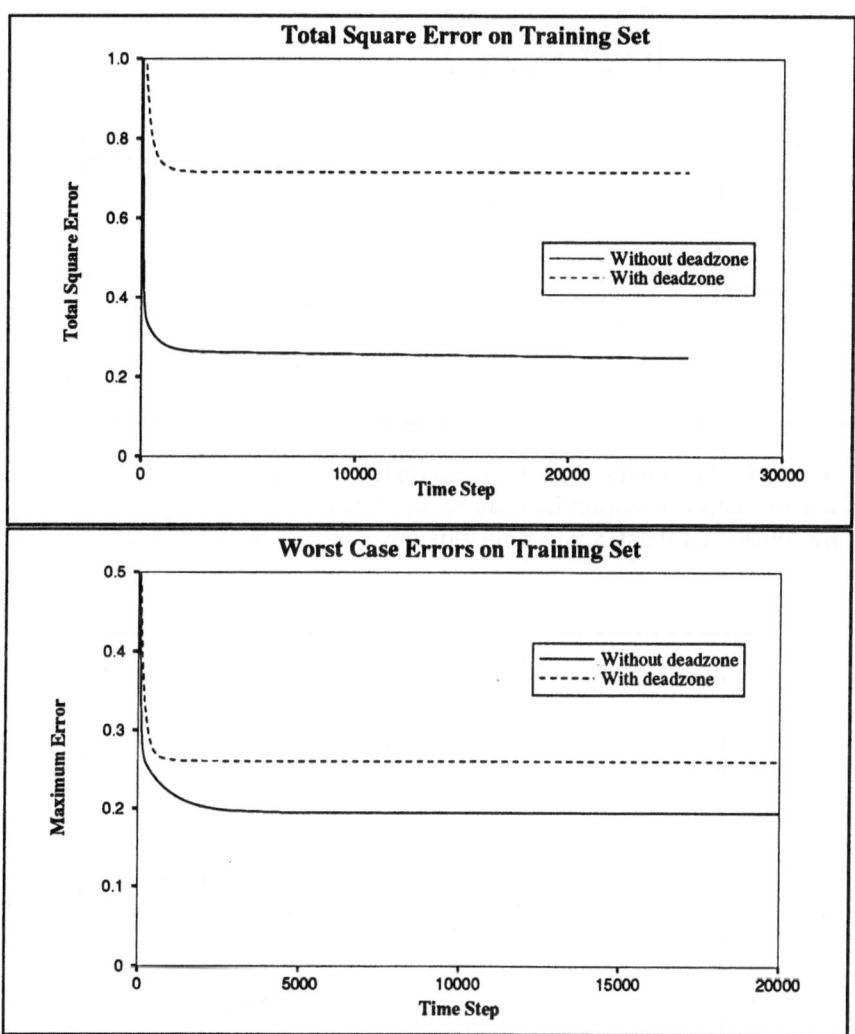

Figure 6.1: Total square error and worst case error on training set as a function of training iterations. A deadzone robustness modification results in slightly worse performance on the training set.

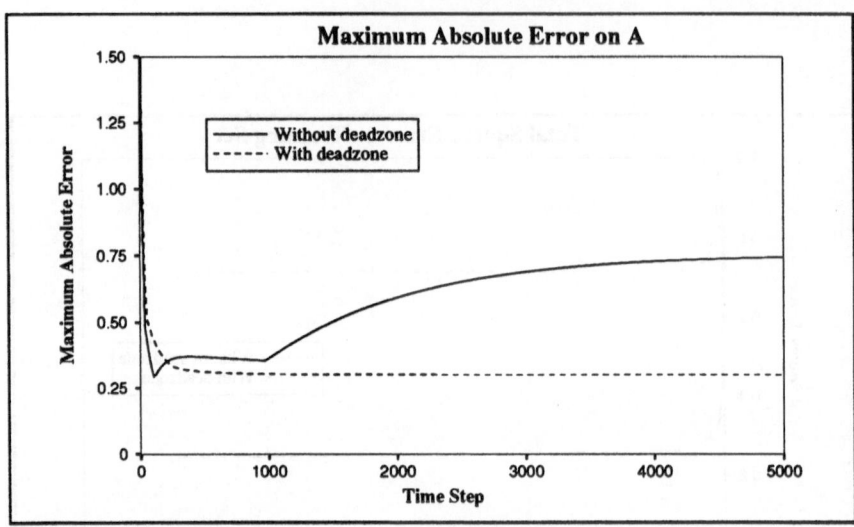

Figure 6.2: Worst case errors recorded on the compact set $A = [-2, 2]$ in the Gaussian network approximation to the function underlying the training data Without robust adaptation this error begins to increase as training continues.

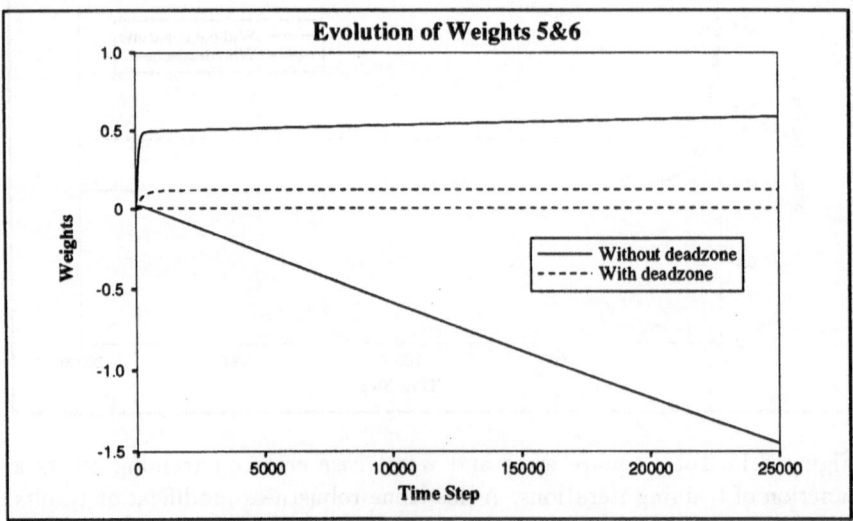

Figure 6.3: Evolution of network output weights. Without the robustness modification, the drift of the output weight of the node centered at -4Δ causes the worsening generalization errors seen in Figure 3.

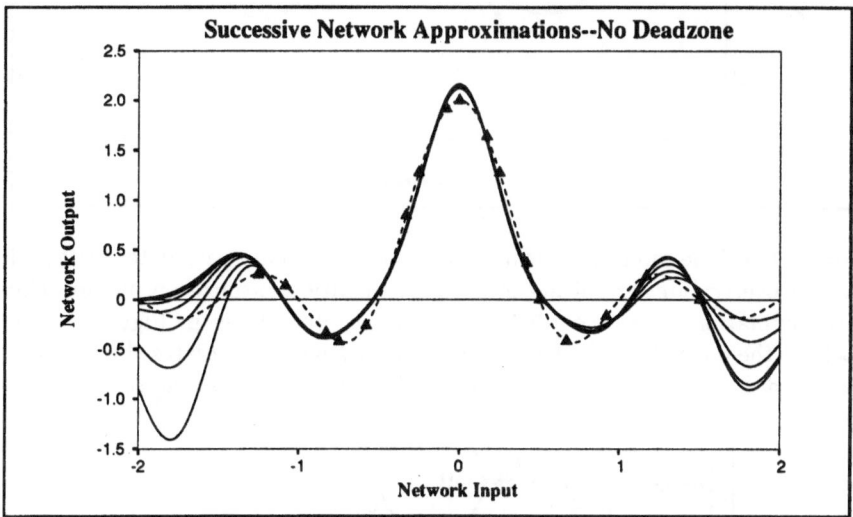

Figure 6.4: The drift in the output weights causes the overall fit to the function underlying the data to become much worse as training increases, although behavior near the training points is almost constant. The different approximations were recorded after 400, 800, 1600, 3200, 6400, 12800, and 25600 training cycles. The dashed line is the actual function f which generates the training set data, and the filled triangles are the points in the training set.

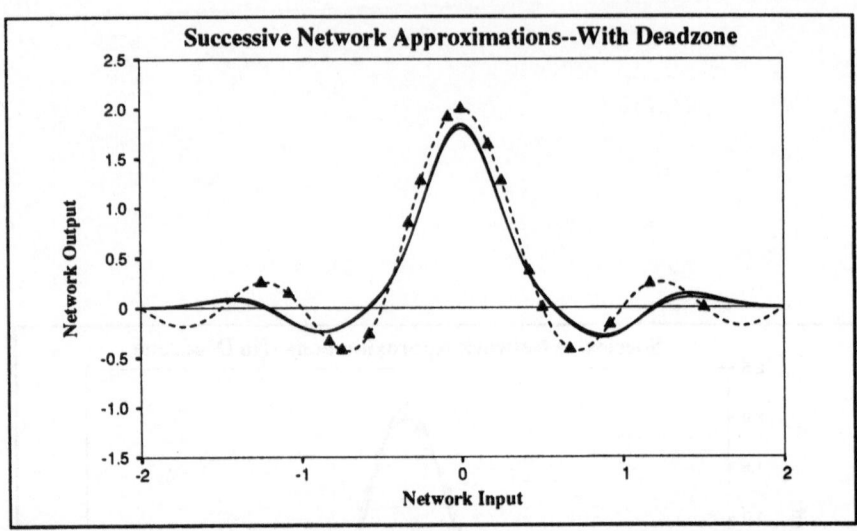

Figure 6.5: The modified learning mechanism prevents parameter drift and avoids overtraining, resulting in a higher quality identification model. The approximations were recorded at the same intervals as in Figure 5. As before, the dashed line is the actual function f which generates the training set data, and the filled triangles are the points in the training set.

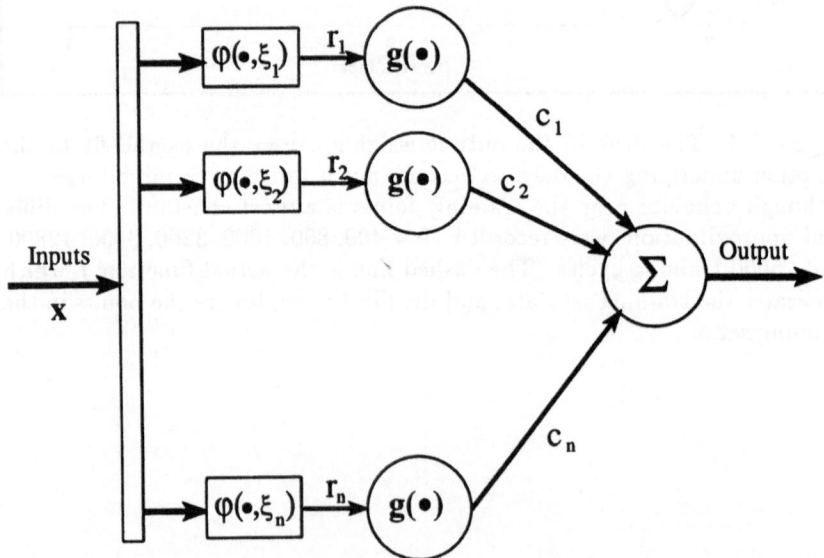

Figure 6.6: Generalized representation of the static, feedforward networks used in this paper.

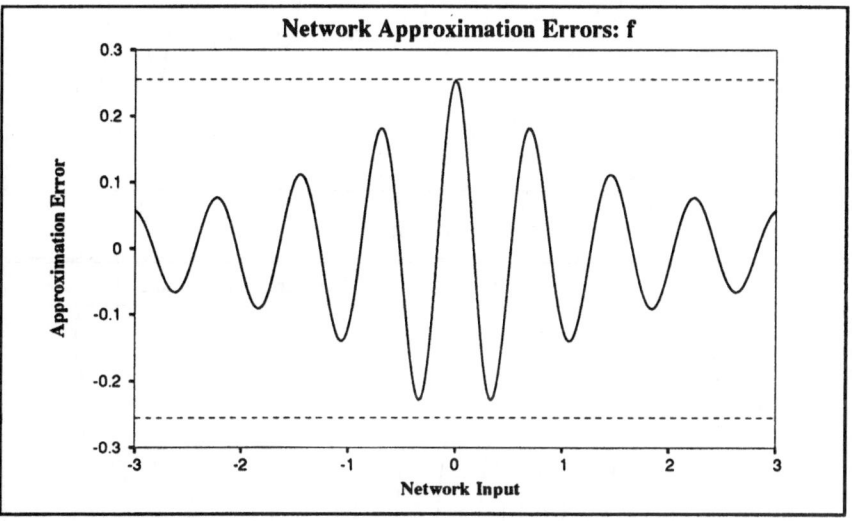

Figure 6.7: Error approximating f on A using the Gaussian network construction in Section 4.3.

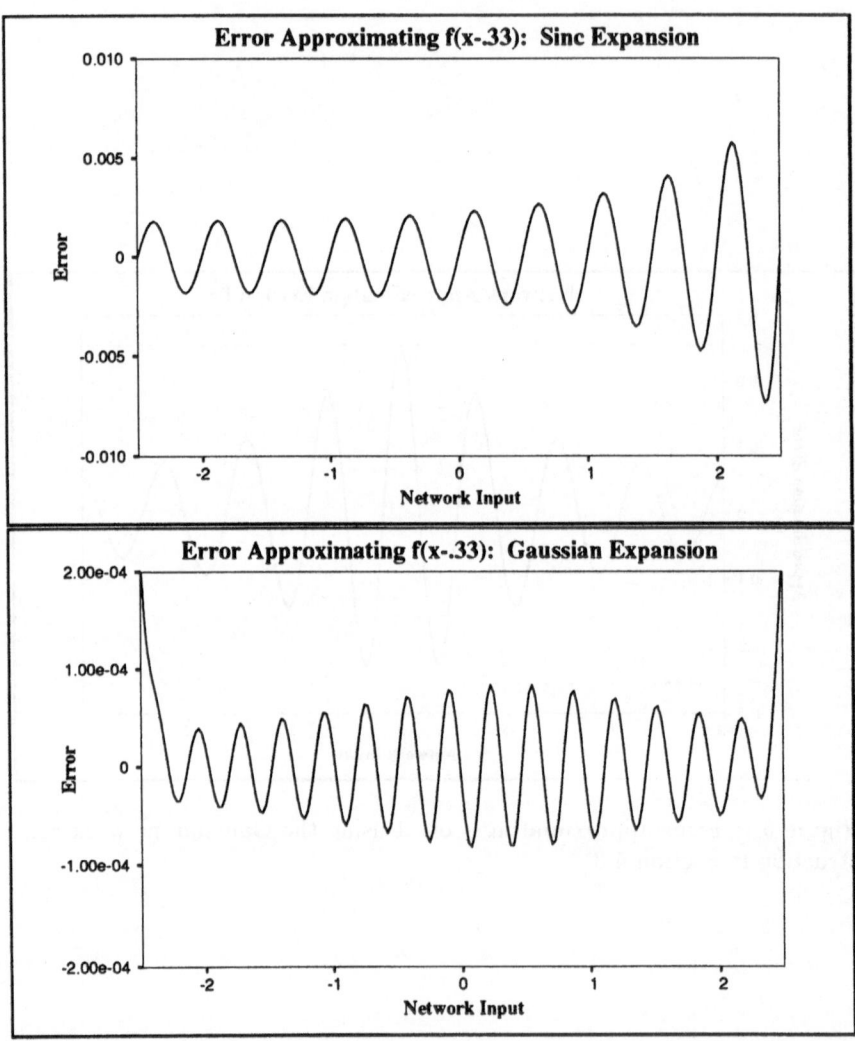

Figure 6.8: Comparison of a truncated sinc and a truncated Gaussian approximation to $f(x - .33)$; the Gaussian expansion achieves much greater accuracy with the same number of nodes.

Progress in Systems and Control Theory

Series Editor

Christopher I. Byrnes
Department of Systems Science and Mathematics
Washington University
Campus P.O. 1040
One Brookings Drive
St. Louis, MO 63130-4899

Progress in Systems and Control Theory is designed for the publication of workshops and conference proceedings, sponsored by various research centers in all areas of systems and control theory, and lecture notes arising from ongoing research in theory and applications control.

We encourage preparation of manuscripts in such forms as LATEX or AMS TEX for delivery in camera-ready copy which leads to rapid publication, or in electronic form for interfacing with laser printers.

Proposals should be sent directly to the editor or to: Birkhäuser Boston, 675 Massachusetts Avenue, Cambridge, MA 02139, U.S.A.